AF809726

Green Chemistry and Sustainable Technology

Series Editors

Liang-Nian He
State Key Lab of Elemento-Organic Chemistry, Nankai University, Tianjin, China

Pietro Tundo
Department of Environmental Sciences, Informatics and Statistics, Ca' Foscari University of Venice, Venice, Italy

Z. Conrad Zhang
Dalian Institute of Chemical Physics, Chinese Academy of Sciences, Dalian, China

Aims and Scope

The series Green Chemistry and Sustainable Technology aims to present cutting-edge research and important advances in green chemistry, green chemical engineering and sustainable industrial technology. The scope of coverage includes (but is not limited to):

- Environmentally benign chemical synthesis and processes (green catalysis, green solvents and reagents, atom-economy synthetic methods etc.)
- Green chemicals and energy produced from renewable resources (biomass, carbon dioxide etc.)
- Novel materials and technologies for energy production and storage (bio-fuels and bioenergies, hydrogen, fuel cells, solar cells, lithium-ion batteries etc.)
- Green chemical engineering processes (process integration, materials diversity, energy saving, waste minimization, efficient separation processes etc.)
- Green technologies for environmental sustainability (carbon dioxide capture, waste and harmful chemicals treatment, pollution prevention, environmental redemption etc.)

The series Green Chemistry and Sustainable Technology is intended to provide an accessible reference resource for postgraduate students, academic researchers and industrial professionals who are interested in green chemistry and technologies for sustainable development.

Pradeep Verma · Bikash Kumar
Editors

Sustainable Radiation Technologies in Waste-biomass Valorization

Greener Energy and High-Value Product Generation

Springer

Editors
Pradeep Verma
Bioprocess and Bioenergy Laboratory
Department of Microbiology
Central University of Rajasthan
Kishangarh, Ajmer, Rajasthan, India

Bikash Kumar
Department of Biosciences and Biomedical Engineering
Indian Institute of Technology Indore
Indore, Madhya Pradesh, India

ISSN 2196-6982 ISSN 2196-6990 (electronic)
Green Chemistry and Sustainable Technology
ISBN 978-3-031-63940-1 ISBN 978-3-031-63941-8 (eBook)
https://doi.org/10.1007/978-3-031-63941-8

© The Editor(s) (if applicable) and The Author(s), under exclusive license to Springer Nature Switzerland AG 2024

This work is subject to copyright. All rights are solely and exclusively licensed by the Publisher, whether the whole or part of the material is concerned, specifically the rights of translation, reprinting, reuse of illustrations, recitation, broadcasting, reproduction on microfilms or in any other physical way, and transmission or information storage and retrieval, electronic adaptation, computer software, or by similar or dissimilar methodology now known or hereafter developed.
The use of general descriptive names, registered names, trademarks, service marks, etc. in this publication does not imply, even in the absence of a specific statement, that such names are exempt from the relevant protective laws and regulations and therefore free for general use.
The publisher, the authors and the editors are safe to assume that the advice and information in this book are believed to be true and accurate at the date of publication. Neither the publisher nor the authors or the editors give a warranty, expressed or implied, with respect to the material contained herein or for any errors or omissions that may have been made. The publisher remains neutral with regard to jurisdictional claims in published maps and institutional affiliations.

This Springer imprint is published by the registered company Springer Nature Switzerland AG
The registered company address is: Gewerbestrasse 11, 6330 Cham, Switzerland

If disposing of this product, please recycle the paper.

Preface

The growing energy demand and depleting fossil-based fuels are promoting the scientific community for searching sustainable alternatives to meet the needs. Energy security can be supported by the utilization of renewable sources such as waste biomass. This waste biomass is a result of anthropogenic activities and requires advanced approaches for its remediation. Thus, the development of approaches for its valorization is a sustainable approach and serves the dual purpose of value addition and waste remediation. However, several limitations such as the need for new technologies for overcoming lignocellulosic biomass recalcitrance, the need for a sustainable approach for the breakdown of algal biomass, and municipal waste valorization for recovery of high-value compounds. Several radiation-based technologies such as microwave, ultrasound, gamma, X-ray, and laser electric field have come to the forefront of helping bioenergy scientists tackle the limitation of biomass recalcitrance and easy breakdown of biomass to desired high-value compounds. However, these technologies are in the laboratory-scale stage or pre-industry stages but have great potential to be applied in large-scale industries. Some of the physicochemical, thermochemical (pyrolysis, gasification), and catalysis-based approaches can be integrated with radiation-based technologies that have shown great results in the generation of high-value compounds from waste in different biomass-based biorefinery processes in a much faster and effective manner. This high-value product can be the result of a multistep process or single-step direct conversion. Thus, helping in biomass lysis for its conversion to energy (bioethanol, H_2, etc.) and value-added compounds (oligosaccharides, sugars, phenolics, etc.). The book focused on understanding the mechanism, instrumentation, and recent technical advances in the fields of microwave, gamma, ultrasound, electric field, and lased-based radiation-assisted biomass valorization approaches. The book also presents the application of radiation-based technologies for various industrial, human health, and environmental aspects. The book also covers bottlenecks and opportunities along with highlighting techno-economic and environmental impact studies. The book is an attempt to provide a dedicated book on radiation-based technologies which otherwise is lacking in the literature, to understand its potential in waste valorization. Further, through this book, we are making an attempt to provide compiled literature useful to stakeholders working in this area. This book can provide one-stop solutions to researchers, students, academicians, scientists, and professionals working in the area of biorefinery and waste valorization.

Ajmer, India Pradeep Verma
Indore, India Bikash Kumar

Acknowledgments

First of all, we would like to convey our gratitude toward Springer and Series editors of the book series *Green Chemistry and Sustainable Technology* for providing us the opportunity to submit this book entitled *Sustainable Radiation Technologies in Waste-biomass Valorization: Greener Energy and High-Value Product Generation* under the book series. Dr. Bikash is thankful to Prof. Pradeep Verma for motivating and providing an opportunity to be part of this book as editor. The current book is the result of the constant support of all the researchers and academicians who contributed to the book and provided all needed support during review and revisions. Therefore the editors are thankful for their contribution and constant support. Professor Pradeep Verma and Dr. Bikash Kumar are thankful to their respective institutes, i.e., "Central University of Rajasthan (CURAJ), Ajmer, India" and the Indian Institute of Technology, Indore, respectively for providing infrastructural support and a suitable teaching and research environment. The teaching and research experience has provided the necessary understanding of the needs of academicians, students, and researchers in a book that was greatly helpful during the development of the book. Professor Verma is also thankful to the Department of Biotechnology for providing me funds through sponsored projects (Grant No. BT/304/NE/TBP/2012 and BT/PR7333/PBD/26/373/2012), for setting up my laboratory "Bioprocess and Bioenergy Laboratory". Dr. Bikash Kumar is thankful to DST-SERB for the Postdoctoral fellowship (NPDF/2022/001781) at IIT Indore. Dr. Bikash Kumar is also thankful to his current NPDF Mentor Prof. Kiran Bala for motivating and supporting him in his endeavors. We are also thankful to god and parents for their blessings. We would also like to express our deep sense of gratitude toward our families for their support during the development of the book and in life. Dr. Kumar would like to especially mention his newborn daughter "Shrinika" for bringing joy to their life.

Contents

Editors and Contributors

About the Editors

Prof. Pradeep Verma completed his Ph.D. at Sardar Patel University in Gujarat, India in 2002. In the same year, he was selected as a UNESCO fellow and joined the Czech Academy of Sciences in Prague, Czech Republic. He later moved to Charles University, Prague to work as a Post-Doctoral Fellow. In 2004 he joined as a visiting scientist at UFZ Centre for Environmental Research, Halle, Germany. He was awarded a DFG fellowship which provided him another opportunity to work as a Post-Doctoral Fellow at Gottingen University, Germany. He moved to India in 2007 where he joined Reliance Life Sciences, Mumbai, and worked extensively on bio-butanol production which attributed a few patents to his name. Later he was awarded the JSPS Post-Doctoral Fellowship Program and joined the Laboratory of Biomass Conversion, Research Institute of Sustainable Humanosphere (RISH), Kyoto University, Japan. He is also a recipient of various prestigious awards such as the Ron-Cockcroft Award by the Swedish society, and UNESCO Fellow ASCR Prague. He has been awarded with Fellow of Mycological Society of India (MSI-2020), Prof. P. C. Jain Memorial Award, Mycological Society of India 2020, and Fellow of Biotech Research Society, India (2021).

Professor Verma began his independent academic career in 2009 as a Reader and Founder Head at the Department of Microbiology at Assam University. In 2011 he moved to the Department of Biotechnology at Guru Ghasidas Vishwavidyalaya (A Central University),

Bilaspur, and served as an Associate Professor till 2013. He works as a Professor at the Department of Microbiology, CURAJ. He is working as the Dean, School of Sport Sciences and acted as the former Head and Dean of the School of Life Sciences. He is a member of various national and international societies/academies. He has completed two collaborated projects worth 150 million INR in microbial diversity and bioenergy.

Professor Verma is a Group leader of the Bioprocess and Bioenergy Laboratory at the Department of Microbiology, School of Life Sciences, CURAJ. His area of expertise involves Microbial Diversity, Bioremediation, Bioprocess Development, and Lignocellulosic and Algal Biomass based Biorefinery. He also holds several international patents in microwave-assisted biomass pretreatment and bio-butanol production. He has more than 94+ research/review articles in peer-reviewed international journals and contributed to several 62+ book chapters in different edited books and having citations 6972+, h-index 44 and i-index-106 and also consider in 2% top scientists in subject expert list by Stanford university. He is a Fellow of the Mycological Society of India, the Biotech Research Society of India, and the Association of Microbiological Society (AMI). He has also edited 20+ books in international publishers such as Springer, Taylor & Francis, De-Gruyter, CRC Press, and Elsevier. He is a guest editor to several journals such as *Biomass Conversion and Biorefinery* (Springer), *Frontier in Nanotechnology* (Frontiers), and *International Journal of Environmental Research and Public Health* (mdpi). He is also an editorial board member for the *Journal of Current Nanomedicine* (Bentham Sciences) and *Biotechnology and Genetic Engineering Reviews* (T&F). He is acting as a reviewer for more than 50 journals in different publication houses such as Springer, Elsevier, RSC, ACS, Nature, Frontiers, and mdpi. He has more than 23 years of research and teaching experience in bioprocess and microbial biotechnology and industrial microbiology.

Dr. Bikash Kumar is currently working as a National Post-Doctoral Fellow, at the Indian Institute of Technology, Indore, He received his B.Sc. (Hons-2012) and M.Sc. (Distinction-2014) in Biotechnology and Bioinformatics from Guru Ghasidas University, Bilaspur (A Central University) and Pondicherry Central University, respectively. Later, in 2014 he worked as Junior Research Fellow at the Department of Microbiology, Central University of Rajasthan, India. In 2015 he enrolled for Ph.D. and started his research at the Bioprocess and Bioenergy Laboratory under the research supervision of Prof. Pradeep Verma from the Department of Microbiology, Central University of Rajasthan, India. Post Ph.D., he worked as a Research Associate at the Department of Biosciences and Bioengineering, Indian Institute of Technology, Guwahati. His work is focused on the development of biomass pretreatment technologies, and microbial enzyme production (cellulases, laccase, and xylanases) for greener energy and environment. The work has resulted in the publication of more than 35+ research articles, review papers, and book chapters. He has been acknowledged for the 11-editorial book. His work has attracted significant impact which is evident from 2154+ citations with h-index—22, i-1—index-27. He is the recipient of several awards AMI-Young Scientist Award, INSA-NASI-SRF, CSIR-SRF, JNMF scholarship, DST-SERB NPDF, etc. He is a life member of the Biotech Research Society of India (BRSI); the Association of Microbiologists of India (AMI); the Mycological Society of India (MSI) and the Association of Carbohydrate Chemists and Technologists, India (ACCTI). He is a reviewer of 20+ peer-reviewed research journals for Elsevier, Wiley, MDPI, Springer, Taylor, and Francis. Dr. Bikash Kumar is working as Associate Editor in *Frontiers in Energy Research*, Review editor for *Frontiers in Bioengineering and Biotechnology & Frontiers in Chemistry*, Guest editor for a Special Issue in *Frontiers in Bioengineering and Biotechnology* (*Green Synthesis of Catalyst for Sustainable Industrial Technologies*), & *Biotechnology for Sustainable Materials (BMC)* (*Green Carbon Revolution: Lignocellulose-based Carbon Materials powering future Biomaterial solutions)*. He has provided editorial assistance to Prof. Pradeep Verma in compiling 11 books for CRC Press, Elsevier, and Springer.

Contributors

Thamer Adnan Abdullah Applied Sciences Department, University of Technology, Baghdad, Iraq

Hiba N. Abu Tayeh Institute of Agricultural Engineering, Agricultural Research Organization (ARO), Volcani Institute, Rishon LeZion, Israel

Hassan Azaizeh Department of Environmental Science, Water Science and Biotechnology, Tel Hai College, Upper Galilee, Israel

Paramjeet Baghel Clean Energy Laboratory, Centre for Rural Development and Technology, Indian Institute of Technology Delhi, Delhi, India

Latika Bhatia Department of Microbiology and Bioinformatics, Atal Bihari Vajpayee University, Bilaspur, Chhattisgarh, India

Arghya Datta Chemical Conversion Division, Sardar Swaran Singh National Institute of Bioenergy, Kapurthala, Punjab, India

Xingyu Fan School of Food and Biological Engineering, Jiangsu University, Zhenjiang, China

Yoram Gerchman Oranim Academic College, Tivon, Israel;
Faculty of Natural Sciences, University of Haifa, Tivon, Israel

Amit Ghosh P.K. Sinha Centre for Bioenergy and Renewables, Indian Institute of Technology Kharagpur, Kharagpur, West Bengal, India;
School of Energy Science and Engineering, Indian Institute of Technology Kharagpur, Kharagpur, West Bengal, India

Geethika Gudapati Department of Chemical Engineering, College of Engineering, Andhra University, Visakhapatnam, India

Husam Talib Hamzah Department of Chemical Engineering, College of Engineering, Andhra University, Visakhapatnam, India

Hasan J. Hasan Faculty of Basic Education, Sumer University, Dhi-Qar, Iraq

Sulfath Hakkim Hazeena Department of Plant Science, School of Biological Sciences, Central University of Kerala, Kasaragod, Kerala, India

Koloina Heriniaina Department of Chemical Engineering, College of Engineering, Andhra University, Visakhapatnam, India

Tejaswini Jaiswal Department of Microbiology and Bioinformatics, Atal Bihari Vajpayee University, Bilaspur, Chhattisgarh, India

Priyanka Kaushal Clean Energy Laboratory, Centre for Rural Development and Technology, Indian Institute of Technology Delhi, Delhi, India

Haoxin Li School of Food and Biological Engineering, Jiangsu University, Zhenjiang, China

Jiakang Liang School of Food and Biological Engineering, Jiangsu University, Zhenjiang, China

Subhajit Majumder Department of Chemical Engineering, National Institute of Technology (NIT) Hamirpur, Hamirpur, India

Aryasree Makkakode Department of Plant Science, School of Biological Sciences, Central University of Kerala, Kasaragod, Kerala, India

Chithra Manisseri Department of Plant Science, School of Biological Sciences, Central University of Kerala, Kasaragod, Kerala, India

Abha Mishra Biomolecular Engineering Laboratory, School of Biochemical Engineering, Indian Institute of Technology (BHU), Varanasi, India

Navneet Kumar Mishra Department of Fuel, Minerals and Metallurgical Engineering, IIT (ISM) Dhanbad, Dhanbad, India

S. Nikhil Department of Chemical Engineering, College of Engineering, Andhra University, Visakhapatnam, India

Venkata Rao Poiba Department of Chemical Engineering, College of Engineering, Andhra University, Visakhapatnam, India

B. Rajasekhar Reddy Department of Fuel, Minerals and Metallurgical Engineering, IIT (ISM) Dhanbad, Dhanbad, India

Dibakar Rakshit Department of Energy Science and Engineering, Indian Institute of Technology Delhi, Delhi, India

Kaviraj Ramesh Department of Plant Science, School of Biological Sciences, Central University of Kerala, Kasaragod, Kerala, India

Shruti S. Raut Biomolecular Engineering Laboratory, School of Biochemical Engineering, Indian Institute of Technology (BHU), Varanasi, India

Somdutt Ravindran P.K. Sinha Centre for Bioenergy and Renewables, Indian Institute of Technology Kharagpur, Kharagpur, West Bengal, India

Ayusmita Ray P.K. Sinha Centre for Bioenergy and Renewables, Indian Institute of Technology Kharagpur, Kharagpur, West Bengal, India

Manni Ren School of Food and Biological Engineering, Jiangsu University, Zhenjiang, China

Dilip Kumar Sahu Department of Microbiology and Bioinformatics, Atal Bihari Vajpayee University, Bilaspur, Chhattisgarh, India

Geddada S. G. D. Saketh Department of Chemical Engineering, College of Engineering, Andhra University, Visakhapatnam, India

Anil Kumar Sarma Chemical Conversion Division, Sardar Swaran Singh National Institute of Bioenergy, Kapurthala, Punjab, India;
Electrochemical Conversion Division, Sardar Swaran Singh National Institute of Bioenergy, Wadala Kalan, Kapurthala, Punjab, India

Shruti Singh Department of Microbiology and Bioinformatics, Atal Bihari Vajpayee University, Bilaspur, Chhattisgarh, India

Aruna Thakur Department of Mechanical and Manufacturing Engineering, National Institute of Advanced Manufacturing Technology (NIAMT), Ranchi, India

Pooja Thakur Department of Chemical Engineering, National Institute of Technology (NIT) Hamirpur, Hamirpur, India

Sridevi Veluru Department of Chemical Engineering, College of Engineering, Andhra University, Visakhapatnam, India

Jampana Vishnu Varma Department of Chemical Engineering, College of Engineering, Andhra University, Visakhapatnam, India

Lei Wu School of Chemistry and Chemical Engineering, Xi'an University of Architecture and Technology, Xi'an, China

Lei Zhang School of Food and Biological Engineering, Jiangsu University, Zhenjiang, China

Cunshan Zhou School of Food and Biological Engineering, Jiangsu University, Zhenjiang, China

Jun Zhou School of Chemistry and Chemical Engineering, Xi'an University of Architecture and Technology, Xi'an, China

Man Zhou School of Food and Biological Engineering, Jiangsu University, Zhenjiang, China

Chapter 1
Introduction to Radiation-Based Technologies for Biomass Waste: Understanding Mechanism and Its Applications in Future Technologies

Husam Talib Hamzah, Sridevi Veluru, Hasan J. Hasan, Thamer Adnan Abdullah, Geethika Gudapati, and Koloina Heriniaina

Abstract This book chapter is an introductory exploration into the innovative use of radiation-based technologies for addressing biomass waste challenges in future technologies. Biomass waste, including agricultural residues, forestry waste, and organic byproducts, represents a significant environmental concern. Radiation-based technologies, encompassing various radiation sources and techniques, offer sustainable solutions for biomass waste conversion, energy production, and environmental remediation. This chapter provides a foundational overview of radiation-based processes, their potential applications in biomass waste management, and their pivotal role in shaping environmentally responsible future technologies.

Keywords Radiation-based technologies · Biomass waste · Waste conversion · Energy production · Environmental remediation · Radiation sources

1.1 Introduction

Human civilizations' energy needs have expanded tremendously in the last century. As fossil resources become scarcer and their influence on global warming becomes clearer, green energy must be used to manufacture chemicals and biofuels. Biomass, the world's largest organic carbon source, offers this potential [4]. To satisfy the

H. T. Hamzah (✉) · S. Veluru · G. Gudapati · K. Heriniaina
Department of Chemical Engineering, College of Engineering, Andhra University, Visakhapatnam, India
e-mail: alwatanaliraqi@gmail.com

H. J. Hasan
Faculty of Basic Education, Sumer University, Dhi-Qar, Iraq

T. A. Abdullah
Applied Sciences Department, University of Technology, Baghdad, Iraq

© The Author(s), under exclusive license to Springer Nature Switzerland AG 2024

P. Verma and B. Kumar (eds.), *Sustainable Radiation Technologies in Waste-biomass Valorization*, Green Chemistry and Sustainable Technology,
https://doi.org/10.1007/978-3-031-63941-8_1

ever-increasing need for energy while also containing the effects of global warming, it appears to have incredible potential for the future [2].

Radiation technology has shown potential in biomass waste treatment and monetization. Because of the growing need to address waste management and promote sustainable development globally, radiation-based technologies offer unique and practical techniques to recycle biomass waste into valuable products. This chapter introduces radiation-based biomass waste treatment technology. To comprehend how biological systems are impacted by radio and microwave radiation, one must be familiar with basic principles. Maxwell's equations control radio and microwave waves; these equations represent experiment outcomes mathematically. These four electromagnetic rules can analytically and precisely infer all macroscopic electromagnetic events [70]. Radiation-based technologies use gamma, X-ray, microwaves, and electron beams to convert biomass waste physically and chemically. These technologies employ radiation to sterilize waste, break down complex organic compounds, and make biofuels, biochemicals, and other valuable products. Radiation-based technologies offer higher efficiency, shorter processing times, and less environmental effects than previous methods [14, 77]. Some ways radiation is used to treat biomass waste are pyrolysis, decomposition, modification, gasification, combustion, and hydrothermal liquefaction. These technologies use radiation to eliminate impurities and pollutants and make energy and valuable chemicals. It is essential to use radiation-based technology to kill germs in biomass garbage. Biomass waste is often sterilized with radiation to eliminate germs, viruses, and fungi. Pathogens can harm medicine, food preparation, and health care, so this method is essential. By exposing biomass waste to controlled radiation, sterilization improves the safety and quality of the result. Besides deterioration, radiation-based technologies can change biomass waste properties. Radiation modifies biomass waste's physical, chemical, or biological qualities to increase its value. Radiation crosslinks biomass waste polymers, improving their mechanical properties and versatility. Radiation can enhance biomass waste functions and can convert biomass waste into biogas, biofuels, and syngas, lowering greenhouse gas emissions and encouraging sustainable energy. Radiation-based technology can clear soil, break down organic contaminants, purify wastewater, preserve food, control pests, and increase fertilizer production [56]. Transforming waste into mass and energy requires more efficient methods. Ionizing radiation irreversibly changes materials' molecular structure and chemical composition. Basic physical and chemical properties change significantly. The unique features of ionizing radiation have inspired various new material processing methods. Initial efficiency and reliability studies suggest they could transform food processing and biotechnology [24, 36]. Reduced energy usage, fewer production steps, and elimination of potentially harmful chemical treatments are all possible outcomes of radiation-induced cellulose degradation [98]. The method is financially feasible because of its low energy use and other advantages. As a result, high-energy radiations represent an alternative energy source that has the potential to facilitate biofuel manufacturing processes that are both economical and kind to the environment [89]. We must know how to pretreat the biomass with high-energy irradiation to help the enzymes hydrolyze the cellulose and convert it to glucose [13, 51] (Table 1.1). The sugar

Table 1.1 Evaluation of the various physical pre-treatment approaches

Radiation type	Feedstocks	Condition	Yield	References
Microwave Ultrasound	Eucalyptus regnans sawdust	MWAH at 180 °C, 30 min	Sugar: 3.5 times higher than LHW	[3]
	Corn meal	MW power: 80 W	Ethanol: Improved by 13%	[79]
	Sawdust	Frequency: 20 kHz, power: 20 W	Glucose: 22–53%	[88]
	Cassava chips	Pulses of 10 and 30 s at an amplitude of 80, 160, and 320 μm_{pp}	Ethanol: 2.7 fold	[80]
Gamma-ray irradiation	Corncob reed stover, Eucalyptus, Camellia oleifera	Gamma irradiation: 800 kGy	Hemicellulose: ~84 g/L, cellulose: ~251 g/L	[1, 62]
Electron beam irradiation	Korean Miscanthussinensis	1 meV, 500 kGy, energy: 7.4 mA	Glucose: 1.26 fold	[111]
	Rice straw	1 meV at 80 kGy, energy of 0.12 mA	Ethanol: 9.3 g/L, recovery: 57.0%	[10]
Pulsed electric energy	Ley crop silage	Field strength: 48–96 kV/cm	Biogas: 16%	[42]
Autohydrolysis	Wood barks	Field strength: 20 kV/cm	Polyphenol recovery: 8–tenfold	[16]

is fermented into ethanol and other valuable byproducts, including a widely used commercial biofuel, throughout the fermentation process.

1.2 Mechanism of Radiation-Based Technologies

1.2.1 Mechanism of Microwave Pre-treatment

A dipole, defined as the space between two charges perpendicular to one other and spaced at a constant distance apart, is the source of low-frequency non-ionizing waves like microwaves [73]. The thermal and athermal effects account for the heating caused by MW irradiation [39]. When the temperature and pressure rise, the thermal action breaks down the lignin and cell wall components. When dielectric materials relax and polarize in response to an applied electromagnetic field, athermal or non-thermal effect forces are produced [30, 53, 93, 99]. Hydrolysis is enhanced when polar molecules realign, which disrupts hydrogen bonds [93, 99], destroys components of cell walls (such as cellulose crystallinity degradation), and realigns the molecules themselves. Consequently, when exposed to microwave radiation, a material's dielectric characteristics are the most crucial determinant of its heating property [96]. Following are other microwave heating factors: (a) Type and size of input biomass, (b) molecular dynamics of dipole molecules, (c) Joule heating due to magnetic material induction current and electrolyte ionic conduction, (d) heating rate and Residence time, (e) moistness of material, (f) MW power output, and (g) depth of penetration [73]. The groundbreaking study that used microwaves to pretreat lignocellulose was published in the early 1980s by Azuma et al. [7, 8] and Ooshima et al. [82]. To demonstrate microwave-assisted auto-hydrolysis (MWAH), Azuma and colleagues utilized a twin magnetrons system with a multimode microwave (model: TMB-3210, Make: Toshiba Co).

1.2.2 Mechanism of Ultrasound Pre-Treatment

Frequencies of sound waves more significant than 20 kHz, which are inaudible to humans, are referred to as ultrasound. Wood and Loomis showed ultrasound to cause floc to break apart or even collapse when applied to cellular biomass [38]. Supersonic pre-treatment is based on the "cavitational effect" principle. During this procedure, the application of ultrasound helps to disrupt the β-O-4 and αO-4 bonds in lignin, leading to cavitation bubbles forming that separate the biomass's polysaccharides and lignin fractions [50]. When these bubbles get to a specific size and then suddenly break, there is a sharp increase in temperature (2000–5000 K) and pressure (1800 atm) [55]. According to More and Ghangrekar [72] and Chandrapala et al. [19], used ultrasound can cause a brief but localized rise in temperature (500–15,000 K) and pressure (100–5000 atm) that can dramatically alter the substrate's morphology and speed up chemical reactions. According to Yu et al. [112], the cellulase enzyme can access more cellulose fibers because cavitation moves the intermediate layer and ruptures the cell wall. Ultrasound irradiation may reduce the degree of polymerization of cellulose, according to Wang et al. [104] and Yu et al. [114]. According to Wang et al. [103] and Khanal et al. [45], a longer duration of high-power ultrasound can raise the pore volume, which in turn reduces the crystallinity of cellulose. Using a shorter reaction time and high-energy ultrasound (400 W), Kunaver et al. [55] showed how to extract valuable compounds from lignocellulosic wood waste by liquefying materials like spruce sawdust, wheat straw, particle boards, and plywood.

1.2.3 Mechanism Gamma-Ray Pre-treatment

Gamma rays (γ) are released into the atmosphere when radioisotopes like cobalt-60 or cesium-137 decay radioactively; a radioactive cesium isotope known as cesium-137 is created when uranium-235 and other fissionable isotopes undergo nuclear fission. Ionizing radiation of the cobalt-60 variety decays at a rate of around 1% per month and, as a result of its inherent instability, releases two gamma rays with an energy of 1.17 and 1.33 meV [18, 81]. These ionizing radiations change lignin and decrease cellulose crystallinity by penetrating biomass and transferring their energy to the atoms of lignocellulosic components [41, 85]. These alterations in structure are the outcome of free radical production, which facilitates the hydrolysis of polysaccharides and the lysis of cell walls by the scission of glycosidic linkages [11]. After experimenting with various dosages of Co-60 γ-radiation, Yang et al. [109] found that irradiating wheat straw at 500 kGy significantly altered the size of the biomass particles, leading to an improvement in glucose yield. A total of 90.3 mg/g of sugar (glucose and xylose) was produced by the irradiation straw [102]. When rapeseed straw was exposed to γ (^{60}Co) radiation, Zhang et al. [113] noted alterations in various physicochemical characteristics. Liu et al. [63] used γ-irradiation to evaluate the bioconversion efficiency of microcrystalline cellulose in terms of sugar yield and ethanol generation. Various chemicals produced by irradiation can eliminate their harmful effects during ethanol synthesis if the right amount of γ-irradiation is used. A specialized chamber can be used to conduct the irradiation system, limiting the biomass used and the accompanying safety measures [85].

1.2.4 Mechanism of Electron Beam Irradiation (EBI) Pre-treatment

An electron beam, a highly charged stream of ionizing radiation, is generated by a linear electron accelerator [65]. The cell wall polymers are destroyed when an accelerated electron stream is directed onto biomass, which destroys the polymers and triggers free radical production by transferring energy to the biomass atoms. It reduces the degree of polymerization and induces chain scission, which leads to de-crystallization [22, 66]. It does this by breaking the hydrogen connection between the cellulose chains. Bak proved that by introducing EBI into rice straw, its digestibility and bioconversion capacity were both enhanced. The formation of any inhibitory chemicals was also noted in his findings [10]. At 500 kGy, the EBI of Korean *Miscanthus sinensis* increased sugar production [111]. The de-polymerization of cellulose, hemicelluloses, lignin, and increased output of phenolics are effects of exposing sugar maple to EBI at 1000 kGy [69]. The pre-treatment of bamboo chips, which involved low electron beam power levels (0.5–50 kGy), significantly reduced the amount of cellulose while retaining the hemicelluloses [44]. The literature supports that lower-power EBI pre-treatment is cellulose-specific and that beam power is a critical EBI parameter [66, 69]. The main benefits of EBI pre-treatment include being an efficient, clean, and cost-effective procedure that requires minimal chemical or solvents and no changes to temperature or pressure [25].

1.2.5 Mechanism of Pulsed Electric Energy Pre-treatment

The application of pulsed electric energy (PEE) was used to treat a wide variety of biological materials [76, 101]. PEE pre-treatment involves applying non-thermal voltage pulses to biomass or bio-suspensions in a simple device with two electrodes. The pulse amplitude ranges from 0.1 to 100 kV/cm, and the treatment only takes nano to milliseconds. Because the membranes of exposed biomass undergo structural changes and disturbance, they become less semi-permeable, allowing hydrolyzing enzymes to enter and hydrolyze polysaccharides and lignin [50]. Switchgrass and wood chip samples were pretreated with PEE by Kumar et al. [54] using electric field intensities ranging from 1–10 kV/cm. Researchers examined the absorption of colored dye in PEE-treated substrates and discovered that tissue porosity increased, leading them to conclude that PEE may improve cellulose hydrolysis. Compared to other thermal or chemical treatments, PEE-assisted treatment has a low energy requirement and does not alter biological material in any way that would be undesired [29]. As a result, PEE has further applications beyond biorefineries, such as enhanced porosity of LCB [68] and high-temperature-based extraction of value-added compounds [15]. However, this method's development is still in its early stages.

1.3 Principles of Radiation and Biomass Interaction

Recent interest in radiation-based technologies has grown because of their potential to solve environmental and energy issues. Radiation-based biomass waste treatment and valorization show potential. Biomass waste—including agricultural, forestry, and organic municipal garbage—is an environmental issue and resource source. Radiation and biomass interaction principles are essential to understanding and using radiation-based technologies for biomass waste [59].

1.3.1 Radiation Basics

Basic radiation knowledge is needed to understand radiation and biomass interaction. Radiation is the energy released as electromagnetic waves or particles. It is divided into ionizing and non-ionizing radiation. Gamma, X, and high-energy ultraviolet (UV) rays can remove firmly bound electrons from atoms or molecules, forming ions [21]. de Moraes et al. [27] Used gamma radiation to pretreat cells, allowing lipid extraction and astaxanthin extraction from *Haematococcus pluvialis* biomass. The experimental design included two factors, which were cell pretreatment approaches: biomass with (chloroform: methanol) under ultrasound (BCMU), biomass with (chloroform: methanol) using γ radiation (BCMR), dried biomass under γ radiation (DBR). Chen et al. [20] studied treatment with gamma irradiation to break down Laminaria japonica material for dark fermentation hydrogen generation. We found that 20 kGy radiation enhanced hydrogen synthesis by 71.4%. Gamma irradiation at 10–30 kGy destroys macroalgae structure and increases soluble organic components including soluble chemical oxygen demand (SCOD), protein, and polysaccharides, making more hydrogenable organic compounds available. However, 30 kGy of irradiation may decrease hydrogen generation due to biodegradable organic compound destruction and affect microbial structure, as indicated by lactic acid buildup. Shvets et al. [90] conducted a study on the effects of microwave power on sawdust pyrolysis properties. The highest yield of flammable gases rose in direct proportion to the power input, and the time required to release these gases accounted for a sizable chunk of the overall pyrolysis duration. Increasing the power had a non-linear and scale-dependent effect on the yield of specific gases. The average CO and CO_2 production was almost unaffected by increasing the microwave power from 840 to 1760 W. Photocatalysis can promote spontaneous ($\Delta G < 0$) and non-spontaneous processes ($\Delta G > 0$). Photocatalysis is thought to entail hole oxidation and electron reduction processes (Fig. 1.1). A semiconductor is typically chosen as the photocatalyst. When light energy exceeds the photocatalyst band gap, an electron (e−) is excited from the valence band (VB) to the conduction band (CB), creating a hole (h+). Redox reactions are effective when photogenerated electrons and holes activate them, while bulk recombination is nonproductive and dissipates heat or emits light. In certain reaction conditions, electron–hole pairs can contribute to the generation of reactive oxygen species (ROS), including superoxide radicals (O2·−), hydroperoxyl radicals (HOO·), and hydroxyl radicals (·OH) [83]. Figure 1.1 shows photocatalyst band gap topologies and redox couple potentials in water (pH = 7). Photocatalysts like TiO_2, $SiTiO_3$, ZnO, and WO_3 can produce ·OH radicals due to their higher VB position (+2.27), while CdS, CdSe, MoS_2, and C_3N_4 cannot. Engineering band gap structure can improve selectivity and activity [49].

1.3.2 Biomass Composition

Organic components like cellulose, hemicellulose, lignin, proteins, lipids, and carbohydrates compose biomass waste. Chemical structure and radiation susceptibility differ among these components. Biomass's physical and chemical qualities depend on its main components: cellulose, hemicellulose, and lignin. Plant cell walls are strengthened by cellulose, a linear glucose polymer. Hemicellulose, a branching polymer of sugar units, establishes cellulose fibers. Complex aromatic polymer lignin gives plant tissues stiffness and protection [51, 84].

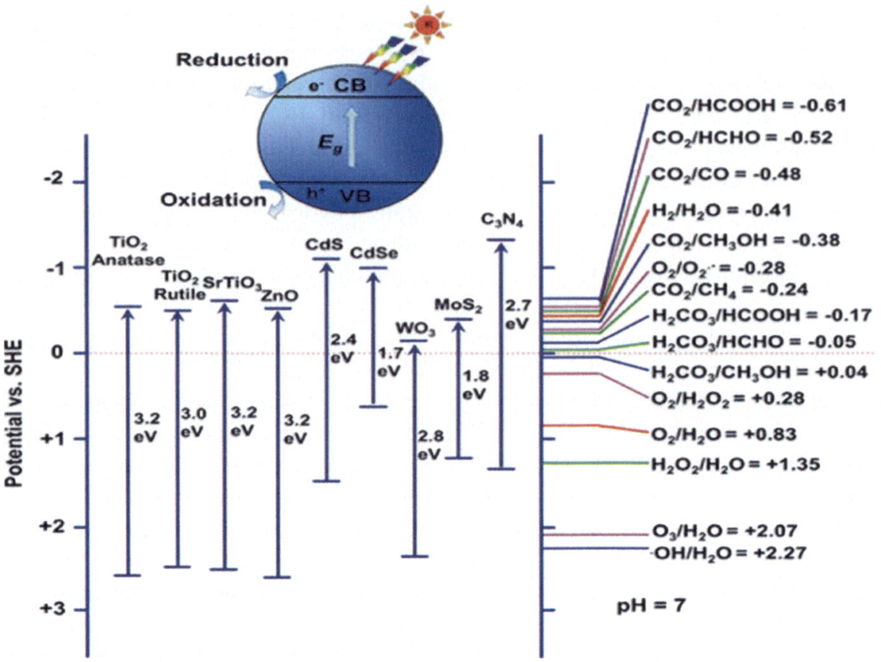

Fig. 1.1 Potentials for various redox couples in water (pH 7) and the band-edge positions of semiconductor photocatalysts (Kou et al. [49], *Creative Commons License*)

Various LCB components, including lignin, heteropolymeric hemicellulose, and homopolymeric cellulose, can be seen in Fig. 1.2. The complicated structure of LCF makes it resistant to enzymatic hydrolysis, making biofuel conversion difficult [12]. Pre-treatment disorganizes lignin and breaks down LCB to overcome its intrinsic recalcitrance. Thus, pre-treatment increases hydrolyzing enzyme accessibility to cellulosic and hemicellulosic material, generating simple sugars fermented to make biofuels [75, 95]. In addition to constructing economically significant compounds, the hydrolysate of polyphenolic lignin and other substances is transformed into chemicals with additional value [106]. The rationale behind developing a wide range of pre-treatment tactics, some of which include physical, chemical, physicochemical, and biological techniques. From small-scale pilot projects to large-scale biofuel production, this article delves into various biomass pre-treatment processes.

1.3.3 Radiation Interactions with Biomass

The biomass waste undergoes several transformations due to radiation's interactions with it. Two main types of radiation interactions with biomass are direct and indirect. Radiation causes chemical and physical changes in biomass components

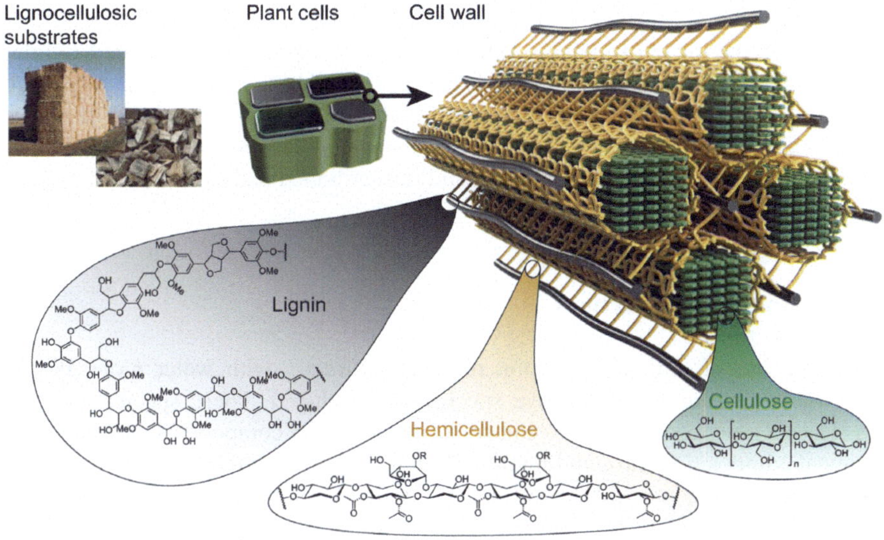

Fig. 1.2 Composition of lignin, hemicelluloses, and cellulose (Brethauer et al. [17]. *Creative commons license*)

through direct interactions. In indirect interactions, water molecules in the biomass are exposed to radiation, which causes the creation of reactive species. These species then react with the components of the biomass [100]. Cellular water gamma radiation quickly produces hydroxyl radical (•OH) and ionized water (H_2O+), reactive oxygen species (ROS), and the less studied reductants hydrogen radical (H·) and hydrated electrons (eaq−). In a span of one picosecond (tenths to twelve seconds), the intermediate reactive oxygen species (ROS) products of infrared include superoxide ($O_2\cdot-$) and hydrogen peroxide (H_2O_2) [87]. Later on, these reactive species' intracellular stoichiometry is impacted by chemical cascades, which in turn produce more compounds that harm cells. Take metal catalysis by intracellular ferrous and/or cuprous ions as an example. It turns $O_2\cdot-$ and H_2O_2 into more ·OH [23].

1.3.3.1 Direct Interactions

In direct contact, ionizing radiation breaks biomass chemical bonds, forming free radicals. Highly reactive free radicals with unpaired electrons can start chemical reactions. Biomass free radicals can degrade complex organic molecules into simpler ones. A treatment known as radiolysis is being performed. Ionizing radiation not only creates differences in the structure of biomass but also induces the components of biomass to crosslink, which results in the formation of new chemical linkages [57]. Direct ionization and the reactivity of high-energy species resulting from water radiolysis (indirect effects of ionization) alter cellular macromolecules, impacting around 2000 primary ionization events (Ward et al. 1994). From chemical processes that happen as quickly as 0.01 ps following IR to significant cellular impacts that take place in the range of minutes to hours, the time aspects of IR-induced cellular damage cover a wide spectrum [87]. When S–H, O–H, N–H, and C–H bonds are broken, direct radiation damage begins within the 10^{-14}–10^{-12} s range. Within 1 ps (10^{-12} s), the radiolytic byproducts of water start to cause extensive biomolecular damage, including the depletion of thiol and further bond breakage, such as C–C and C–N bonds [87].

1.3.3.2 Indirect Interactions

In the process of ionizing radiation reacting indirectly with water molecules in biomass, two instances of reactive species are produced. These species are OH· and H_2O_2. These reactive entities can combine with other components and alter their chemical composition through biomass. Lignin, cellulose, and hemicellulose are all susceptible to being broken down by hydroxyl radicals, particularly reactive hydroxyl radicals. Radiation's indirect interactions with water molecules are the cause of the composting of waste from biomass [57].

1.4 Radiation Effects on Biomass Waste

The radiation that is effective on waste biomass could be utilized in a variety of different ways. Through the use of radio waves, the complex organic chemicals that are found in waste biomass can be broken down into their more fundamental forms. Additionally, radiation can alter the chemical components that are found in various regions of biomass, which can result in the creation of new products that are of more excellent value. For the production of biofuels and biochemicals, it is possible to employ radiation to decompose lignin into aromatic compounds [74]. The use of radiation can improve the treatability of biomass components. Biomass becomes more amenable to enzymatic hydrolysis, a crucial step in biofuel synthesis, when exposed to radiation [52]. Radiation impacts the intricate structure of biomass in a way that enhances biofuel generation and enzyme breakdown [5]. Emitted high-energy radiation transmits energy within the materials by traveling into the biomass and biofiber components. Depolymerization of cellulose and the formation of a carbonyl group as a consequence of oxidation are two of the many chemical and thermal reactions that begin in the biomass when it is heated. There have been reports of crosslinking of biomass components when subjected to irradiation beams [65]. The mechanical strength of biomass that was subjected to radiation was shown to be reduced. Cellulose may become less crystalline and more amorphous if disrupted by the hydrogen bond between its chains [44] (Fig. 1.3).

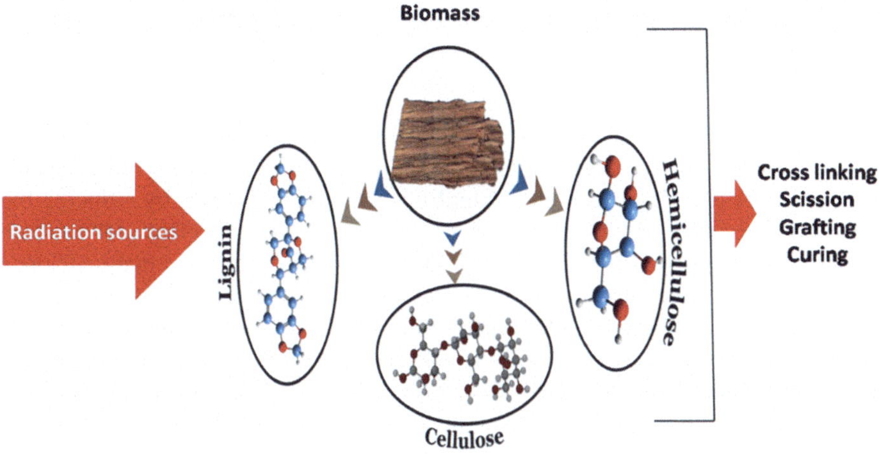

Fig. 1.3 Biomass irradiation

1.4.1 Radiation-Assisted Pyrolysis of Biomass Waste

Radiation pyrolysis is promising for biomass waste treatment and valorization. High-temperature pyrolysis decomposes organic molecules without oxygen. This method turns biomass waste into usable products more effectively with radiation. Radiation pyrolysis produces biofuels, that can be liquid biofuels that replace fossil fuels like bio-oil or pyrolysis oil. Bio-oil may be processed into petrol, diesel, and jet fuel due to its high energy content. It can generate heat and power. One of the many valuable byproducts of radiation pyrolysis is biochar. The acronym "syngas" stands for "synthesis gas," a class of gases that includes carbon monoxide, hydrogen, and others. Ammonia, synthetic natural gas, and methanol are all products of its synthesis. Syngas can be converted into heat and power through gasification or combustion. The process of pyrolysis generates biochar, bio-oil, and syngas by thermally breaking down biomass waste. By heating waste biomass in an oxygen-free environment, the method decomposes organic materials into smaller molecules. An intriguing new way to increase efficiency is radiation-assisted pyrolysis. Solar pyrolysis of rice and poultry husks generates highly calorific gases, which can be used as fuel for motors and electricity for buildings [37, 86, 97, 105]. Developed a novel pyrolysis method using focused solar radiation to create biochar and investigated its nitrate adsorption capability. Biochar from biomass pyrolysis improves soil quality, reduces greenhouse gas emissions, and metal toxicity, and increases phosphorus and potassium usage [58].

1.4.2 Radiation-Based Degradation of Biomass Waste and Its Mechanisms

Radiation has demonstrated promising results in the breakdown of waste biomass. Biomass waste is broken down into its parts using gamma rays, electron beams, or microwaves [97]. This technique simplifies waste management by reducing the volume and toxicity of biomass waste [2, 86]. Radiation can break down waste biomass in many ways. Energy from high-energy radiation destroys chemical bonds in organic waste molecules, releasing free radicals. These reactive free radicals can start a chain reaction that degrades organic compounds. Carbon splintering is necessary for radiation breakdown. Organic materials lose dimensionality when high-energy radiation dissolves carbon–carbon bonds. Because of complex polymers like cellulose and lignin, biomass waste is ideal for decomposition. Radioactive decay can also occur through the oxidation of organic molecules. Oxygen radicals and other reactive oxygen species can be produced by high-energy radiation, these radicals can then combine with organic substances to cause their oxidation. Degradation of organic molecules into more straightforward and less harmful chemicals can occur due to this oxidation process [60]. Many variables affect biomass waste radiation degradation. The degree of degradation depends on the waste's radiation dose. Increasing radiation exposure usually increases organic molecule breakdown. Optimizing the radiation dose prevents harmful byproducts and severe degradation. Also, radiation type affects the deterioration rate. Remediating biomass waste with gamma rays and electron beams is common. Gamma rays can efficiently irradiate massive amounts of garbage due to their broad penetration. Electron beams, on the other hand, may expose waste to more radiation but penetrate less. The properties and decomposition level of biomass waste dictate the radiation type. Radioactive decay may be affected by biomass waste composition. Lignocellulosic, hemicellulose, and cellulose concentrations vary among biomass waste kinds. Due to their radiation resistance and complexity, lignin can complicate breakdown. Thus, biomass waste composition must be considered while planning radiation degradation [33, 47]. Reactive organisms from biomass waste can break down organic contaminants and reduce heavy metals when irradiated. This procedure removes contaminants and restores contaminated places. Degradation and carbon and nitrogen elimination were studied using (d, p) reactions. Using PIXE-PIGME, the compositions of biomass, feed, and processed materials were investigated concurrently. Protons at 2.5 meV were measured using 25 nA currents, respectively, in the PIXE-PIGME experiments. At the same time, X-ray and γ-ray data were archived. Comparisons using calibration standards yielded absolute concentrations. The biomass concentrates Cl, Ti, Ni, and Fe, removing 54–58% carbon and 45–50% nitrogen from wastes [31]. Bioplastics can be biodegraded in less than 60 days using gamma radiation and ethylene oxide while retaining all of their mechanical properties. Solar pyrolysis can convert various waste biomass materials, such as straw, sewage sludge, and waste wood, into bio-oil, gas, and char. One potential long-term usage for the high-quality bio-oil made by pyrolyzing wood waste is as a solid fuel [92].

1.4.3 Radiation-Assisted Gasification of Biomass Waste

Gasification is a promising method for turning biomass waste into syngas, charcoal, and bio-oil. It involves biomass partial oxidation at high temperatures with regulated oxygen or steam. The process produces carbon monoxide (CO) and hydrogen (H_2), which can be utilized as fuel. The unique use of ionizing radiation in radiation-assisted gasification improves biomass waste conversion efficiency. This method irradiates biomass feedstock before or during gasification with microwaves, gamma, electrons, or X-rays [2].

1.4.4 Radiation-Assisted Combustion of Biomass Waste

Radiation-assisted combustion exploits biomass waste combustion energy in a promising way. Radiation improves biomass waste combustion and energy generation. This method uses radiation to manage biomass waste efficiently and generate renewable energy. Solar gasification decreases flame damage and increases heat load, energy, biomass conversion, and waste management [91]. Recently developed biomass pellets can replace fossil fuels in electricity generation, they are efficiently heated and transported. Solid biomass fuel must be evaluated quickly to maximize combustion efficiency. Lu et al. [64] conducted a study on 66 wood pellet samples to determine the gross calorific value, carbon content, volatile matter content, and ash content all at once using laser-induced breakdown spectroscopy (LIBS) in combination with partial least squares (PLS). The PLS model that utilized spectra-integrated baseline correction and Z-score standardization yielded the most favorable quantitative analysis outcomes. Liu et al. [61] distinguished Biomass pellets using chemometrics techniques and laser-induced breakdown spectroscopy (LIBS). Before further analysis using LIBS spectra, the components of lignocellulosic material were identified. To statistically differentiate biomass pellets, we used principal component analysis (PCA), partial least squares discrimination analysis (PLS-DA), support vector machines (SVM), radial basis function neural network (RBFNN), and extreme learning machine (ELM).

1.4.5 Radiation-Assisted Nanocomposite Formation

A combination of nanoparticles and radiation is required to produce radiation-assisted nanocomposite bioplastics [46]. The capacity of radiation therapy to alter the surface characteristics of waste biomass enhances the interactions and distribution of nanoparticles. This method produces bioplastics with improved mechanical, thermal, and barrier properties. Bioplastics with nanoparticles like clay, graphene, or cellulose nanocrystals are more substantial, stiffer, and flame-resistant [9]. Radiation-assisted nanocomposite production may enable biomass waste-based high-performance bioplastics. Here, laser ablation in liquid (LAL), a promising nanoparticle production method, is employed to coat calcium lignosulfonate (CLS), an abundant biowaste, with palladium nanoparticles. The Pd/CLS nanocomposite, synthesized by stirring, is used for hydrogen storage and environmental cleanup research. Its 5.8 C g^{-1} hydrogen storage capacity proved that Pd NPs are hydrogen adsorption sites [71]. Radiation-assisted bioplastics can also help the circular economy by reusing biomass waste and minimizing the environmental impact of plastic waste. In radiation on starch-based bioplastics, the correct energy range of ionizing radiation improves their mechanical and barrier properties, making them more practical [78].

1.4.6 Radiation-Assisted Production of Biochemicals from Biomass Waste

Pharmaceuticals, agriculture, and materials science use biochemicals, a vast collection of substances obtained from biological sources. Biochemicals from biomass waste are sustainably becoming a more popular alternative to fossil fuels and petrochemicals. Biochemicals can be produced from biomass waste using radiation-based methods, simplifying complicated biomass molecules into more valuable compounds. Biomass conversion techniques can yield chemicals and fuels; pyrolysis is the most effective method for making char and liquid products [28, 37, 97].

1.4.7 Radiation-Induced Biomass De-Polymerization

Biochemistry from biomass waste requires de-polymerization, which simplifies biomass molecules. Gamma rays or electron beams destroy biomass molecules' chemical bonds in radiation-induced biomass de-polymerization. Agricultural and forestry waste can be depolymerized into fermentable sugars by radiation. These sugars make biofuels, bioplastics, and other valuable compounds [32, 94]. Radiation can depolymerize food, algae, and biomass waste into organic acids, amino acids, and enzymes [26, 43]. Biochar, a carbon-rich soil additive or activated carbon precursor, can be made from biomass waste via radiation-induced de-polymerization. Accelerating biomass waste de-polymerization by radiation requires catalysts or specific radiation doses. Microwave-assisted peracetic acid pre-treatment delignifies biomass best, simplifying decomposition and increasing product yield [6]. Microwave selectivity against more polar compounds causes unequal heating of the material interior, causing microexplosions that degrade lignocellulose. Radiation generates internal heat, causing molecular explosions. Microwaves' non-thermal effect causes polar bond vibrations, which disrupts them and accelerates chemical and physical processes. High radiation energy affects the lignocellulose biomass's structure, increases its specific surface, reduces cellulose polymerization and crystallinity, and accelerates hemicellulose and lignin hydrolysis [115].

1.4.8 Radiation-Assisted Catalytic Conversion

The flexible catalytic conversion process produces biochemicals from biomass waste. Radiation boosts catalyst activity and selectivity, making biomass molecule conversion into biochemicals more efficient and selective [37, 97]. Radiation can make catalysts like metal nanoparticles or zeolites more active and selective. Free radicals and excited states from radiation can react with biomass molecules and convert them into biochemicals. Various types of biomass waste, such as lignocellulosic biomass, food waste, and algae, can be converted using radiation-assisted catalytic conversion [108]. Many other types of biochemicals, including platform chemicals, fine chemicals, and specialized chemicals, can be made using this method [108]. These biochemicals find usage in fields as diverse as materials science, cosmetics, and pharmaceuticals [48, 86]. An interesting strategy for rapid and effective biomass conversion is using microwaves to aid in the catalytic liquefaction of cotton stalks with catalysts enhanced with iron, specifically HZSM-5. It generates aromatic compounds and significantly improves conversion efficiency [48]. Solar photocatalytic oxidation is an effective method for eliminating ionic liquids, which ultimately results in wastewater that is environmentally friendly and more biodegradable [34].

1.4.9 Applications of Radiation in Biomass Waste Pre-treatment

Radioactive biomass waste treatment has several applications in various fields. These technologies use gamma, X-rays, electron beams, and microwaves to treat and turn biomass waste into valuable goods. Energy generation, valorization, environmental remediation, food production, and agricultural output are all achieved by the use of radiation in the context of treating biomass waste [34, 35, 78, 97]. Yang et al. [110] reported that Hydrogen fermentation of grass waste was initially improved using a combined ionizing radiation-acid pretreatment procedure, which proved to have a synergistic effect. Maceiras et al. [67] analyzed the efficacy of alkaline pretreatment of algal waste with microwave assistance. The product was hydrolyzed with sulfuric acid after pretreatment. Researchers looked at how different values of microwave power, irradiation period, solid–liquid ratio, and NaOH concentration affected the results. With *Saccharomyces Cerevisiae*, the fermentable sugars were transformed into cellulosic bioethanol under optimal circumstances. Huang et al. [40] suggested an innovative approach to manufacture starch-phenolic complexes using electron beam irradiation (EBI) as a pretreatment. The native and EBI-pretreated TB starch were complexed with four structurally identical monomeric phenolic acids found naturally in Tartary buckwheat (TB) seeds: gallic acid (GA), syringic acid (SA), and vanillic acid (VA). Wu et al. [107] examined the physicochemical composition and enzymatic efficiency of rice straw, maize stalk, and rapeseed straw that had been exposed to gamma illumination. Milling energy consumption was reduced by 64.40%, 66.62%, and 71.68% for rice straw, corn stalk, and rapeseed straw irradiated with 600 kGy, respectively, compared to unirradiated samples, and particle size was significantly smaller as the absorbed dosage of irradiation rose.

1.5 Conclusion

It is necessary to have a comprehensive understanding of the interaction between radiation and biomass to incorporate radiation-based technology into applications involving waste from biomass. Radiation can destroy, modify, and make more accessible biomass than it would otherwise be. Utilize these effects for the treatment of waste biomass, the valorization of waste, and the rehabilitation of environmental damage. Technology that is based on radiation has the potential to improve our capacity to manage waste biomass and put it to use for energy and environmental purposes. Radiation degradation appears to be a promising strategy when it comes to the treatment of waste biomass. The volume and toxicity of biomass waste are both reduced in an efficient and environmentally responsible manner by this method, which makes the waste appropriate for use in the generation of energy, the cleanup of the environment, and the valorization of waste. Radiation degradation is complicated, but scientists are improving it and solving financial and legal issues to make it extensively employed in future technology. Radiation-assisted hydrothermal liquefaction can turn biomass into chemicals. Radiation and heat treatments improve liquefaction efficiency and yield, increasing bio-oil yields, feedstock flexibility, and waste valorization. To make this technology broadly available, we must address radiation safety, process optimization, and financial sustainability. Radiation-assisted hydrothermal liquefaction may make biomass waste conversion for electricity and chemical manufacture more sustainable with more research. Radiation-assisted bioplastic synthesis from biomass waste is efficient and green. Radiation changes biomass waste's chemical structure, allowing bioplastics with desired qualities. Bioplastics from biomass waste aid the environment and circular economy. Radiation-assisted bioplastics production requires more study and development to overcome its challenges and fulfill its potential in future technologies. Biofuels from biomass waste using radiation are an intriguing renewable energy solution.

Conflicts of Interest The authors declare that there is no conflict of interest.

Declaration on Non-usage of the AI Tools The authors clarify that they did not use artificial intelligence tools in preparing this chapter.

References

1. Abdel Rahman RO, Hung YT (2019) Application of ionizing radiation in wastewater treatment: an overview. Water 12(1):19. https://doi.org/10.3390/w12010019
2. Al Gharib S, Mostafavi M, Belloni J (2022) Radiation-assisted hydrolysis of lignocellulosic biomass. Mechanistic study. Waste Biomass Valorization 14(4):1113–1122. https://doi.org/10.1007/s12649-022-01933-4
3. Amini N, Haritos VS, Tanksale A (2018) Microwave-assisted pretreatment of eucalyptus sawdust enhances enzymatic saccharification and maximizes fermentable sugar yield. Renew Energy 127:653–660. https://doi.org/10.1016/j.renene.2018.05.001
4. Arumugam A, Malolan VV, Ponnusami V (2020) Contemporary pretreatment strategies for bioethanol production from corncobs: a comprehensive review. Waste Biomass Valorization 12(2):577–612. https://doi.org/10.1007/s12649-020-00983-w
5. Astuti W, Sulistyaningsih T, Prastiyanto D, Rusiyanto L, Riayanti FI, Wulandari DA (2023) Influence of lignocellulosic composition in biomass waste on the microstructure and dye adsorption characteristics of microwave-assisted $ZnCl_2$ activated carbon. Biomass Convers Biorefinery. https://doi.org/10.1007/s13399-023-04281-y
6. Azelee NIW, Adnan SAM, Manas NHA, Dailin DJ, Ramli ANM, Illias RM (2019) Assessment of microwave-assisted pretreatments for enhancing pineapple waste delignification. In: Proceedings of the 2nd international conference on biosciences and medical engineering (ICBME2019): towards innovative research and cross-disciplinary collaborations. https://doi.org/10.1063/1.5125507
7. Azuma J, Higashino J, Isaka M, Koshijima T (1985) Microwave irradiation of lignocellulosic materials: IV. Enhancement of enzymatic susceptibility of microwave irradiated softwoods. Wood Res Tech Notes 71:13–24
8. Azuma J, Tanaka F, Koshijima T (1984) Enhancement of enzymatic susceptibility of lignocellulosic wastes [sugar cane bagasse, rice straw and rice hulls] by microwave irradiation. J Ferment Technol 62:377–384
9. Bahrami M, Abenojar J, Martínez MN (2020) Recent progress in hybrid biocomposites: mechanical properties, water absorption, and flame retardancy. Materials 13(22):5145. https://doi.org/10.3390/ma13225145
10. Bak JS (2014) Electron beam irradiation enhances the digestibility and fermentation yield of water-soaked lignocellulosic biomass. Biotechnol Rep 4:30–33. https://doi.org/10.1016/j.btre.2014.07.006
11. Betiku E, Adetunji OA, Ojumu TV, Solomon BO (2009) A comparative study of the hydrolysis of gamma irradiated lignocelluloses. Braz J Chem Eng 26(2):251–255. https://doi.org/10.1590/s0104-66322009000200002
12. Bhardwaj N, Kumar B, Verma P (2019) A detailed overview of xylanases: an emerging biomolecule for current and future prospective. Bioresour Bioprocess 6:40. https://doi.org/10.1186/s40643-019-0276-2
13. Bhardwaj N, Agrawal K, Kumar B, Verma P (2021) Role of enzymes in deconstruction of waste biomass for sustainable generation of value-added products. In: Thatoi H, Mohapatra S, Das SK (eds) Bioprospecting of enzymes in industry, healthcare and sustainable environment, pp 219–250. Springer, Singapore. https://doi.org/10.1007/978-981-33-4195-1_1
14. Bhardwaj N, Kumar B, Verma P (2022) Microwave-assisted pretreatment using alkali metal salt in combination with orthophosphoric acid for generation of enhanced sugar and bioethanol. Biomass Convers Biorefinery 12:1069–1076. https://doi.org/10.1007/s13399-020-00640-1
15. Bouras M, Grimi N, Bals O, Vorobiev E (2016) Impact of pulsed electric fields on polyphenols extraction from Norway spruce bark. Ind Crops Prod 80:50–58. https://doi.org/10.1016/j.indcrop.2015.10.051
16. Bouras M, Grimi N, Bals O, Girard F, Vorobiev E (2013) Impact of electrical treatment on the extraction of polyphenol from Norway spruce (*Picea abies*). In: 10th International bioelectrics symposium (BIOELECTRICS 2013). Karlsruhe Institute of Technology (KIT), Karlsruhe

17. Brethauer S, Shahab RL, Studer MH (2020) Impacts of biofilms on the conversion of cellulose. Appl Microbiol Biotechnol 104(12):5201–5212. https://doi.org/10.1007/s00253-020-10595-y
18. Brewer I, Negro M, Striebig N, Kierans C, Caputo R, Leys R, Perkins J (2021) Developing the future of gamma-ray astrophysics with monolithic silicon pixels. Nucl Instrum Methods Phys Res Sect A 1019:165795. https://doi.org/10.1016/j.nima.2021.165795
19. Chandrapala J, Oliver C, Kentish S, Ashokkumar M (2012) Ultrasonics in food processing. Ultrason Sonochem 19(5):975–983. https://doi.org/10.1016/j.ultsonch.2012.01.010
20. Chen Y, Yin Y, Wang J (2022) Biohydrogen production using macroalgal biomass of *Laminaria japonica* pretreated by gamma irradiation as substrate. Fuel 309:122179. https://doi.org/10.1016/j.fuel.2021.122179
21. Chen Z, Chen H, Xu Y, Hu M, Hu Z, Wang J, Pan Z (2023) Reactor for biomass conversion and waste treatment in supercritical water: a review. Renew Sustain Energy Rev 171:113031. https://doi.org/10.1016/j.rser.2022.113031
22. Cheng K, Barber VA, Driscoll MS, Winter WT, Stipanovic AJ (2013) Reducing woody biomass recalcitrance by electron beams, biodelignification and hot-water extraction. J Bioprocess Eng Biorefinery 2(2):143–152. https://doi.org/10.1166/jbeb.2013.1048
23. Chiu SM, Xue LY, Friedman LR, Oleinick NL (1993) Copper ion-mediated sensitization of nuclear matrix attachment sites to ionizing radiation. Biochemistry 32:6214–6219
24. Chmielewski AG (2017) Future developments in radiation processing, 1st edn. In: Applications of ionizing radiation in materials processing. Institute of Nuclear Chemistry and Technology, pp 501–516
25. Clough R (2001) High-energy radiation and polymers: a review of commercial processes and emerging applications. Nucl Instrum Methods Phys Res Sect B 185(1–4):8–33. https://doi.org/10.1016/s0168-583x(01)00966-1
26. Das H, Singh SK (2004) Useful byproducts from cellulosic wastes of agriculture and food industry—a critical appraisal. Crit Rev Food Sci Nutr 44(2):77–89. https://doi.org/10.1080/10408690490424630
27. de Moraes LB, Malafaia CB, da Silva TD, Mota GC, Marinho YF, de Albuquerque Melo AM, de Oliveira AF, de Macêdo Dantas DM, Olivera Gálvez A, de Souza Bezerra R (2022) Gamma Radiation as a Pretreatment for Co-extraction of Lipids and Astaxanthin in *Haematococcus pluvialis*. BioEnergy Res 16(3):1841–1850. https://doi.org/10.1007/s12155-022-10531-0
28. Demirbaş A (2001) Biomass resource facilities and biomass conversion processing for fuels and chemicals. Energy Convers Manage 42(11):1357–1378. https://doi.org/10.1016/s0196-8904(00)00137-0
29. Donsì F, Ferrari G, Pataro G (2010) Applications of pulsed electric field treatments for the enhancement of mass transfer from vegetable tissue. Food Eng Rev 2(2):109–130. https://doi.org/10.1007/s12393-010-9015-3
30. Eskicioglu C, Terzian N, Kennedy K, Droste R, Hamoda M (2007) A thermal microwave effects for enhancing digestibility of waste activated sludge. Water Res 41(11):2457–2466. https://doi.org/10.1016/j.watres.2007.03.008
31. Fakhouri HG, Dyson NA (1989) (d, p) Reactions and the combination of PIXE-PIGME techniques for the analysis of biomass samples. J Radioanal Nucl Chem Art 129(1):163–170. https://doi.org/10.1007/bf02037579
32. Gao S, Song W, Guo M (2020) The integral role of bioproducts in the growing bioeconomy. Ind Biotechnol 16(1):13–25. https://doi.org/10.1089/ind.2019.0033
33. Gayathiri M, Pulingam T, Lee K, Sudesh K (2022) Activated carbon from biomass waste precursors: factors affecting production and adsorption mechanism. Chemosphere 294:133764. https://doi.org/10.1016/j.chemosphere.2022.133764
34. Gomez Herrero E, Tobajas M, Rodriguez J, Mohedano A (2020) Ionic liquids removal by sequential photocatalytic and biological oxidation. J Chem Technol Biotechnol 95(7):1926–1935. https://doi.org/10.1002/jctb.6306
35. González ME, Lunati C, Floccari M, Salmoral EM (2009) Effects of sterilizing agents on the biodegradation of a bioplastic material. Int J Polym Mater 58(3):129–140. https://doi.org/10.1080/00914030802583726

36. Güven O (2017) Established and emerging applications of radiation-induced graft polymerization. Appl Ioniz Radiat Mater Process 2:355–373
37. Hamzah HT, Sridevi V, Surya DV, Ramesh P, Rao CS, Palla S, Abdullah TA (2024) Synergistic effects and product yields in microwave-assisted in-situ co-pyrolysis of rice straw and paraffin wax. Process Saf Environ Prot 182:45–55. https://doi.org/10.1016/j.psep.2023.11.048
38. Hassan SS, Williams GA, Jaiswal AK (2018) Emerging technologies for the pretreatment of lignocellulosic biomass. Biores Technol 262:310–318. https://doi.org/10.1016/j.biortech.2018.04.099
39. Hong SM, Park JK, Lee Y (2004) Mechanisms of microwave irradiation involved in the destruction of fecal coliforms from biosolids. Water Res 38(6):1615–1625. https://doi.org/10.1016/j.watres.2003.12.011
40. Huang Y, Liu Q, Yu J, Zhang Q, He C, Han L, Wang M (2023) Electron beam irradiation pretreatment enhances the formation of granular starch-phenolics complexes. Food Res Int 163:112288. https://doi.org/10.1016/j.foodres.2022.112288
41. Hyun Hong S, Taek Lee J, Lee S, Gon Wi S, Ju Cho E, Singh S, Yeoup Chung B (2014) Improved enzymatic hydrolysis of wheat straw by combined use of gamma ray and dilute acid for bioethanol production. Radiat Phys Chem 94:231–235. https://doi.org/10.1016/j.radphyschem.2013.05.056
42. Janositz A, Semrau J, Knorr D (2011) Impact of PEF treatment on quality parameters of white asparagus (*Asparagus officinalis* L.). Innov Food Sci & Emerg Technol 12(3):269–274. https://doi.org/10.1016/j.ifset.2011.02.003
43. Kant Bhatia S, Ahuja V, Chandel N, Gurav R, Kant Bhatia R, Govarthanan M, Yang YH (2022) Advances in algal biomass pretreatment and its valorisation into biochemical and bioenergy by the microbial processes. Biores Technol 358:127437. https://doi.org/10.1016/j.biortech.2022.127437
44. Kassim MA, Khalil HA, Serri NA, Kassim MHM, Syakir MI, Aprila NS, Dungani R (2016) Irradiation pretreatment of tropical biomass and biofiber for biofuel production. Radiat Eff Mater. https://doi.org/10.5772/62728
45. Khanal SK, Takara D, Nitayavardhana S, Lamsal BP, Shrestha P (2010) 14 Ultrasound applications in biorenewables for enhanced bioenergy and biofuel production. Green Chem Environ Sustain 303.
46. Kharissova OV, Kharisov BI, Oliva González CM, Méndez YP, López I (2019) Greener synthesis of chemical compounds and materials. R Soc Open Sci 6(11):191378. https://doi.org/10.1098/rsos.191378
47. Koh LM, Khor SM (2022) Biodegradation process: basics, factors affecting, and industrial applications. Handb Biodegrad Mater 1–39. https://doi.org/10.1007/978-3-030-83783-9_66-1
48. Kong X, Li X, Wu S, Zhang X, Liu J (2016) Efficient conversion of cotton stalks over a Fe modified HZSM-5 catalyst under microwave irradiation. RSC Adv 6(34):28532–28537. https://doi.org/10.1039/c5ra26918k
49. Kou J, Lu C, Wang J, Chen Y, Xu Z, Varma RS (2017) Selectivity enhancement in heterogeneous photocatalytic transformations. Chem Rev 117(3):1445–1514. https://doi.org/10.1021/acs.chemrev.6b00396
50. Kumar AK, Sharma S (2017) Recent updates on different methods of pretreatment of lignocellulosic feedstocks: a review. Bioresour Bioprocess 4(1). https://doi.org/10.1186/s40643-017-0137-9
51. Kumar B, Bhardwaj N, Verma P (2020) Microwave assisted transition metal salt and orthophosphoric acid pretreatment systems: generation of bioethanol and xylo-oligosaccharides. Renew Energy 158:574–584. https://doi.org/10.1016/j.renene.2020.05.006
52. Kumar B, Bhardwaj N, Agrawal K, Chaturvedi V, Verma P (2020) Current perspective on pretreatment technologies using lignocellulosic biomass: an emerging biorefinery concept. Fuel Process Technol 199:106244. https://doi.org/10.1016/j.fuproc.2019.106244
53. Kumar B, Bhardwaj N, Verma P (2019) Pretreatment of rice straw using microwave assisted $FeCl_3$-H_3PO_4 system for ethanol and oligosaccharides generation. Bioresour Technol Rep 7(2019):100295. https://doi.org/10.1016/j.biteb.2019.100295

54. Kumar P, Barrett DM, Delwiche MJ, Stroeve P (2011) Pulsed electric field pretreatment of switchgrass and wood chip species for biofuel production. Ind Eng Chem Res 50(19):10996–11001. https://doi.org/10.1021/ie200555u
55. Kunaver M, Jasiukaitytė E, Čuk N (2012) Ultrasonically assisted liquefaction of lignocellulosic materials. Biores Technol 103(1):360–366. https://doi.org/10.1016/j.biortech.2011.09.051
56. Leroy C, Rancoita PG (2014) Principles of radiation interaction in matter and detection. https://doi.org/10.1142/9167
57. Li J, Han Z, Surapipith V, Fan W, Thongboonchoo N, Wu J, Zhang R (2022) Direct and indirect effects and feedbacks of biomass burning aerosols over Mainland Southeast Asia and South China in springtime. Sci Total Environ 842:156949. https://doi.org/10.1016/j.scitotenv.2022.156949
58. Li S, Galoustian T, Trejo H (2023) Biochar pyrolyzed with concentrated solar radiation for enhanced nitrate adsorption. J Anal Appl Pyrol 174:106131. https://doi.org/10.1016/j.jaap.2023.106131
59. Liauw SL, Connell PP, Weichselbaum RR (2013) New paradigms and future challenges in radiation oncology: an update of biological targets and technology. Sci Transl Med 5(173). https://doi.org/10.1126/scitranslmed.3005148
60. List R, Gonzalez-Lopez L, Ashfaq A, Zaouak A, Driscoll M, Al-Sheikhly M (2023) On the mechanism of the ionizing radiation-induced degradation and recycling of cellulose. Polymers 15(23):4483. https://doi.org/10.3390/polym15234483
61. Liu X, Feng X, He Y (2019) Rapid discrimination of the categories of the biomass pellets using laser-induced breakdown spectroscopy. Renew Energy 143:176–182. https://doi.org/10.1016/j.renene.2019.04.137
62. Liu Y, Guo L, Wang L, Zhan W, Zhou H (2017) Irradiation pretreatment facilitates the achievement of high total sugars concentration from lignocellulose biomass. Biores Technol 232:270–277. https://doi.org/10.1016/j.biortech.2017.01.061
63. Liu Y, Zhou H, Wang S, Wang K, Su X (2015) Comparison of γ-irradiation with other pretreatments followed with simultaneous saccharification and fermentation on bioconversion of microcrystalline cellulose for bioethanol production. Biores Technol 182:289–295. https://doi.org/10.1016/j.biortech.2015.02.009
64. Lu Z, Chen X, Yao S, Qin H, Zhang L, Yao X, Lu J (2019) Feasibility study of gross calorific value, carbon content, volatile matter content and ash content of solid biomass fuel using laser-induced breakdown spectroscopy. Fuel 258:116150. https://doi.org/10.1016/j.fuel.2019.116150
65. Ma X, Zheng X, Zhang M, Yang X, Chen L, Huang L (2014) Electron beam irradiation of bamboo chips: degradation of cellulose and hemicelluloses. Cellulose 21(6):3865–3870
66. Ma X, Zheng X, Zhang M, Yang X, Chen L, Huang L, Cao S (2014) Electron beam irradiation of bamboo chips: degradation of cellulose and hemicelluloses. Cellulose 21(6):3865–3870. https://doi.org/10.1007/s10570-014-0402-4
67. Maceiras R, Alfonsín V, Seguí L, González JF (2021) Microwave assisted alkaline pretreatment of algae waste in the production of cellulosic bioethanol. Energies 14(18):5891. https://doi.org/10.3390/en14185891
68. Mahnič Kalamiza S, Vorobiev E, Miklavčič D (2014) Electroporation in food processing and biorefinery. J Membr Biol 247(12):1279–1304. https://doi.org/10.1007/s00232-014-9737-x
69. Mante OD, Amidon TE, Stipanovic A, Babu SP (2014) Integration of biomass pretreatment with fast pyrolysis: an evaluation of electron beam (EB) irradiation and hot-water extraction (HWE). J Anal Appl Pyrol 110:44–54. https://doi.org/10.1016/j.jaap.2014.08.004
70. Michaelson SM, Lin JC (1987) Physical description of radio and microwave radiation. Biol Eff Health Implic Radiofreq Radiat 11–45. https://doi.org/10.1007/978-1-4757-4614-3_2
71. Mohazzab BF, Jaleh B, Nasrollahzadeh M, Khazalpour S, Sajjadi M, Varma RS (2020) Upgraded valorization of biowaste: laser-assisted synthesis of Pd/calcium lignosulfonate nanocomposite for hydrogen storage and environmental remediation. ACS Omega 5(11):5888–5899. https://doi.org/10.1021/acsomega.9b04149

72. More T, Ghangrekar M (2010) Improving performance of microbial fuel cell with ultrasonication pre-treatment of mixed anaerobic inoculum sludge. Biores Technol 101(2):562–567. https://doi.org/10.1016/j.biortech.2009.08.045
73. Motasemi F, Afzal MT (2013) A review on the microwave-assisted pyrolysis technique. Renew Sustain Energy Rev 28:317–330
74. Mousavi Rabeti SA, Khoshgoftar Manesh MH, Amidpour M (2023) An innovative optimal 4E solar-biomass waste polygeneration system for power, methanol, and freshwater production. J Clean Prod 412:137267. https://doi.org/10.1016/j.jclepro.2023.137267
75. Nair LG, Agrawal K, Verma P (2022a) An insight into the principles of lignocellulosic biomass-based zero-waste biorefineries: a green leap towards imperishable energy-based future. Biotechnol Genet Eng Rev 1–51. https://doi.org/10.1080/02648725.2022.2082223
76. Nair LG, Agrawal K, Verma P (2022b) An overview of sustainable approaches for bioenergy production from agro-industrial wastes. Energy Nexus. https://doi.org/10.1016/j.nexus.2022.100086
77. Nair LG, Agrawal K, Verma P (2023) Organosolv pretreatment: an in-depth purview of mechanics of the system. Bioresour Bioprocess, Springer 10:50(1–29). https://doi.org/10.1186/s40643-023-00673-0
78. Nasir NN, Othman SA (2019) Effect of radiation treatment on starch bioplastic- a review. Int J Adv Life Sci Res 02(04):1–7. https://doi.org/10.31632/ijalsr.2019v02i04.001
79. Nikolic S, Mojovic L, Rakin M, Pejin D, Savic D (2008) A microwave-assisted liquefaction as a pretreatment for the bioethanol production by the simultaneous saccharification and fermentation of corn meal. Chem Ind Chem Eng Q 14(4):231–234. https://doi.org/10.2298/ciceq0804231n
80. Nitayavardhana S, Shrestha P, Rasmussen ML, Lamsal BP, van Leeuwen JH, Khanal SK (2010) Ultrasound improved ethanol fermentation from cassava chips in cassava-based ethanol plants. Biores Technol 101(8):2741–2747. https://doi.org/10.1016/j.biortech.2009.10.075
81. Oglat AA (2023) Gamma ray, beta and alpha particles as a sources and detection. J Radiat Res Appl Sci 16(1):100503. https://doi.org/10.1016/j.jrras.2022.100503
82. Ooshima H, Aso K, Harano Y, Yamamoto T (1984) Microwave treatment of cellulosic materials for their enzymatic hydrolysis. Biotech Lett 6(5):289–294. https://doi.org/10.1007/bf00129056
83. Parrino F, Bellardita M, García-López EI, Marcì G, Loddo V, Palmisano L (2018) Heterogeneous photocatalysis for selective formation of high-value-added molecules: some chemical and engineering aspects. ACS Catal 8(12):11191–11225. https://doi.org/10.1021/acscatal.8b03093
84. Pérez Vila S, Fenelon MA, O'Mahony JA, Gómez-Mascaraque LG (2022) Extraction of plant protein from green leaves: biomass composition and processing considerations. Food Hydrocoll 133:107902. https://doi.org/10.1016/j.foodhyd.2022.107902
85. Postek MT, Poster DL, Vládar AE, Driscoll MS, LaVerne JA, Tsinas Z, Al-Sheikhly MI (2018) Ionizing radiation processing and its potential in advancing biorefining and nanocellulose composite materials manufacturing. Radiat Phys Chem 143:47–52. https://doi.org/10.1016/j.radphyschem.2017.09.015
86. Potnuri R, Surya DV, Rao CS, Yadav A, Sridevi V, Remya N (2023) A review on analysis of biochar produced from microwave-assisted pyrolysis of agricultural waste biomass. J Anal Appl Pyrol 173:106094. https://doi.org/10.1016/j.jaap.2023.106094
87. Reisz JA, Bansal N, Qian J, Zhao W, Furdui CM (2014) Effects of ionizing radiation on biological molecules—mechanisms of damage and emerging methods of detection. Antioxid Redox Signal 21(2):260–292. https://doi.org/10.1089/ars.2013.5489
88. Rezania S, Ye Z, Berson RE (2009) Enzymatic saccharification and viscosity of sawdust slurries following ultrasonic particle size reduction. Appl Biochem Biotechnol 153(1–3):103–115. https://doi.org/10.1007/s12010-008-8467-x
89. Saini A, Aggarwal N, Sharma A, Yadav A (2015) Prospects for irradiation in cellulosic ethanol production. Biotechnol Res Int 2015:1–13. https://doi.org/10.1155/2015/157139

90. Shvets A, Vershinina K, Vinogrodskiy K, Kuznetsov G (2023) Microwave pyrolysis of woody biomass: influence of radiation power on the composition of conversion products. Appl Sci 13(13):7926. https://doi.org/10.3390/app13137926
91. Singh A, Kaundal A, Jha RK, Powar S, Dhar A (2018) Solar assisted gasification. Energy, Environ, Sustain 551–575. https://doi.org/10.1007/978-981-13-3302-6_18
92. Sobek S, Werle S (2021) Solar pyrolysis of waste biomass: a comparative study of products distribution, in situ heating behavior, and application of model-free kinetic predictions. Fuel 292:120365. https://doi.org/10.1016/j.fuel.2021.120365
93. Sólyom K, Mato RB, Pérez-Elvira SI, Cocero MJ (2011) The influence of the energy absorbed from microwave pretreatment on biogas production from secondary wastewater sludge. Biores Technol 102(23):10849–10854. https://doi.org/10.1016/j.biortech.2011.09.052
94. Srivastava RK, Shetti NP, Reddy KR, Kwon EE, Nadagouda MN, Aminabhavi TM (2021) Biomass utilization and production of biofuels from carbon neutral materials. Environ Pollut 276:116731. https://doi.org/10.1016/j.envpol.2021.116731
95. Sun S, Sun S, Cao X, Sun R (2016) The role of pretreatment in improving the enzymatic hydrolysis of lignocellulosic materials. Biores Technol 199:49–58. https://doi.org/10.1016/j.biortech.2015.08.061
96. Surati MA, Jauhari S, Desai KR (2012) A brief review: microwave assisted organic reaction. Arch Appl Sci Res 4:645–661
97. Talib Hamzah H, Sridevi V, Seereddi M, Suriapparao DV, Ramesh P, Sankar Rao C, Pritam K (2022) The role of solvent soaking and pretreatment temperature in microwave-assisted pyrolysis of waste tea powder: analysis of products, synergy, pyrolysis index, and reaction mechanism. Biores Technol 363:127913. https://doi.org/10.1016/j.biortech.2022.127913
98. Tissot C, Grdanovska S, Barkatt A, Silverman J, Al-Sheikhly M (2013) On the mechanisms of the radiation-induced degradation of cellulosic substances. Radiat Phys Chem 84:185–190
99. Tsubaki S, Azuma JI, Yoshimura T, Maitani M, Suzuki E, Fujii S, Wada Y (2016) Microwave-induced biomass fractionation. Biomass Fractionation Technol Lignocellul Feed Based Biorefinery 103–126. https://doi.org/10.1016/b978-0-12-802323-5.00005-0
100. Verma A, Tripathi M (2023) Interaction of Mustard Husk with microwave radiation: a study on dielectric properties and its variation with frequency. J Anal Appl Pyrol 173:106090. https://doi.org/10.1016/j.jaap.2023.106090
101. Vorobiev E, Lebovka N (2016) Applications of pulsed electric energy for biomass pretreatment in biorefinery. Biomass Fractionation Technol Lignocellul Feed Based Biorefinery 151–168. https://doi.org/10.1016/b978-0-12-802323-5.00007-4
102. Wang KQ, Xiong XY, Chen JP, Chen L, Su X, Liu Y (2012) Comparison of gamma irradiation and steam explosion pretreatment for ethanol production from agricultural residues. Biomass Bioenerg 46:301–308. https://doi.org/10.1016/j.biombioe.2012.08.013
103. Wang X, Fang G, Hu C, Du T (2008) Application of ultrasonic waves in activation of microcrystalline cellulose. J Appl Polym Sci 109(5):2762–2767. https://doi.org/10.1002/app.27975
104. Ward JF (1994) DNA damage as the cause of ionizing radiation-induced gene activation. Radiat Res 138(1):S85. https://doi.org/10.2307/3578769
105. Weldekidan H, Strezov V, Li R, Kan T, Town G, Kumar R, Flamant G (2020) Distribution of solar pyrolysis products and product gas composition produced from agricultural residues and animal wastes at different operating parameters. Renew Energy 151:1102–1109. https://doi.org/10.1016/j.renene.2019.11.107
106. Werpy T, Petersen G (2004) Top value added chemicals from biomass, vol I. Results of Screening for Potential Candidates from Sugars and Synthesis Gas. https://doi.org/10.2172/15008859
107. Wu X, Chen L, He W, Qi H, Zhang Y, Zhou Y, Wang K (2020) Characterize the physico-chemical structure and enzymatic efficiency of agricultural residues exposed to γ-irradiation pretreatment. Ind Crops Prod 150:112228. https://doi.org/10.1016/j.indcrop.2020.112228
108. Xiong X, Yu IK, Tsang DC, Bolan NS, Sik Ok Y, Igalavithana AD, Vikrant K (2019) Value-added chemicals from food supply chain wastes: state-of-the-art review and future prospects. Chem Eng J 375:121983. https://doi.org/10.1016/j.cej.2019.121983

109. Yang C, Shen Z, Yu G, Wang J (2008) Effect and aftereffect of γ radiation pretreatment on enzymatic hydrolysis of wheat straw. Biores Technol 99(14):6240–6245. https://doi.org/10.1016/j.biortech.2007.12.008
110. Yang G, Wang J (2018) Pretreatment of grass waste using combined ionizing radiation-acid treatment for enhancing fermentative hydrogen production. Biores Technol 255:7–15. https://doi.org/10.1016/j.biortech.2018.01.105
111. Yang SJ, Yoo HY, Choi HS, Lee JH, Park C, Kim SW (2015) Enhancement of enzymatic digestibility of *Miscanthus* by electron beam irradiation and chemical combined treatments for bioethanol production. Chem Eng J 275:227–234. https://doi.org/10.1016/j.cej.2015.04.056
112. Yu J, Zhang J, He J, Liu Z, Yu Z (2009) Combinations of mild physical or chemical pretreatment with biological pretreatment for enzymatic hydrolysis of rice hull. Biores Technol 100(2):903–908. https://doi.org/10.1016/j.biortech.2008.07.025
113. Zhang C, Su X, Xiong X, Hu Q, Amartey S, Tan X, Qin W (2016) 60 Co-γ radiation-induced changes in the physical and chemical properties of rapeseed straw. Biomass Bioenerg 85:207–214. https://doi.org/10.1016/j.biombioe.2015.11.022
114. Zhang PF, Pei ZJ, Wang DH, Wu XR, Cong WL, Zhang M, Deines T (2011) Ultrasonic vibration-assisted pelleting of cellulosic biomass for biofuel manufacturing. J Manuf Sci Eng 133(1). https://doi.org/10.1115/1.4003475
115. Zieliński M, Nowicka A, Dębowski M (2017) Hydrothermal depolymerization of Virginia fanpetals (*Sida Hermaphrodita*) biomass with the use of microwave radiation as a potential method for substrate pre-treatment before the process of methane fermentation. Energy Procedia 105:694–699. https://doi.org/10.1016/j.egypro.2017.03.377

Chapter 2
State-of-the-Art Irradiation Technologies for the Waste Biomass Pretreatment: Potential and Challenges

Sulfath Hakkim Hazeena, Kaviraj Ramesh, Aryasree Makkakode, and Chithra Manisseri

Abstract Increasing fossil energy dependency on fuels and chemicals leads to escalated environmental impacts, including global climate changes. The search for renewable alternatives for petroleum feedstock is gaining much attention from the scientific community worldwide. On the other side, a huge amount of waste biomass generated from agricultural, forestry, municipal, and industrial activities raises serious issues of pollution due to lack of proper management. This waste biomass could be considered as a cost-effective feedstock for generating cleaner and sustainable energy. However, the complex biomass structure must be degraded with efficient pretreatment techniques for its conversion into high-value products. Various physical, chemical, and biological pretreatment methods have been widely studied for industrially relevant product generation. Irradiation technologies are one of the major physical pretreatment methods for degrading complex recalcitrant biomass. Irradiation methods including microwave, ultrasonication, electron-beam, and gamma irradiation are applied in transforming the chemical structure of biomass, making the further steps of the bioprocess more effective. This chapter reviews the state-of-the-art of various irradiation techniques for the valorization of waste biomass into biofuels and chemicals. It explains specific irradiation methods as a stand-alone as well as a co-pretreatment technique with special emphasis on lignocellulosic biomass. It also dives into various biomass, products, the structural changes occurring to biomass during radiation reaction, potentialities, and challenges in upscaling the techniques to industrial application.

Keywords Biomass · Biofuels · Pretreatment · Radiation techniques · Value-added chemicals

---

Contributed equally to this chapter—Sulfath Hakkim Hazeena, Kaviraj Ramesh and Aryasree Makkakode

---

S. H. Hazeena · K. Ramesh · A. Makkakode · C. Manisseri (✉)
Department of Plant Science, School of Biological Sciences, Central University of Kerala, Tejaswini Hills, Kasaragod, Kerala 671320, India
e-mail: mchithra@cukerala.ac.in

© The Author(s), under exclusive license to Springer Nature Switzerland AG 2024

P. Verma and B. Kumar (eds.), *Sustainable Radiation Technologies in Waste-biomass Valorization*, Green Chemistry and Sustainable Technology,
https://doi.org/10.1007/978-3-031-63941-8_2

2.1 Introduction

The modern concepts of an integrated biorefinery aim at a circular bioeconomy by utilizing the waste biomass from diverse sources. Increased population growth and energy demands accelerate the accumulation of various types of waste [88]. Huge amounts of waste biomass generated from agricultural, forestry, municipal, and industrial activities raise serious issues of pollution due to the lack of proper management [85]. Hence, the valorization of waste biomass in a sustainable way becomes a highly multidisciplinary problem. Physical, chemical, thermochemical, or biological processes can be applied for the economic conversion of the waste biomass to high-value products [66]. This chapter focuses on converting lignocellulosic waste to fuel and other value-added products through different irradiation techniques. Currently, most of our energy and high-value product needs are based on petroleum refineries. Excessive usage of fossil fuels leads to various environmental issues like increased greenhouse gases and severe climate changes [50]. In this scenario, the search for renewable alternatives for petroleum feedstock is gaining increased attention from the scientific community worldwide [64]. It is possible to reduce carbon footprint by depending on renewable sources for energy needs and it is being implemented by most of the countries. Renewables contribute 20–30% of the world's primary energy, and reports indicate that a 100% renewable energy system is theoretically feasible by 2050 [43]. The lignocellulosic waste biomass could be considered a cost-effective feedstock for generating cleaner and sustainable energy. Also, the second-generation biorefinery concept will avoid the competition between food and fuels.

There are several strategies (physical, chemical, and biological) for the conversion of waste biomass to fuels and chemicals (Fig. 2.1). Each method has its effectiveness and limitations. For example, lesser product yield, and higher process time are the limitations of biological processes. The high heat and chemical requirements of the conventional pretreatment techniques limit their applicability as well as make them hazardous. So, an integrated strategy could mitigate the limitations of each pretreatment process. A detailed account of different categories of waste biomass, structural characteristics of lignocelluloses, and widely adopted irradiation technologies for the conversion of lignocelluloses to fuels, biomaterials, and chemicals are included in this chapter. It also highlights the challenges associated with each irradiation technology. Further, the advances and future potential of the irradiation technologies are also discussed.

2.2 Waste Biomass as Feedstocks for High-Value Products

Waste biomass valorization is a process of converting waste to high-value products or energy materials such as bioethanol, biogas, and chemicals [86]. From an energy perspective, the term biomass refers to any organic substrate that can be converted

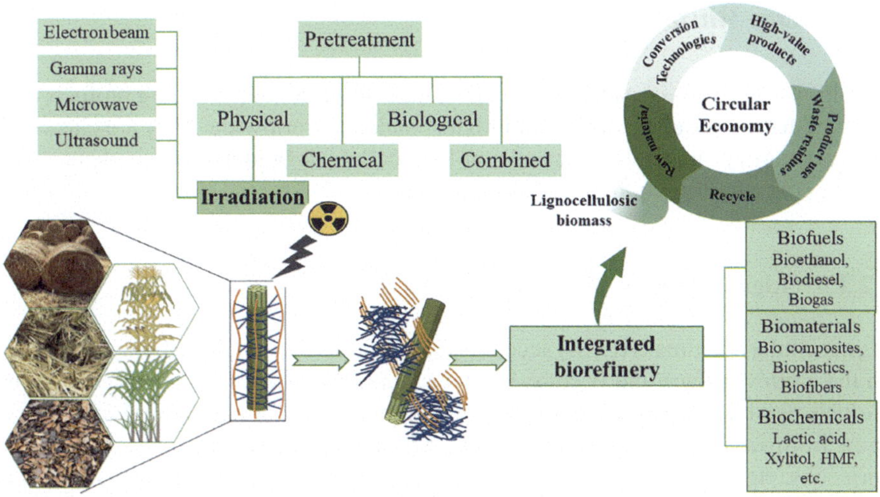

Fig. 2.1 Conversion of lignocellulosic biomass to value-added products

into energy through different methods. In general, the wastes can be categorized into two, namely, dry waste and wet waste based on their physical state.

2.2.1 Wet Waste

Water waste generated from various industries such as sugar, paper, and food are useful resources for value-added products. Molasses, a byproduct of sugar production, is converted into ethanol through fermentation and the remaining distillery spent wash composed of melanoidins, caramels, phenolics, and furfurals can be used as a substrate for the production of organic acids (acetic acid, butyric acid, etc.) and enzymes (cellulase, laccase, etc.) [35, 81]. Similarly, the spent sulfite liquor is a byproduct of the sulfite pulping process from the paper and pulp mill industry and it is formed during the delignification process of wood chips. Lignosulfonates and sugars constitute the major components in spent sulfite liquors. However, the sugars (hexose or pentose) liberated will differ depending on the wood type and can act as a substrate for the production of bioethanol, bioplastic (polyhydroxyalkanoate, PHA, polyhydroxybutyrate, PHB), and other value-added chemicals. Apart from that, it also contains extractives, methanol, furfural, etc. [5]. The huge quantities of waste generated from soft drink industries with high sugar content can also serve as a feedstock for high-value products [46].

2.2.2 *Dry Waste*

The dry waste biomass stands out as a potential substrate for producing biofuels and value-added compounds, due to their high abundance worldwide. They are classified into two namely, lignocellulosic and non-lignocellulosic biomass. In this chapter, we will be concentrating on lignocellulosic biomass.

2.2.2.1 Lignocellulosic Biomass

Lignocellulosic biomass (LCB) accounts for major portions of dry biomass available on the earth's surface and is considered a suitable feedstock for second-generation biofuels and bio-based chemicals [14]. It mainly consists of agricultural residues—straw from various food crops (rice straw, wheat straw, sugarcane tops, corn stover, corn cobs, etc.), industrial residues—bagasse, peanut shells, husk, bran, solid waste from food industries, oil refinery waste (palm oil, sunflower, soybean, rapeseed, and castor oil), forestry residues—natural waste (dead trees, leaves, bark, etc.) and man-made waste (sawdust, rejected portions of trees, etc.) [45, 156]. Out of 181.5 billion tons of LCB produced worldwide annually, only 8.2 billion tons are explored for various applications [29]. LCB is mainly composed of three polymers, namely, cellulose, hemicellulose, and lignin and the composition may vary in different feedstocks [154] (Table 2.1).

In LCB, cellulose is the major structural constituent, made up of glucose units, which are packed together by hydrogen and van der Waals bonds to form microfibrils, and subsequently assigned together to form cellulose fibers [91, 108]. Cellulose exists in crystalline and amorphous forms in biomass. The crystalline cellulose is formed by evenly organized chains while the amorphous form is composed of minor amounts of unorganized chains. Hence, amorphous cellulose is more accessible to microbes for enzymatic hydrolysis compared to crystalline. The hemicellulose and lignin form a complex assembly around microfibrils making it more recalcitrant [100].

Table 2.1 Lignocellulosic composition of different biomass

Biomass	Cellulose (%)	Hemicellulose (%)	Lignin (%)	References
Rice straw	27.9–44.6	15.3–25.5	14.9–28	[21]
Sugarcane straw	31.5–53.8	15–34.7	13.2–25.8	[28, 94]
Bagasse	32.7–47	16–35.2	18–27	[17, 94]
Wheat straw	28–39	23–24	16–25	[4]
Corn stover fractions	31.3–39.4	21.2–27.3	17–20.1	[73]
Perennial grass	36.3–56.5	11.9–25.9	11.9–20.2	[96, 102]
Sawdust	44	26	26	[110]
Softwood (spruce)	33–42	22–40	27–32	[129, 134]
Hardwood (poplar)	38–51	17–38	21–31	[129, 134]

Unlike cellulose, hemicellulose constitutes heteropolysaccharides and its monomer composition may vary in different biomass. For example, in *Gramineae*, arabinose and glucuronic acids are attached to the xylan backbone (xylose chain) to form glucuronoarabinoxylan. Xyloglucan is the major hemicellulose in the primary cell walls of dicots. In hardwood, methylated glucuronic acid and a few acetic acid molecules are attached to the xylan backbone forming O-acetyl-4-O-methylglucuronoxylan. In softwood, galactose attached to mannan (mannose chain) forms O-acetyl galactomannan [41]. Different types of hemicellulose are represented in Fig. 2.2.

Lignin is the second most abundant biopolymer after cellulose and accounts for 15–40% of biomass dry weight [101]. It is primarily a non-linear three-dimensional amorphous polymer, composed of three monomers syringyl (S), guaiacyl (G), and p-hydroxyphenyl (H) alcohol via free radical coupling reaction. The composition of lignin may vary in different species and tissues. Hardwood lignin is mainly composed of G and S monolignols, with only a small content of H units. On the other hand, softwood lignin is largely made up of G units. In contrast, herbaceous plants have significant quantities of all three units G, S, and H, but in varying proportions [19]. The structure of lignin monomers is represented in Fig. 2.3. Lignin acts as a physical barrier to cellulase action. It is integrated with hemicellulose through a covalent bond, thus increasing the mechanical strength of biomass. The relationship between lignin content and enzymatic hydrolysis was studied and found that high sugar yield can be achieved from biomass with low lignin content [149]. Apart from lignin content, the composition of monomers (S/G ratio), distribution, and linkages in lignin-carbohydrate complexes (LCCs) can also influence enzymatic saccharification [120].

2.3 Pretreatment of Biomass to Biofuels and Other Value-Added Products

The presence of cellulose (35–55%) and hemicellulose (20–40%) in LCB makes them an important precursor for biofuels and platform chemicals [92]. However, the compact arrangement of cell wall components restricts sugar release from biomass and becomes a major obstacle in its valorization [154]. Pretreatment is a crucial step for overcoming biomass recalcitrance, increasing solubility, reducing particle size, and increasing surface area, thus improving the conversion into commodity chemicals. So, the composition and complexity of biomass and the efficiency of pretreatment should be taken into consideration when developing a new or altering an existing pretreatment strategy. There are various pretreatment techniques including physical, chemical, biological, and combined methods widely studied for transforming LCB [80, 106].

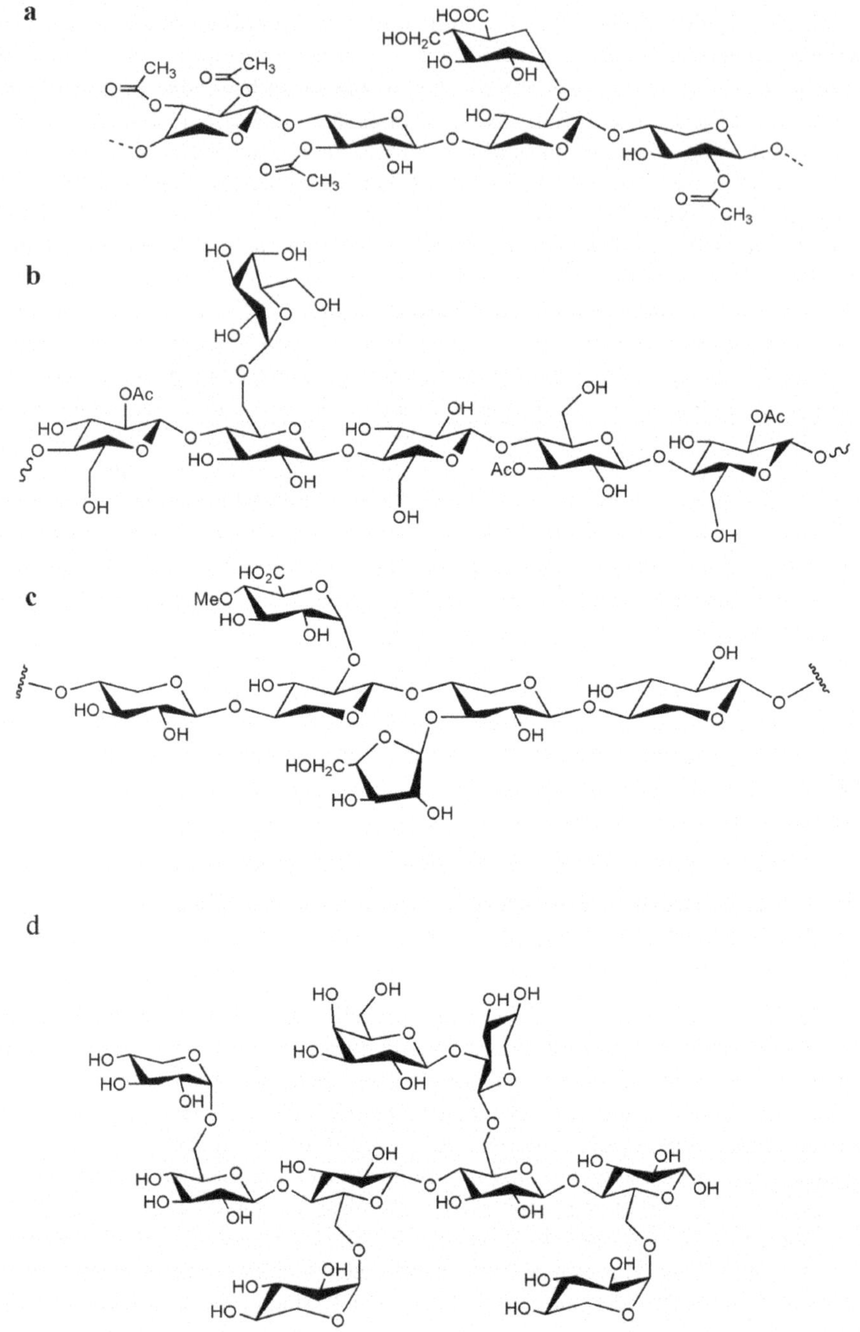

Fig. 2.2 The major hemicelluloses in lignocellulosic biomass **a** O-acetyl-4-O-methylglucuronoxylan, **b** O-acetyl-galactoglucomannan, **c** Glucuronoarabinoxylan, and **d** Xyloglucan

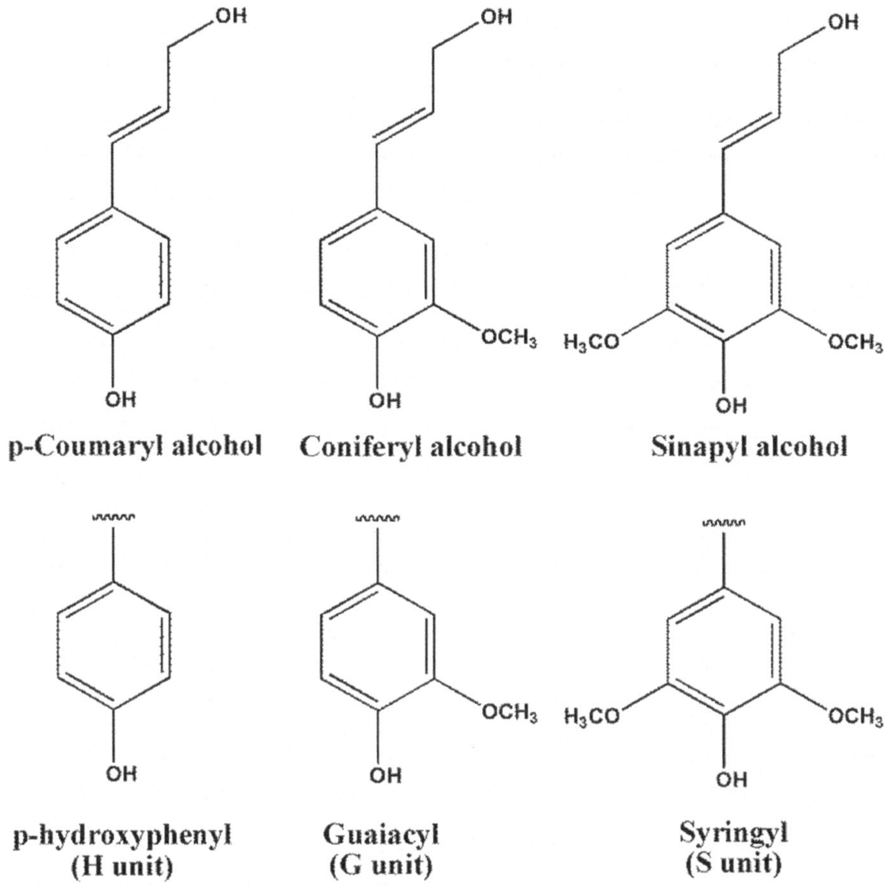

Fig. 2.3 Monolignols and their structural lignin units in biomass

2.3.1 Chemical Pretreatment

Chemical pretreatment involves the utilization of various alkalis, acids, solvents, etc., for the breakdown of complex LCB. Acid pretreatment using H_2SO_4, HCl, CH_3COOH, etc., is a widely employed technique, which breaks the lignin and carbohydrate bonds resulting in hemicellulose dissolution. Similarly, alkali pretreatment using NaOH, ammonium hydroxide, sulfite, etc., is considered to be the most reliable techniques that effectively cleave ester and ether bonds in lignin [80, 95, 106]. The optimum pretreatment condition and their efficiency may vary depending on the type and complexity of biomass. The formation of inhibitors and secondary pollutants during the process, which interferes with fermentation and requires additional separation steps, is counted as one of the major disadvantages of chemical pretreatment.

Recently, the pretreatment of LCB using green solvents has been gaining increased attention because of their eco-friendly nature. The properties of ionic liquids (ILs) such as high thermal stability, and less vapor pressure make them a promising green alternative to conventional methods. Ionic Liquids (ILs) formed by cations (imidazolium, ammonium, phosphonium, and pyridinium) and anions (phosphonate, chloride, and acetate) attached with alkyl/allyl side chains have been applied for LCB pretreatment [16, 133, 137, 152]. Deep eutectic solvents (DESs) are advanced green solvents with greater potential for LCB degradation. They are composed of two or more ionic substances that function as hydrogen bond donors (HBD) or hydrogen bond acceptors (HBA) [119]. Characteristics such as easy to synthesize, low cost, biocompatibility, and low toxicity promote their use in LCB pretreatment [2, 24].

2.3.2 Biological Pretreatment

The process of biomass degradation using microbes and their enzymes is collectively known as biological pretreatment. Several bacteria and fungi produce lignocellulolytic enzymes to break down LCC and release the polymers [3]. Lignin can be converted into various chemicals such as phenolic acids, vanillin, vanillic acids, syringaldehyde, and ferulic acid [98]. Additionally, the microbial consortia that can produce an array of enzymes are also used for LCB pretreatment [37]. They are predominantly used in the anaerobic digestion of biomass. Low energy requirements, simple equipment and process conditions, and less inhibitor production make them attractive for biomass pretreatment. However, its industrial application is limited due to the requirement of efficient microbes, extended duration, etc. [151].

2.3.3 Physical Pretreatment

Physical pretreatment methods include mechanical and irradiation techniques causing alterations in the size, shape, and structure of LCB. Size reduction through mechanical treatment is a crucial step to decrease cellulose crystallinity and increase the surface area for enzymatic saccharification. There are several methods to reduce the particle size such as cutting/crushing (m to cm), coarse milling (cm to mm), intermediate micronization (cm to 100 mm), fine grinding (<100 mm), ultra-fine grinding (<30 mm), and nano grinding (<1 mm) [12]. The advantages of this method include increasing the calorific value of biomass, providing solutions for sample storage difficulties, simplifying the supply chain of raw materials, and reducing energy input during enzymatic hydrolysis. Even though it has several benefits, there are factors such as high energy consumption, low efficiency, and high energy cost that limit its usage [106]. Woody biomass utilizes more energy than herbaceous biomass for size reduction. For example, poplar and pine wood utilize 85.4 and 118.5 kWh/metric ton respectively, whereas corn stover and switchgrass require only 11.0 and 27.6

kWh/metric ton respectively, for size reduction to a definite metric [103]. Besides mechanical methods, irradiation techniques are also widely applied for biomass pretreatment.

2.4 Irradiation Techniques for the Lignocellulosic Biomass Valorization: Potential and Challenges

Irradiation pretreatment is an emerging method used for various applications, including biofuel and biogas production from waste biomass, such as agricultural residues, forestry waste, and food waste. Four different irradiation methods are commonly used for biomass pretreatment; gamma-ray irradiation, electron-beam irradiation (EBI), microwave (MW), and ultra-sonication (US) [58]. These technologies can be used alone as well as in combination with other chemical and/or biological pretreatments to enhance the efficiency of biomass degradation [76] (Table 2.2). High energy radiations such as electron beams, X-rays, or gamma rays can produce ions and/or radicals that can result in cleavage of chemical bonds in the biomass [126]. Short and long-lived radicals can initiate secondary degradation via polymer scission and cross-linking. It is noted that the radicals generated in the crystalline region of the biomass decay after a certain period, while in amorphous regions they extinguish quickly [132]. The rate of degradation of chemical bonds depends on polymeric structure and radiation dose and by integrating with other physical/chemical pretreatment strategies, the radiation dose can be reduced accordingly.

2.4.1 Ionizing Radiation-Based Lignocellulosic Biomass Valorization: Potential and Challenges

The Gamma rays and electron beams are high energy ionizing radiations (IR), which can displace electrons from atoms or molecules, creating ions. This type of radiation is highly energetic and can cause damage to biological tissues and genetic material (DNA). Common sources of IR include X-rays, gamma rays, and certain types of nuclear radiation (e.g., alpha and beta particles).

2.4.1.1 Gamma Radiation-Based Lignocellulosic Biomass Valorization

Gamma rays are produced by the unstable nucleus of an atom during radioactive decay. ^{60}Co and ^{137}Cs are the widely used sources of gamma radiation. This IR penetrates the biomass and transfers its energy to atoms of LCB resulting in the reduction of cellulose crystallinity and lignin modifications [66]. The radicals formed by the

Table 2.2 Irradiation pretreatments in different lignocellulosic biomass

Irradiation method	Biomass	Experimental condition	Product and observation	References
Gamma irradiation	Wheat straw	1000 kGy + 3% (w/w) H_2SO_4	Glucose yield increased to 69.7%, Xylose content was entirely removed	[42]
	Wheat straw	100 kGy followed by 2% NaOH for 1 h	Reduction in crystallinity index	[147]
	Rice straw Cornstalk Rapeseed straw	800 kGy 96 h incubation, 7.5% solid consistency and 15 FPU/g of enzyme loading	72.42%, 69.79%, and 72.68% of glucose yield from rice straw, corn stalk, and rapeseed straw, respectively	[140]
	Sugarcane bagasse	2000 kGy	Significant reduction in hemicellulose and cellulose Reduction in cellulose crystallinity	[54]
	Empty fruit bunch fibers	200 kGy followed by 5% Na_2CO_3 at 150 °C for 2 h	65.03% reducing sugar yield, Reduction in degree of crystallinity	[26]
	Sugarcane bagasse	Delignified sample, 400 kGy + 1% APS	Enhanced the yield of TRS from 33360 μg/ml (MW alone) to 122237 μg/ml (MW + 1% APS)	[55]
	Hybrid poplar sawdust	300 kGy + sludge protein 2 mg/ml	Reducing sugar yield increased up to 519 mg/g	[141]
	Poplar bark	1000 kGy + 3% (w/w) H_2SO_4	83.0% (w/w) fermentable sugar yield, Reduction in crystallinity index	[27]
	Brachypodium distachyon	250 Gy and 300 Gy	139.4% and 213.7% glucose yield from 250 and 300 Gy, respectively	[63]
	Macroalgae *Laminaria japonica*	20 kGy	Biohydrogen production (71%)	[23]
	Senna tora seeds	400 Gy	Increased glucan, lignin, and total phenolic content	[62]

(continued)

Table 2.2 (continued)

Irradiation method	Biomass	Experimental condition	Product and observation	References
	Microcrystalline cellulose (MCC)	800 kGy	Glucose yield (82.3%)	[74]
	Microcrystalline cellulose (MCC)	891 kGy + IL [BMIM][Cl]	Ethanol yield (67%)	[75]
Electron beam	Rice straw	80 kGy at 0.12 mA and 1 meV, Enzymatic hydrolysis with 60 FPU of cellulase and 30 CBU of b-glucosidase for 24 and 132 h, respectively	43.1% (24 h) and 52.1% (132 h) glucose yield	[9]
	Rice straw	500 kGy by autoclaving with 3% NaOH	Cellulose content increases from 39.5% to 71.1%	[61]
	Rice straw	200 kGy and 3% dilute H_2SO_4	80% total sugar yield at 72 h	[68]
	Corn cob	270 kGy	Transformation of cellulose I to cellulose II Increase the glucose yield (44.3%) and reducing sugar yield (267.9 mg/g)	[38]
	Sugarcane bagasse	50 kGy, followed by dilute acid (0.1% H_2SO_4) and hydrothermal treatment (180 °C) for 40 and 60 min	30% enhancement in enzymatic saccharification 71.55% glucose yield	[33]
	Oil palm empty fruit bunches	100 kGy + NaOH (10%)	5.36% ethanol concentration	[30]
	Sugar maple wood chips	500 kGy and 750 kGy	Reduces 80% of the biomass recalcitrance, increases the rate and the yield of glucose	[25]

(continued)

Table 2.2 (continued)

Irradiation method	Biomass	Experimental condition	Product and observation	References
	Hardwood (Birch) and softwood (Pine)	Sequential electron beam-steam explosion (EB-SE)	Conversion of hemicelluloses into water-soluble oligomers and enhanced the overall enzymatic saccharification	[71]
	Napier grass	150 g biomass + 300 ml of either 1% H_2SO_4, or 1% NaOH separately + 150 kGy	40% increase in glucose yield Softwood	[57]
	Kenaf core (*Hibiscus cannabinus*)	500 kGy + 3% NaOH	72.4% total sugar recovery	[51]
Microwave	Wheat straw fibers	MW (270 W) + Choline chloride: Formic acid (1:4 Mol), 2 min	TRS yield of 619 mg/g biomass (99% glucose, 85% xylose)	[44]
	Corn straw	Multimode-US (20 kHz (60 W) + 40 kHz (60 W) for 30 min followed by MW assisted ChCl: OA: Gly (2: 2: 1), 120 °C for 1 min	Achieved removal of 61.5% lignin, 90% hemicellulose, and 76% cellulose recovery	[142]
	Sugarcane bagasse	MW + [BMIM][Cl], 90 °C, 2 h followed by surfactant (PEG-8000) assisted saccharification	Sugar yield 326.76 mg/g biomass ethanol yield 146.9 mg/g biomass	[118]
	Rice straw	MW + (681 W) NaOH (0.54 M) for 3 min	255 mg/g of TRS	[49]
	Sugarcane bagasse	Dilute H_2SO_4 (0.005 M) and MW irradiation (180 °C for 30 min)	87% hemicellulose dissolution along with minimum furfural production	[22]
	Rice straw	DES and MW Choline chloride-Formic acid (ChCl-FA), MW 50 Hz, 130 °C for 10 min	214 mg ethanol/g biomass	[116]

(continued)

Table 2.2 (continued)

Irradiation method	Biomass	Experimental condition	Product and observation	References
	Corncob	MW with autohydrolysis 180 °C, 5 min and 2% (v/v) HCl catalyst	37% furfural yield	[111]
	Japanese cedar (Softwood), *Eucalyptus globulus* (hardwood)	MW + CuO/H_2O_2 + 2N NaOH at 200 °C for 20 min	Vanillin and vanillic acid (total 6.5% of wood weight) in Japanese cedar wood vanillin, syringaldehyde, and syringic acid (total 10.6% of wood weight) in *E. globulus*	[99]
	Tall wheatgrass hybrid Szarvasi-1 (*Agropyron elongatum*)	180 °C for 20 min at 600 rpm stirring, alkaline catalyst (0.6 M NaOH)	Lignin removal (96%), 16-fold higher saccharification than untreated	[47]
	Kenaf (*Hibiscus cannabinus*)	MW + (^{30}Ce/MgO-48)/H_2O_2 (170 °C for 20 min)	Yielded high vanillin (3.7%) with less amount of 4-vinylguaiacol, syringol, and syringaldehyde	[59]
	Kenaf stalks	MW + H_2O_2/CeO_2-Nps-400	Vanillin yield (3.8%)	[104]
	Brewer's spent grain	MW-assisted hydrothermal pretreatment (192.7 °C and 5.4 min)	Biobutanol yield of 8.3 g/L	[77]
	Brewer's spent grain	MW heating at 172 °C with 0.38 M NaOH	17 g/L arabinoxylans and 9.9 g/L butanol	[78]
	Tomato	MW-assisted, catalyst-free aerobic oxidation process, 170 °C for 30 min	Vanillin 8.2% and syringaldehyde 9.0% along with traces of p-hydroxyl benzaldehyde, acetovanillone, and organic acids	[128]

(continued)

Table 2.2 (continued)

Irradiation method	Biomass	Experimental condition	Product and observation	References
	Marine macroalgae (*Ulva reticulate*)	Acidic—hydrogen peroxide (H_2O_2) induced MW (AHMW) pretreatment (pH 5, 360 W power for 10 min)	92.5 mL H_2/g COD	[65]
Ultrasound radiation	Wheat straw	US (20 kHz) + 2% NaOH	70% delignification, 84.5% hydrolysis, 90% glucose, and 65% xylose yield. Increased PHB yield 0.441 g/g of reducing sugar	[113]
	Cornstalk	US + Ammonium bicarbonate solution. (liquid/solid ratio 12:1 temperature 42 °C and time 11 min at 100 W)	82.6% of the saccharification rate	[52]
	Rice straw	US-assisted alkali (2 M NaOH) pretreatment at 500 W, 90 °C for 30 min, stirring up to 1.5 h	Yielded 84.7% lignin	[136]
	Bagasse	US + [Bmim]OAc 40 kHz/30 min	65.72% of HMF was produced	[72]
	Rice straw	Choline hydroxide ([Ch][OH]) with US at 300 W/40 kHz	96.22% yield of total reducing sugar	[143]
	Corn stover	US + Sodium percarbonate (2 g biomass, Na_2CO_3: 0.4 mol/L; H_2O_2: 0.6 mol/L, US 20 kHz/50 W)	Increased lignin removal, efficient saccharification of hemicellulose into xylose without any furfural formation	[87]
	Corn stover Sorghum stalk	US (20 kHz) and pelleting followed by dilute acid pretreatment (2%)	93.1 and 92.8% sugar yield obtained from corn stover and sorghum stalk, respectively	[153]

(continued)

Table 2.2 (continued)

Irradiation method	Biomass	Experimental condition	Product and observation	References
	Corn meal	US-assisted liquefaction (40 kHz, 5 min, 60 °C)	Increased glucose concentration after liquefaction, ethanol yields 3.02 g/L/h	[89]
	Corn slurry	274 ± 5 W	30% increase in glucose yield	[60]
	Kenaf powder	US (24 kHz) + [EMIM][Ac] pretreatment for 15 min at 25 °C	86% cellulose saccharification	[91]
	Grass biomass	US (260 W) + 1% HCl, 100 °C for 30 min	Improve the biomass solubilization, 42.2 mL H_2/g-dry grass	[144]
	Miscanthus sinensis	US assisted-alkali-urea, US 20 kHz, 15.5 min, 2.1% NaOH, 1.7% urea	Recovery of 47.8% cellulose and 27.5% lignin	[122]
	Saccharum arundinaceum *Arundo donax* *Typha angustifolia* *Ipomoea carnea*	US (400W/24 kHz) + 0.5% NaOH	Maximum ethanol concentration was observed as 3.5 ± 1.1%, 4.7 ± 0.9%, 4.1 ± 0.8%, and 5.6 ± 1.1% (w/v) for *Saccharum arundinaceum, Arundo donax, Typha angustifolia,* and *Ipomoea carnea*, respectively	[84]
	Dictyota dichotoma	40% amplitude for 40 min	Reducing sugar concentration (151.73 mg/g of dried biomass) is 65.3% higher than untreated	[107]
	Green algae *(Ulva rigida)*	40 kHz/120 W	2.53 g/L of reducing sugar yield	[56]
	Microcrystalline cellulose (MCC)	4 mol L^{-1} HNO_3, US for 60 min at 30 °C in a cup horn system at 50% of amplitude	78% of conversion to furfural	[112]

gamma irradiation break the glycosidic bonds leading to degradation of polysaccharides and destruction of the cell wall. These changes can increase the accessibility of cellulose to enzymes and improve the enzymatic hydrolysis efficiency. The dosage of gamma irradiation is an important factor that affects the lignocellulosic degradation. However, the optimal irradiation dose depends on the type and source of biomass. Irradiation at 250 Gy and 300 Gy was reported to increase the glucose yield in *Brachypodium distachyon* [63]. A dosage of 400 Gy gamma irradiation on the seeds of *Senna tora*, an important medicinal plant, was also reported effective in glucan and lignin disintegration [62]. It was reported that irradiation dose at 10–30 kGy disrupts the structure of macroalgae and increases the concentrations of soluble organic compounds, such as soluble chemical oxygen demand (SCOD), protein, and polysaccharides, making more organic compounds available for hydrogen production [23]. High glucose yield was also observed in agri-residues such as rice straw (72.42%), corn stalk (69.79%), and rapeseed straw (72.68%) treated with gamma irradiation [140].

The impacts of gamma irradiation (0–1400 kGy) on the microstructure, heat stability, and derivatives of microcrystalline cellulose (MCC) were studied previously and found that irradiation dose had a significant impact on the properties of MCC, and 800 kGy, resulting in maximum glucose yield (82.3%) after enzymatic hydrolysis [74]. A comparative study on the impact of gamma irradiation combined with other pretreatment methods such as ILs, 1% HCl, 1% H_2SO_4, and acidic aqueous ILs (AA-ILs) was carried out in MCC and found that 891 kGy was optimum for ethanol production (67%) and an irradiation dosage over 1000 kGy can lead to the production of toxic, and unwanted inhibitors [75]. Cellulose and hemicellulose were found to decrease with the high dosage of gamma radiation in *Miscanthus* [124]. Similarly, when the irradiation dose was 1800 kGy or above, glycolytic degradation resulted in a decline in glucose and fructose concentration [93]. Also, sugarcane bagasse exposed to gamma radiation at different doses showed a reduction in cellulose and hemicellulose content at 2000 kGy [54]. Abdelhafez et al. [1] reported the exposure of microorganisms to different doses of IR on the production of bioethanol from sugarcane bagasse and potato peels. Even though gamma irradiation is an efficient pretreatment strategy, its combination with other methods was reported to increase the efficiency or give comparable results with a mild dosage [132]. Combined pretreatment of gamma rays and dilute H_2SO_4 was used for the bioethanol production from poplar bark [27]. Increased glucose yield (69.7%) was observed from wheat straw by gamma rays and dilute acid pretreatment [42]. Similarly, 65.03% reducing sugar yield was observed when empty fruit bunch fibers were treated with 200 kGy gamma radiation and 5% sodium carbonate [26]. Combined urea and gamma irradiation treatment was found to enhance reducing sugar yield in Thai rice straw and corn stalk [11]. Gamma irradiation with 1% ammonium persulfate (APS) resulted in a four-fold increase in reducing sugar yield in sugarcane bagasse (122378 μg/ml) compared to untreated biomass (33360 μg /ml) [55]. Combined pretreatment of gamma rays with IL, 1-butyl-3-methylimidazolium chloride ([BMIM][Cl]) was reported to increase cellulose depolymerization in MCC [39]. Gamma irradiation effectively reduced the limonene (a common fermentation inhibitor) content in orange peels and led to increased ethanol production [7]. Enhanced sugar yield (72.67%) was reported in wheat straw with 100 kGy radiation followed by 2% sodium hydroxide (NaOH) pretreatment for 1 h [147]. In sunflower seed cake gamma irradiation, pretreatment was found to decrease the biomethane production by 1.2–2.4 times compared to untreated samples under batch mesophilic anaerobic digestion, which may be due to the over-acidification and presence of competing cross-linking reactions in the anaerobic environment [32].

In brief, the prime challenge in gamma irradiation pretreatment of plant biomass is to break the recalcitrant lignocellulosic structure. Though a high dose of irradiation (>1000 kGy) could significantly degrade the crystalline structure of cellulose, it generates toxic compounds, which could slow down enzymatic hydrolysis, sugar recovery, and ethanol fermentation. Therefore, the determination of appropriate irradiation dose with minimum inhibitor generation is one of the key challenges observed during gamma irradiation pretreatment. From previous studies, it is evident that the gamma irradiation needs to be optimized for each biomass to obtain the desired end product with minimum input energy. Also, gamma irradiation when combined with other cost-effective chemical or biological pretreatments may help yield the desired high-value products at mild dosage.

2.4.1.2 Electron Beam Irradiation-Based Lignocellulosic Biomass Valorization

Electron beam irradiation (EBI) is an efficient strategy for LCB pretreatment. It does not require any temperature or pressure adjustments and needs only limited chemicals or solvents. Hence, it is considered to be environmentally friendly and affordable. Electron beam (EB) is a highly charged ionizing radiation generated from a linear electron accelerator. The high-energy electrons create radical species that lead to a chain reaction within the cell wall polymers [71]. The dosage of an EB plays a critical role in biomass pretreatment. A lower EBI dosage may not break the bonds effectively, while chain scission occurs at a higher dosage. However excessive dosage may decrease glucose recovery by disintegrating the glucose monomer [76, 79]. When compared to gamma irradiation pretreatment, EBI is usually conducted at a higher dosage. Because EB has limited penetration and deposits its entire energy on thin layers of material, whereas gamma rays can penetrate deeper [33]. The energy of electrons can be controlled by changing the irradiation dose [58]. Ma et al. [79] observed that cellulose degradation can be controlled in bamboo chips by adjusting the EBI dosage. There was a close relationship observed between radiation dose and degree of cellulose degradation. High EBI dosage was associated with drastic cellulose degradation and a sharp decline in the degree of polymerization. At 10 kGy dose of EBI, the degree of polymerization of cellulose was 684, which was decreased to 324 with an increased dosage of 50 kGy [79].

Similar to gamma irradiation, EBI can be used alone or in combination with other pretreatments to enhance the biomass conversion to value-added products. EBI pretreatment was reported to cause a significant increase in the enzymatic digestibility and glucose yield in rice straw, and *Miscanthus sinensis* [9, 145]. The study conducted on sugar maple wood chips showed that a 500 kGy dose of EBI reduced 80% biomass recalcitrance and 750 kGy caused an increase in glucose yield compared to the untreated sample [25]. Removal of hemicellulose and lignin along with degradation of cellulose chain was noticed after EBI pretreatment in corn cob [38]. Rice straw pretreated with EBI based biodegradation resulted in alterations to the surface of the biomass, an increase in digestibility (65.5%), and a high sugar yield (20.7 g) without any inhibitor formation [8].

Combined pretreatment strategies involving EBI can help to reduce process severity, and prevent extreme sugar degradation as well as the formation of toxic/inhibitory by-products during saccharification of lignocelluloses [145]. EBI is used more often in combination with alkali and acid pretreatments. The combined alkaline-EBI method resulted in high total sugar recovery in the kenaf core (*Hibiscus cannabinus*) [51]. EBI followed by alkali pretreatment was reported to enhance sugar yield from rice straw than alkali alone, due to the impact of irradiation on crystalline cellulose and the removal of lignin and hemicellulose [61]. Similarly, pre-soaking in mild acid (1% H_2SO_4) and alkali (1% NaOH) before EBI treatment resulted in the removal of hemicellulose, reduction in crystallinity, and an increase in glucose yield in hybrid Napier grass [57]. EBI pretreatment of rice straw with 3% dilute acid, and subsequent autoclaving was reported to generate a high amount of sugars (80%) [68]. The impact of EB pretreatment in birch and pine wood was analyzed individually and in combination with a steam explosion (SE) previously. It was found that the combined pretreatment increased the sugar yield to three and two-fold in birch and pine wood, respectively [71]. The EBI pretreatment in switchgrass combined with hot water (HW) was found to increase the sugar release [127]. When sugarcane bagasse was subjected to EBI followed by dilute acid and hydrothermal treatments, an enhanced saccharification rate (30%) was observed [33].

In general, most of the combined pretreatments showed positive impacts concerning glucose yield from biomass. On the contrary, some research suggests that the combination of EBI with other pretreatments may not be always recommended. Dahnum et al. [30] reported that the combination of irradiation and NaOH yielded high ethanol but with a significant biomass loss. The combined effect of EBI and imidazolium-based ILs on biomass was reported to cause fixation of nitrogen from the IL into the cellulosic matrix and hence should be avoided [53]. Thus, for efficient pretreatment yields, the type of biomass and pretreatment solvent must be selected accordingly, before its application with another pretreatment strategy. Since the phenolic polymer lignin resists the EBI pretreatment, it could be extracted from the pretreated LCB in its near-native state for multiple applications in lignin-based biorefinery [114]. Also, similar to other irradiation treatments EBI offers many advantages such as less inhibitory compounds, short radiation time, heat control, etc., and an optimized EBI pretreatment may be efficient in reducing biomass recalcitrance and its conversion to value-added products. However, EBI pretreatment is not much explored for multiple product generation in biorefineries and more research needs to be carried out to utilize its complete potential.

2.4.2 Non-Ionizing Radiation based Lignocellulosic Biomass Valorization: Potential and Challenges

Non-ionizing radiation has a lower energy which is not enough to ionize atoms or molecules and includes electromagnetic waves with longer wavelengths, such as radio waves, microwaves, visible light, and ultraviolet radiation. Widely used microwave and ultrasound radiations for the production of value-added products from lignocellulosic biomass are explained in this section.

2.4.2.1 Microwave Radiation-Based Lignocellulosic Biomass Valorization

Microwaves (MW) are low-frequency non-ionizing waves. MW heating may depend on dipolar reorientation and ionic conduction, which are connected to various properties such as reaction time, heating rate, moisture content of biomass, MW power, etc. Cell wall degradation is caused by both thermal and non-thermal effects of MW heating. The thermal effect is associated with the rapid rearrangement of dipoles in a changing electric field of MW radiation. The friction and collision generated by the rearrangement of dipoles generate heat energy in the molecules. The non-thermal effects are generated due to the realignment of the molecules under an electromagnetic field disrupting hydrogen bonds and cell wall degradation [66].

MW pretreatment can alter the structure of cell wall components leading to increased saccharification [97]. The major advantage of MW heating compared to conventional is that it is more uniform and selective temperature distribution can be achieved [13]. However, the penetration depth of material by MW is critical in scaling up the method. MW when used alone as the pretreatment method requires high temperature for efficient biomass disintegration and can lead to the formation of phenolics and furfurals [66]. MW irradiation has been reported in combination with other pretreatment methods such as hot water, acid, and IL to maximize the yield and minimize the inhibitor formation [6, 66, 83]. MW pretreatment of wheatgrass cultivar Szarvasi (*Agropyron elongatum*) biomass resulted in 16-fold higher enzymatic saccharification, lignin removal (96%), and glucose production [47].

MW-assisted pretreatment of corn straw and rice husk with alkaline glycerol resulted in high sugar recovery [31]. Rice straw pretreated with MW-assisted NaOH also yielded high total reducing sugars (255 mg/g TRS) [49]. Similarly, MW-assisted dilute acid pretreatments were found to be effective in stillage of wheat and rye yielding high glucose [82]. Alkaline copper oxide-peroxide/MW pretreatment was used in the production of vanillin and vanillic acid in Japanese cedar wood and vanillin, syringaldehyde, and syringic acid in *E. globulus* [99]. MW-assisted combined pretreatment strategies were also applied for furfural production from agri-residues. Sugarcane bagasse pretreated with dilute sulfuric acid and MW irradiation resulted in 87% hemicellulose dissolution along with minimum furfural production [22]. MW with HCl yielded high furfural (37%) compared to MW alone (10.8%) in corncob [111]. MW was found to be more effective than the US in terms of sugar yield and ethanol production in corn meal. In addition, MW pretreatment was found to reduce the time of simultaneous saccharification and fermentation (SSF) [90].

MW heating is often combined with green solvents including IL and DES for biomass pretreatment. [123] studied the effect of ILs (1-butyl-3-methylimidazolium hydrogen sulfate, [BMIM][HSO_4] and [BMIM][Cl] combined with MW on sugarcane bagasse and efficient delignification was observed. Similarly, MW-[BMIM][Cl] and PEG-8000 treatment yielded high TRS (326.7 mg/g) from sugarcane bagasse [118]. MW/DES pretreatment in rice straw also resulted in high TRS and ethanol production (214 mg ethanol/g biomass) [116].

Studies have also been conducted on MW-assisted chemical pretreatments such as MW/dilute ammonia and MW/$FeCl_3$ [34]. Tomato plant biomass on MW-assisted, catalyst-free aerobic oxidation pretreatment produced vanillin (8.2%) and syringaldehyde (9.0%) with a minor amount of p-hydroxyl benzaldehyde, acetovanillone, and organic acids [128]. Kenaf (*Hibiscus cannabinus*) biomass and kenaf lignin with hydrogen peroxide (H_2O_2, oxidizing agent) and ^{30}Cerium/Magnesium oxide (^{30}Ce/MgO-48, catalyst) pretreatment produced significant vanillin yield. Some industrially important compounds such as 4-vinylguaiacol, syringol, and syringaldehyde were also generated during the same process [59]. MW pretreatment with H_2O_2 and CeO_2-Nps-400 catalyst was used for vanillin production in kenaf stalks [104]. MW-assisted hydrothermal pretreatment was performed in brewer's spent grain for fermentable sugar release followed by biobutanol production [77]. In a related study on biobutanol fermentation from brewer's spent grain, after MW-assisted NaOH pretreatment arabinoxylans were also produced [78]. MW pretreatment of sorghum leaves also generated high sugars, furfural, acetic acid, 5-hydroxymethylfurfural (HMF), and phenol [109].

MW pretreatment was also applied in wet microalgal biomass for high recovery of biofuel precursors such as sugars and lipids [105]. H_2O_2-induced MW pretreatment in acidic condition of marine macroalgae (*Ulva reticulate*) was reported to significantly reduce the pretreatment time and increase the liquefaction rate for biohydrogen production (92.5 mL H_2/g COD) [65]. Kumar et al. [67] reported that surfactant-induced MW pretreatment is more energy efficient than MW pretreatment alone in macroalgae for biohydrogen production. The wide applications of MW-assisted pretreatment, having the advantages of short reaction time and high pretreatment speed, are limited by certain crucial factors. The need for specific reactor design for biomass pretreatment, product chain establishment, and modeling for diverse origins of biomass could be considered as some of the challenges to be faced for scaling up at the commercial level. However, an optimized MW pretreatment with low energy requirements or in combination with cost-effective chemical/biological pretreatments with minimum inhibitor generation can be effectively utilized for LCB degradation.

2.4.2.2 Ultrasound Radiation-Based Lignocellulosic Biomass Valorization

Ultrasound (US) can be defined as sound waves with a frequency higher than the human hearing limit (20 kHz). US-assisted pretreatment works on the principle of cavitational effect, which leads to the formation of cavitational bubbles. These bubbles enlarge to a critical size and collapse to cause an increase in pressure (1800 atm) and temperature (2000–5000 K) [66]. This region or condition is called a hot spot [18]. The cavitational bubbles disrupt LCCs by breaking down β-O-4 and α-O-4 linkage in lignin. The factors such as US frequency, sonication duration, power, and temperature influence the US pretreatment. US irradiation at high power for a longer duration can effectively reduce the cellulose crystallinity by increasing the pore volume [138]. Also, the choice of solvent is critical in determining optimum pretreatment conditions. US irradiation of eucalyptus wood in different solutions, i.e., aqueous soda, distilled water, and aqueous acetic acid was reported to cause changes in their physiochemical structures resulting in increased crystallinity [40]. Also, this pretreatment was applied to a wide range of LCB and fibers, like kenaf biomass, corn biomass, and empty fruit bunches for their conversion into biofuels [58]. A higher degree of cell wall disintegration and increased biogas production (384 ml/g) were observed in filamentous algae by US pretreatment [69]. Similarly, US-treated *Dictyota dichotoma* showed 65.3% enhanced sugar release (151.73 mg/g) compared to untreated biomass [107]. US pretreatment followed by acid hydrolysis on oil palm empty fruit bunch resulted in an increased xylose yield of 58% while it was 22% untreated [150]. Though it reduces the production of impurities during the process, the US alone is less effective in hydrolyzing complex polysaccharides into simple monomeric sugars. However, it can be used in combination with other treatments to enhance the disruption of lignocellulosic and non-lignocellulosic linkages [155].

US-assisted acid pretreatment was used in grass, wood waste, and sugarcane straw for furfural production [15]. Santos et al. [112] reported 78% furfural conversion from MCC using US-assisted acid pretreatment. US-assisted HCl pretreatment was reported to improve the degradation of grass biomass by altering the molecular morphology and enhancing biohydrogen production. The hydrogen yield was 42.2 mL/g of dry biomass which was increased to 311.7 and 190.0% in the US and US with acid pretreatment, respectively compared to the control [144].

US power required for pretreatment is lowered when it is used in combination with alkali. It enables lignin solubilization, increases cellulose digestibility, and reduces pretreatment time. US alkaline pretreatment of chili post-harvest residue, grass clipping, and Chinese fir showed increased sugar yield without any inhibitor production [121, 139]. The US-assisted alkali pretreatment of sugarcane bagasse showed decreased sugar degradation and inhibitor formation than thermal/acid pretreatments with a substantial reduction in pretreatment temperature and time [135]. But to upscale US pretreatment a well-designed reactor and energy optimization are essential. The US-assisted alkaline pretreatment showed enhanced delignification, high carbohydrate release, and lower inhibitor generation which led to high bioethanol production from *Saccharum arundinaceum* (3.5%), *Arundo donax* (4.7%), *Typha angustifolia* (4.1%), *and Ipomoea carnea* (5.6%) [84]. Xylitol production from sugarcane bagasse via US-assisted alkali pretreatment and US enhanced fermentation was reported previously [36, 130, 131]. Also, high cellulose and lignin recovery was reported by US assisted-alkali-urea (UAAU) pretreatment in *Miscanthus*. US in combination with organosolv pretreatment was also reported for biomass degradation. Combined US-organic solvent (ethanol, methanol, and dioxane) pretreatment was reported to cause high delignification (96.2%) and hemicellulose dissolution (75.5%) in dewaxed poplar wood [148]. Sun et al. [125] noticed comparable results in wheat straw pretreated with US and organic solvents.

US can be used in combination with ILs for efficient pretreatment of biomass. IL/Water system reduces the cavitation value of the US wave and promotes its mass transfer. Surfactant (SDS) assisted US-IL [BMIM][Cl] pretreatment was reported to cause high cellulose recovery (58.74%) and delignification (21.01%) in water hyacinth [20]. US-surfactant-assisted IL pretreatment of sugarcane bagasse also showed enhanced sugar yield (16%) [117]. Probe sonication-assisted [BMIM][Cl] treatment ($CrCl_3$ as catalyst) was used for HMF production from glucose, cellulose, and bamboo biomass [115]. The specific mechanism of ultrasonic irradiation and its integration with other pretreatment methods have to be understood well for efficient biomass degradation. Ultrasonic reactor design and optimization of pretreatment parameters are crucial for its applications at the industrial scale. In the up-scale process, dissipation of power was found to be one of the main obstacles due to energy loss from reactor walls, surrounding air, and bulk liquid [48, 70]. To improve the ultrasonic field efficiency, innovative designs were created, including power transducers and self-focusing reactors [146]. Energy efficiency and cost-effectiveness are the other major challenges that may be resolved by optimizing ultrasound frequency, combination with other pretreatments requiring cheaper solvents and mild treatment conditions.

2.5 Economic Perspectives for Irradiation-Assisted Lignocellulosic Biomass Valorization

Economic assessment of any bioprocess is very important to understand the long-term enduringness of the same. It evaluates the cost of production and the income generated through the sale of products. There are two major expenses, i.e., capital expenses (CAPEX) and operational expenses (OPEX) that have to be considered to assess the economic viability of any technology. CAPEX is the expenses required to start the facility and related infrastructure. Operational expenses include raw materials, labor costs, utilities, maintenance, and general plant costs. Technology with low CAPEX/OPEX able to sustain itself to generate profit is considered to be economically viable. Pretreatment is an important step of bioethanol production and leads to the fractionation of biomass and these fractions can be derivatized to produce different value-added products. Hence, a well-designed biorefinery is essential for making the process cost-effective and helps in efficient waste management and other environmental issues associated with the process. There are fundamental disadvantages in decentralized biomass pretreatment when it is about achieving full economic potential. Consequently, additional investigation into biomass processing and pretreatment is necessary to ensure its financial and economic sustainability.

2.6 Future Prospects

Biomass pretreatment is the major challenge faced by the biorefinery sector in achieving a circular economy. Despite the advantages of irradiation technologies for the pretreatment of biomass, real-time usage is limited due to several reasons. The major reasons include high cost, energy requirements limited understanding of the mechanism of irradiation technologies, and the need for replacement of the current plant design. Optimization of operating parameters and management of energy expenses have to be studied well to get the maximum product recovery. Optimization and economic feasibility studies could reduce the environmental impacts and make the system sustainable. Using radiation heating in combination with physicochemical approaches provides several advantages for integration into biomass pretreatment, but still, many issues need to be completely understood and resolved. As a result, there are difficulties in pretreating lignocellulosic biomass for the production of biofuels/biochemicals on a commercial and industrial scale. More research has to be carried out to learn better irradiation techniques for use in industrial biorefineries. Furthermore, while designing and developing innovative methods and procedures, energy input should be carefully taken into account. In brief, techno-economic assessments are crucial to the success of biorefineries.

2.7 Conclusions

Radiation technology is used to convert waste biomass into fuels, chemicals, and other high-value products. Most of the radiation-assisted processes are laboratory level and require scale-up thorough techno-economic evaluation. The major disadvantage of irradiation technology lies in the difficulty of incorporating it into existing engineering and the replacement of current infrastructure requiring huge investment. Another limitation is the constant energy input which may be very high for long pretreatment periods. Though irradiation pretreatments have shown promising results, the applicability is comparatively low for their wide use. However, irradiation methods in combination with other pretreatments have been reported to be more efficient. There was a noticeable influence of temperature, irradiation time, radiation power, and biomass loading on the radiation-assisted physicochemical pretreatment of biomass. These findings will help to make effective use of radiation pretreatment strategies for converting LCB into green energy and value-added products.

Declaration The authors declare that AI tools are not used for writing this chapter.

References

1. Abdelhafez AA, El-Tayeb TS, El-Mongi TM, Moussa MM (2014) Effect of gama irradiation of bioethanol producing microorganisms on bioethanol formation from sugarcane bagasse and potato peels. Egypt J Microbiol 49:55–79
2. Ai B, Li W, Woomer J, Li M, Pu Y, Sheng Z, Zheng L, Adedeji A, Ragauskas AJ, Shi J (2020) Natural deep eutectic solvent mediated extrusion for continuous high-solid pretreatment of lignocellulosic biomass. Green Chem 22:6372–6383
3. Agrawal K, Bhardwaj N, Kumar B, Chaturvedi V, Verma P (2019) Process optimization, purification and characterization of alkaline stable white laccasse from *Myrothecium verrucaria* ITCC-8447 and its application in delignification of agroresidues. Int J Biol Macromol 125:1042–1055
4. Álvarez C, Mullen AM, Pojić M, Hadnađev TD, Papageorgiou M (2021) Classification and target compounds. In: Food waste recovery, pp 21–49. Elsevier
5. Amândio MST, Pereira JM, Rocha JMS, Serafim LS, Xavier AMRB (2022) Getting value from pulp and paper industry wastes: on the way to sustainability and circular economy. Energies 15:4105
6. Anoopkumar AN, Reshmy R, Aneesh EM, Madhavan A, Kuriakose LL, Awasthi MK, Pandey A, Binod P, Sindhu R (2023) Progress and challenges of Microwave-assisted pretreatment of lignocellulosic biomass from circular bioeconomy perspectives. Bioresour Technol 369:128459
7. Bai HW, Hong SH, Park CH, Jang DM, Kim TH, Chung BY (2014) Degradation of limonene by gamma radiation for improving bioethanol production. J Korean Soc Appl Biol Chem 57:1–4
8. Bak JS (2014) Process evaluation of electron beam irradiation-based biodegradation relevant to lignocellulose bioconversion. Springerplus 3:487
9. Bak JS, Ko JK, Han YH, Lee BC, Choi IG, Kim KH (2009) Improved enzymatic hydrolysis yield of rice straw using electron beam irradiation pretreatment. Bioresour Technol 100:1285–1290
10. Baksi S, Saha D, Saha S, Sarkar U, Basu D, Kuniyal JC (2023) Pre-treatment of lignocellulosic biomass: review of various physico-chemical and biological methods influencing the extent of biomass depolymerization. Int J Environ Sci Technol 20:13895–13922
11. Banchorndhevakul S (2002) Effect of urea and urea–gamma treatments on cellulose degradation of Thai rice straw and corn stalk. Radiat Phys Chem 64:417–422
12. Barakat A, Mayer-Laigle C, Solhy A, Arancon RAD, De Vries H, Luque R (2014) Mechanical pretreatments of lignocellulosic biomass: towards facile and environmentally sound technologies for biofuels production. RSC Adv 4:48109–48127
13. Bhardwaj N, Kumar B, Verma P (2020) Microwave-assisted pretreatment using alkali metal salt in combination with orthophosphoric acid for generation of enhanced sugar and bioethanol. Biomass Convers Biorefinery. https://doi.org/10.1007/s13399-020-00640-1
14. Bhardwaj N, Kumar B, Agrawal K, Verma P (2020) Bioconversion of rice straw by synergistic effect of in-house produced ligno-hemicellulolytic enzymes for enhanced bioethanol production. Bioresour Technol Rep 10:100352
15. Bizzi CA, Santos D, Sieben TC, Motta GV, Mello PA, Flores EMM (2019) Furfural production from lignocellulosic biomass by ultrasound-assisted acid hydrolysis. Ultrason Sonochem 51:332–339

16. Brandt A, Ray MJ, To TQ, Leak DJ, Murphy RJ, Welton T (2011) Ionic liquid pretreatment of lignocellulosic biomass with ionic liquid–water mixtures. Green Chem 13:2489
17. Bu J, Wang YT, Deng MC, Zhu MJ (2021) Enhanced enzymatic hydrolysis and hydrogen production of sugarcane bagasse pretreated by peroxyformic acid. Bioresour Technol 326:124751
18. Bundhoo ZMA, Mohee R (2018) Ultrasound-assisted biological conversion of biomass and waste materials to biofuels: a review. Ultrason Sonochem 40:298–313
19. Cao Y, Chen SS, Zhang S, Ok YS, Matsagar BM, Wu KC-W, Tsang DCW (2019) Advances in lignin valorization towards bio-based chemicals and fuels: Lignin biorefinery. Bioresour Technol 291:121878
20. Chang KL, Han YJ, Wang XQ, Chen XM, Leu SY, Liu J, Peng YP, Liao YL, Potprommanee L (2017) The effect of surfactant-assisted ultrasound-ionic liquid pretreatment on the structure and fermentable sugar production of a water hyacinth. Bioresour Technol 237:27–30
21. Chen C, Deng X, Kong W, Qaseem MF, Zhao S, Li Y, Wu AM (2021) Rice straws with different cell wall components differ on abilities of saccharification. Front Bioeng Biotechnol 8:624314
22. Chen WH, Ye S-C, Sheen HK (2012) Hydrolysis characteristics of sugarcane bagasse pretreated by dilute acid solution in a microwave irradiation environment. Appl Energy 93:237–244
23. Chen Y, Yin Y, Wang J (2022) Biohydrogen production using macroalgal biomass of *Laminaria japonica* pretreated by gamma irradiation as substrate. Fuel 309:122179
24. Chen Z, Bai X, Wan C (2018) High-solid lignocellulose processing enabled by natural deep eutectic solvent for lignin extraction and industrially relevant production of renewable chemicals. ACS Sustain Chem Eng 6:12205–12216
25. Cheng K, Barber VA, Driscoll MS, Winter WT, Stipanovic AJ (2013) Reducing woody biomass recalcitrance by electron beams, biodelignification and hot-water extraction. J Bioprocess Eng Biorefinery 2:143–152
26. Chin SX, Chia CH, Zakaria S, Ahmad SH, Tasirin SM (2017) Combination of gamma irradiation and sodium carbonate pretreatment on oil palm empty fruit bunch (EFB) for high acidic hydrolysis yield. JSM 46:167–173
27. Chung BY, Lee JT, Bai H-W, Kim UJ, Bae HJ, Gon Wi S, Cho JY (2012) Enhanced enzymatic hydrolysis of poplar bark by combined use of gamma ray and dilute acid for bioethanol production. Radiat Phys Chem 81:1003–1007
28. Coelho BCS, Da Silva DMMC, Hermosilla ED, Teixeira RSS, Bon EPS, Diez MC, Ferreira-Leitão VS, Da Silva AS (2022) A prior biological delignification treatment as an aid for the hydrothermal pretreatment of sugarcane straw. Waste Biomass Valorization 13:4881–4895
29. Dahmen N, Lewandowski I, Zibek S, Weidtmann A (2019) Integrated lignocellulosic value chains in a growing bioeconomy: status quo and perspectives. GCB Bioenergy 11:107–117
30. Dahnum D, Barlianti V, Sembiring KC, Kristiani A, Muryanto M, Sudiyani Y (2017) Effect of combining electron beam irradiation and alkaline pretreatments of OPEFB for enzymatic hydrolysis and fermentation of ethanol. J Kim Terap Indones 19:1–10
31. Diaz AB, Moretti MMDS, Bezerra-Bussoli C, Carreira Nunes CDC, Blandino A, Da Silva R, Gomes E (2015) Evaluation of microwave-assisted pretreatment of lignocellulosic biomass immersed in alkaline glycerol for fermentable sugars production. Bioresour Technol 185:316–323
32. Dima AD, Pârvulescu OC, Mateescu C, Lungulescu EM (2021) Influence of gamma-ray irradiation on the biomethane production of sunflower seed cake. U.P.B Sci Bull Ser B 83:59–72
33. Duarte CL, Ribeiro MA, Oikawa H, Mori MN, Napolitano CM, Galvão CA (2012) Electron beam combined with hydrothermal treatment for enhancing the enzymatic convertibility of sugarcane bagasse. Radiat Phys Chem 81:1008–1011
34. Ethaib S, Omar R, Kamal SMM, Biak DRA (2015) Microwave-assisted pretreatment of lignocellulosic biomass: a review. Journal of Engineering Science and Technology. Taylor's University, School of Engineering, pp 97–109

35. Fito J, Tefera N, Van Hulle SWH (2019) Sugarcane biorefineries wastewater: bioremediation technologies for environmental sustainability. Chem Biol Technol Agric 6:6
36. Gai L, Ren EF, Tian W, Niu D, Sun W, Hang F, Li K (2022) Ultrasonic-assisted dual-alkali pretreatment and enzymatic hydrolysis of sugarcane bagasse followed by *Candida tropicalis* fermentation to produce xylitol. Front Nutr 9:913106
37. González C, Wu Y, Zuleta-Correa A, Jaramillo G, Vasco-Correa J (2021) Biomass to value-added products using microbial consortia with white-rot fungi. Bioresour Technol Rep 16:100831
38. Guo X, Zhang T, Shu S, Zheng W, Gao M (2017) Compositional and structural changes of corn cob pretreated by electron beam irradiation. ACS Sustain Chem Eng 5:420–425
39. Hao Y, Peng J, Ao Y, Li J, Zhai M (2012) Radiation effects on microcrystalline cellulose in 1-butyl-3-methylimidazolium chloride ionic liquid. Carbohydr Polym 90:1629–1633
40. He Z, Wang Z, Zhao Z, Yi S, Mu J, Wang X (2017) Influence of ultrasound pretreatment on wood physiochemical structure. Ultrason Sonochem 34:136–141
41. Huang LZ, Ma MG, Ji XX, Choi SE, Si C (2021) Recent developments and applications of hemicellulose from wheat straw: a review. Front Bioeng Biotechnol 9:690773
42. Hyun Hong S, Taek Lee J, Lee S, Gon Wi S, Ju Cho E, Singh S, Sik Lee S, Yeoup Chung B (2014) Improved enzymatic hydrolysis of wheat straw by combined use of gamma ray and dilute acid for bioethanol production. Radiat Phys Chem 94:231–235
43. IEA (2017) https://www.iea.org/reports/world-energy-outlook-2017. Last Accessed 07 Dec 2023
44. Isci A, Erdem GM, Bagder Elmaci S, Sakiyan O, Lamp A, Kaltschmitt M (2020) Effect of microwave-assisted deep eutectic solvent pretreatment on lignocellulosic structure and bioconversion of wheat straw. Cellulose 27:8949–8962
45. Isikgor FH, Becer CR (2015) Lignocellulosic biomass: a sustainable platform for the production of bio-based chemicals and polymers. Polym Chem 6:4497–4559
46. Isla MA, Comelli RN, Seluy LG (2013) Wastewater from the soft drinks industry as a source for bioethanol production. Bioresour Technol 136:140–147
47. Jablonowski ND, Pauly M, Dama M (2022) Microwave assisted pretreatment of Szarvasi (*Agropyron elongatum*) biomass to enhance enzymatic saccharification and direct glucose production. Front Plant Sci 12:767254
48. Jamshidi R, Brenner G (2013) Dissipation of ultrasonic wave propagation in bubbly liquids considering the effect of compressibility to the first order of acoustical Mach number. Ultrasonics 53(4):842–848
49. Jasmine A, Rajendran M, Thirunavukkarasu K, Abinandan S, Vaidyanathan VK, Krishnamurthi T (2023) Microwave-assisted alkali pre-treatment medium for fractionation of rice straw and catalytic conversion to value-added 5-hydroxymethyl furfural and lignin production. Int J Biol Macromol 236:123999
50. Jeswani HK, Chilvers A, Azapagic A (2020) Environmental sustainability of biofuels: a review. Proc R Soc A Mat Phys Eng Sci 476(2243):20200351
51. Jeun JP, Lee BM, Lee JY, Kang PH, Park JK (2015) An irradiation-alkaline pretreatment of Kenaf core for improving the sugar yield. Renew Energy 79:51–55
52. Jiao Y, Guo C, Wang S, Guan S, Pan X, Ma X, Zhang Z, Zhang Q, He C (2020) Enhancement of converting corn stalk into reducing sugar by ultrasonic-assisted ammonium bicarbonate pretreatment. Bioresour Technol 302:122878
53. Jusner P, Sulaeva I, Schiehser S, Potthast K, Tischer A, Barbini S, Potthast A, Rosenau T (2023) β-Irradiation in the presence of 1, 3-dialkylimidazolium ionic liquids causes covalent cellulose derivatization with simultaneous nitrogen incorporation. Cellulose 30:10551–10558
54. Kapoor K, Garg N, Diwan RK, Varshney L, Tyagi AK (2017) Study the effect of gamma radiation pretreatment of sugarcane bagasse on its physcio-chemical morphological and structural properties. Radiat Phys Chem 141:190–195
55. Kapoor K, Tyagi AK, Diwan RK (2020) Effect of gamma irradiation on recovery of total reducing sugars from delignified sugarcane bagasse. Radiat Phys Chem 170:108643

56. Karray R, Hamza M, Sayadi S (2015) Evaluation of ultrasonic, acid, thermo-alkaline and enzymatic pre-treatments on anaerobic digestion of *Ulva rigida* for biogas production. Bioresour Technol 187:205–213
57. Karthika K, Arun AB, Melo JS, Mittal KC, Kumar M, Rekha PD (2013) Hydrolysis of acid and alkali presoaked lignocellulosic biomass exposed to electron beam irradiation. Bioresour Technol 129:646–649
58. Kassim MA, Khalil HPSA, Serri NA, Kassim MHM, Syakir MI, Aprila NAS, Dungani R (2016) Irradiation pretreatment of tropical biomass and biofiber for biofuel production. In: Monteiro WA (ed) Radiation effects in materials
59. Khairul Anuar NASI, Ramli A, Jun Wei L (2021) Synthesis of Ce/MgO catalysts for direct oxidation of *Hibiscus cannabinus* stalks to vanillin. Catalysts 11:1449
60. Khanal SK, Montalbo M, Van Leeuwen J (Hans), Srinivasan G, Grewell D (2007) Ultrasound enhanced glucose release from corn in ethanol plants. Biotechnol Bioeng 98:978–985
61. Kim DY, Lee BM, Lee JY, Kang PH, Jeun JP (2014) Electron beam irradiation and dilute alkali pretreatment for improving saccharification of rice straw. J Korean Soc Appl Biol Chem 57:591–595
62. Kim JE, Gong SH, Jung JT, Lee OR, Lee JW (2018) Effect of gamma ray irradiation on the pretreatment and enzymatic hydrolysis of *Senna tora* stalk. KJMCS 26:127–133
63. Kim JY, Na CS, Kim DS, Kim J-B, Seo YW (2015) The effect of chronic gamma ray irradiation on lignocellulose of *Brachypodium distachyon*. Cellulose 22:2419–2430
64. Kumar B, Bhardwaj N, Verma P (2019) Pretreatment of rice straw using microwave assisted $FeCl_3$-H_3PO_4 system for ethanol and oligosaccharides generation. Bioresour Technol Rep 7(2019):100295. https://doi.org/10.1016/j.biteb.2019.100295
65. Kumar DM, Kaliappan S, Gopikumar S, Zhen G, Rajesh Banu J (2019) Synergetic pretreatment of algal biomass through H_2O_2 induced microwave in acidic condition for biohydrogen production. Fuel 253:833–839
66. Kumar B, Bhardwaj N, Agrawal K, Chaturvedi V, Verma P (2020) Current perspective on pretreatment technologies using lignocellulosic biomass: an emerging biorefinery concept. Fuel Proc Tech 199:106244
67. Kumar DM, Godvin Sharmila V, Kumar G, Park JH, Al-Qaradawi SY, Rajesh Banu J (2022) Surfactant induced microwave disintegration for enhanced biohydrogen production from macroalgae biomass: thermodynamics and energetics. Bioresour Technol 350:126904
68. Lee B-M, Lee J-Y, Kang P-H, Hong S-K, Jeun J-P (2014) Improved pretreatment process using an electron beam for optimization of glucose yield with high selectivity. Appl Biochem Biotech 174:1548–1557
69. Lee K, Chantrasakdakul P, Kim D, Kong M, Park KY (2014) Ultrasound pretreatment of filamentous algal biomass for enhanced biogas production. Waste Manage 34:1035–1040
70. Legay M, Gondrexon N, Le Person S, Boldo P, Bontemps A (2011) Enhancement of heat transfer by ultrasound: review and recent advances. Int J Chem Eng 2011:1–17
71. Leskinen T, Kelley SS, Argyropoulos DS (2017) E-beam irradiation & steam explosion as biomass pretreatment, and the complex role of lignin in substrate recalcitrance. Biomass Bioenergy 103:21–28
72. Li M, Jiang H, Zhang L, Yu X, Liu H, Yagoub AEA, Zhou C (2020) Synthesis of 5-HMF from an ultrasound-ionic liquid pretreated sugarcane bagasse by using a microwave-solid acid/ionic liquid system. Ind Crops Prod 149:112361
73. Li Z, Zhai H, Zhang Y, Yu L (2012) Cell morphology and chemical characteristics of corn stover fractions. Ind Crops Prod 37:130–136
74. Liu Y, Chen J, Wu X, Wang K, Su X, Chen L, Zhou H, Xiong X (2015) Insights into the effects of γ-irradiation on the microstructure, thermal stability and irradiation-derived degradation components of microcrystalline cellulose (MCC). RSC Adv 5:34353–34363
75. Liu Y, Zhou H, Wang S, Wang K, Su X (2015) Comparison of γ-irradiation with other pretreatments followed with simultaneous saccharification and fermentation on bioconversion of microcrystalline cellulose for bioethanol production. Bioresour Technol 182:289–295

76. Loow Y-L, Wu TY, Yang GH, Jahim MJ, Teoh WH, Mohammad AW (2016) Role of energy irradiation in aiding pretreatment of lignocellulosic biomass for improving reducing sugar recovery. Cellulose 23:2761–2789
77. López-Linares JC, García-Cubero MT, Lucas S, González-Benito G, Coca M (2019) Microwave-assisted hydrothermal as greener pretreatment of brewer's spent grains for biobutanol production. Chem Eng J 368:1045–1055
78. López-Linares JC, Lucas S, García-Cubero MT, Jiménez JJ, Coca M (2020) A biorefinery based on brewer's spent grains: arabinoxylans recovery by microwave assisted pretreatment integrated with butanol production. Ind Crops Prod 158:113044
79. Ma X, Zheng X, Zhang M, Yang X, Chen L, Huang L, Cao S (2014) Electron beam irradiation of bamboo chips: degradation of cellulose and hemicelluloses. Cellulose 21:3865–3870
80. Mankar AR, Pandey A, Modak A, Pant KK (2021) Pretreatment of lignocellulosic biomass: a review on recent advances. Bioresour Technol 334:125235
81. Meghana M, Shastri Y (2020) Sustainable valorization of sugar industry waste: Status, opportunities, and challenges. Bioresour Technol 303:122929
82. Mikulski D, Kłosowski G (2020) Microwave-assisted dilute acid pretreatment in bioethanol production from wheat and rye stillages. Biomass Bioenergy 136:105528
83. Mikulski D, Kłosowski G (2023) High-pressure microwave-assisted pretreatment of softwood, hardwood and non-wood biomass using different solvents in the production of cellulosic ethanol. Biotechnol Biofuels 16:19
84. Muthuvelu KS, Rajarathinam R, Kanagaraj LP, Ranganathan RV, Dhanasekaran K, Manickam NK (2019) Evaluation and characterization of novel sources of sustainable lignocellulosic residues for bioethanol production using ultrasound-assisted alkaline pre-treatment. Waste Manage 87:368–374
85. Nair LG, Agrawal K, Verma P (2022a) An insight into the principles of lignocellulosic biomass-based zero-waste biorefineries: a green leap towards imperishable energy-based future. Biotechnol Genet Eng Rev 1–51
86. Nair LG, Agrawal K, Verma P (2022b) An overview of sustainable approaches for bioenergy production from agro-industrial wastes. Energy Nexus 100086
87. Nakashima K, Ebi Y, Kubo M, Shibasaki-Kitakawa N, Yonemoto T (2016) Pretreatment combining ultrasound and sodium percarbonate under mild conditions for efficient degradation of corn stover. Ultrason Sonochem 29:455–460
88. Nanda S, Mohammad J, Reddy SN, Kozinski JA, Dalai AK (2014) Pathways of lignocellulosic biomass conversion to renewable fuels. Biomass Convers Biorefin 4:157–191
89. Nikolić S, Mojović L, Rakin M, Pejin D, Pejin J (2010) Ultrasound-assisted production of bioethanol by simultaneous saccharification and fermentation of corn meal. Food Chem 122:216–222
90. Nikolić S, Mojović L, Rakin M, Pejin D, Pejin J (2011) Utilization of microwave and ultrasound pretreatments in the production of bioethanol from corn. Clean Techn Environ Policy 13:587–594
91. Ninomiya K, Kamide K, Takahashi K, Shimizu N (2012) Enhanced enzymatic saccharification of kenaf powder after ultrasonic pretreatment in ionic liquids at room temperature. Bioresour Technol 103:259–265
92. Okolie JA, Nanda S, Dalai AK, Kozinski JA (2021) Chemistry and specialty industrial applications of lignocellulosic biomass. Waste Biomass Valorization 12:2145–2169
93. Orozco RS, Hernández PB, Ramírez NF, Morales GR, Luna JS, Montoya AJC (2012) Gamma irradiation induced degradation of orange peels. Energies 5:3051–3063
94. Palliprath S, Poolakkalody NJ, Ramesh K, Mangalan SM, Kabekkodu SP, Santiago R, Manisseri C (2023) Pretreatment of sugarcane postharvest leaves by γ-valerolactone/water/$FeCl_3$ system for enhanced glucan and bioethanol production. Ind Crops Prod 197:116571
95. Poddar BJ, Nakhate SP, Gupta RK, Chavan AR, Singh AK, Khardenavis AA, Purohit HJ (2022) A comprehensive review on the pretreatment of lignocellulosic wastes for improved biogas production by anaerobic digestion. Int J Environ Sci Technol 19:3429–3456

96. Poolakkalody NJ, Ramesh K, Palliprath S, Nittoor SN, Santiago R, Kabekkodu SP, Manisseri C (2023) Understanding triethylammonium hydrogen sulfate ([TEA][HSO_4]) pretreatment induced changes in *Pennisetum polystachion* cell wall matrix and its implications on biofuel yield. Renew Energy 209:420–430
97. Priecel P, Lopez-Sanchez JA (2019) Advantages and limitations of microwave reactors: from chemical synthesis to the catalytic valorization of biobased chemicals. ACS Sustain Chem Eng 7:3–21
98. Putro JN, Soetaredjo FE, Lin S-Y, Ju YH, Ismadji S (2016) Pretreatment and conversion of lignocellulose biomass into valuable chemicals. RSC Adv 6:46834–46852
99. Qu C, Kaneko M, Kashimura K, Tanaka K, Ozawa S, Watanabe T (2017) Direct production of vanillin from wood particles by copper oxide-peroxide reaction promoted by electric and magnetic fields of microwaves. ACS Sustain Chem Eng 5:11551–11557
100. Quiroz-Castañeda, R. E., & Folch-Mallol, J. L., 2013. Hydrolysis of biomass mediated by cellulases for the production of sugars. In Sustainable Degradation of Lignocellulosic Biomass—Techniques, Applications and Commercialization. IntechOpen.
101. Ragauskas AJ, Beckham GT, Biddy MJ, Chandra R, Chen F, Davis MF, Davison BH, Dixon RA, Gilna P, Keller M, Langan P, Naskar AK, Saddler JN, Tschaplinski TJ, Tuskan GA, Wyman CE (2014) Lignin valorization: improving lignin processing in the biorefinery. Science 344:1246843
102. Rahaman T, Biswas S, Ghorai S, Bera S, Dey S, Guha S, Maity D, De S, Ganguly J, Das M (2023) Integrated application of morphological, anatomical, biochemical and physico-chemical methods to identify superior, lignocellulosic grass feedstocks for bioenergy purposes. Renew Sust Energ Rev 187:113738
103. Rajendran K, Drielak E, Sudarshan Varma V, Muthusamy S, Kumar G (2018) Updates on the pretreatment of lignocellulosic feedstocks for bioenergy production–a review. Biomass Convers Biorefin 8:471–483
104. Ramli A, Khairul Anuar NASI, Bakhtiar NAA, Mohamad Yunus N, Mohamed AR (2023) Direct oxidation of *Hibiscus cannabinus* stalks to vanillin using CeO_2 nanostructure catalysts. Molecules 28:4963
105. Rana MS, Prajapati SK (2021) Microwave-assisted pretreatment of wet microalgal biomass for recovery of biofuel precursors. Fuel 305:121610
106. Rezania S, Oryani B, Cho J, Talaiekhozani A, Sabbagh F, Hashemi B, Rupani PF, Mohammadi AA (2020) Different pretreatment technologies of lignocellulosic biomass for bioethanol production: an overview. Energy 199:117457
107. Romero-Vargas A, Muñoz I, Marzo C, Díaz AB, Romero-García LI, Blandino A (2023) Ultrasound pretreatment to enhance the enzymatic hydrolysis of *Dictyota dichotoma* for sugars production. Algal Res 71:103083
108. Rongpipi S, Ye D, Gomez ED, Gomez EW (2019) Progress and opportunities in the characterization of cellulose–an important regulator of cell wall growth and mechanics. Front Plant Sci 9:1894
109. Rorke DCS, Suinyuy TN, Gueguim Kana EB (2017) Microwave-assisted chemical pretreatment of waste sorghum leaves: process optimization and development of an intelligent model for determination of volatile compound fractions. Bioresour Technol 224:590–600
110. Rusanen A, Lappalainen K, Kärkkäinen J, Tuuttila T, Mikola M, Lassi U (2019) Selective hemicellulose hydrolysis of Scots pine sawdust. Biomass Convers Biorefin 9:283–291
111. Sánchez C, Serrano L, Andres MA, Labidi J (2013) Furfural production from corn cobs autohydrolysis liquors by microwave technology. Ind Crops Prod 42:513–519
112. Santos D, Silva UF, Duarte FA, Bizzi CA, Flores EMM, Mello PA (2018) Ultrasound-assisted acid hydrolysis of cellulose to chemical building blocks: application to furfural synthesis. Ultrason Sonochem 40:81–88
113. Saratale GD, Saratale RG, Varjani S, Cho SK, Ghodake GS, Kadam A, Mulla SI, Bharagava RN, Kim D-S, Shin HS (2020) Development of ultrasound aided chemical pretreatment methods to enrich saccharification of wheat waste biomass for polyhydroxybutyrate production and its characterization. Ind Crops Prod 150:112425

114. Sarosi OP, Bammer D, Fitz E, Potthast A (2021) Partial exchange of ozone by electron beam irradiation shows better viscosity control and less oxidation in cellulose upgrade scenarios. Carbohydr Polym 265:118037
115. Sarwono A, Man Z, Muhammad N, Khan AS, Hamzah WSW, Rahim AHA, Ullah Z, Wilfred CD (2017) A new approach of probe sonication assisted ionic liquid conversion of glucose, cellulose and biomass into 5-hydroxymethylfurfural. Ultrason Sonochem 37:310–319
116. Sawhney D, Vaid S, Bangotra R, Sharma S, Dutt HC, Kapoor N, Mahajan R, Bajaj BK (2023) Proficient bioconversion of rice straw biomass to bioethanol using a novel combinatorial pretreatment approach based on deep eutectic solvent, microwave irradiation and laccase. Bioresour Technol 375:128791
117. Sharma V, Nargotra P, Bajaj BK (2019) Ultrasound and surfactant assisted ionic liquid pretreatment of sugarcane bagasse for enhancing saccharification using enzymes from an ionic liquid tolerant *Aspergillus assiutensis* VS34. Bioresour Technol 285:121319
118. Sharma V, Nargotra P, Sharma S, Sawhney D, Vaid S, Bangotra R, Dutt HC, Bajaj BK (2023) Microwave irradiation-assisted ionic liquid or deep eutectic solvent pretreatment for effective bioconversion of sugarcane bagasse to bioethanol. Energ Ecol Environ 8:141–156
119. Shen XJ, Wen JL, Mei QQ, Chen X, Sun D, Yuan TQ, Sun RC (2019) Facile fractionation of lignocelluloses by biomass-derived deep eutectic solvent (DES) pretreatment for cellulose enzymatic hydrolysis and lignin valorization. Green Chem 21:275–283
120. Shi J, Pattathil S, Parthasarathi R, Anderson NA, Kim JI, Venketachalam S, Hahn MG, Chapple C, Simmons BA, Singh S (2016) Impact of engineered lignin composition on biomass recalcitrance and ionic liquid pretreatment efficiency. Green Chem 18:4884–4895
121. Sindhu R, Binod P, Mathew AK, Abraham A, Gnansounou E, Ummalyma SB, Thomas L, Pandey A (2017) Development of a novel ultrasound-assisted alkali pretreatment strategy for the production of bioethanol and xylanases from chili post-harvest residue. Bioresour Technol 242:146–151
122. Singh SS, Lim LT, Manickavasagan A (2020) Ultrasound-assisted alkali-urea pre-treatment of *Miscanthus* × *giganteus* for enhanced extraction of cellulose fiber. Carbohydr Polym 247:116758
123. Sorn V, Chang KL, Phitsuwan P, Ratanakhanokchai K, Dong CD (2019) Effect of microwave-assisted ionic liquid/acidic ionic liquid pretreatment on the morphology, structure, and enhanced delignification of rice straw. Bioresour Technol 293:121929
124. Su XJ, Zhang CY, Li WJ, Wang F, Wang KQ, Liu Y, Li QM (2020) Radiation-induced structural changes of *Miscanthus* biomass. Appl Sci 10:1130
125. Sun RC, Sun XF, Ma XH (2002) Effect of ultrasound on the structural and physiochemical properties of organosolv soluble hemicelluloses from wheat straw. Ultrason Sonochem 9:95–101
126. Sun Y, Chmielewski AG, Instytut Chemii Techniki Jądrowej (Warsaw, Poland) (eds) (2017) Applications of ionizing radiation in materials processing. Institute of Nuclear Chemistry and Technology, Warszawa
127. Sundar S, Bergey NS, Salamanca-Cardona L, Stipanovic A, Driscoll M (2014) Electron beam pretreatment of switchgrass to enhance enzymatic hydrolysis to produce sugars for biofuels. Carbohydr Polym 100:195–201
128. Tabasso S, Mariatti F, Grillo G, Boffa L, Tibaldi PS, Cravotto G (2019) Sustainable microwave-assisted aerobic oxidation of tomato plant waste into bioaromatics and organic acids. Ind Eng Chem Res 58:8578–8584
129. Tarasov D, Leitch M, Fatehi P (2018) Lignin–carbohydrate complexes: properties, applications, analyses, and methods of extraction: a review. Biotechnol Biofuels 11:269
130. Thapa SS, Shrestha S, Sadiq MB, Anal AK (2022) Xylitol production from sugarcane bagasse through ultrasound-assisted alkaline pretreatment and enzymatic hydrolysis followed by fermentation. Sugar Tech 24:1135–1146
131. Tizazu BZ, Roy K, Moholkar VS (2018) Ultrasonic enhancement of xylitol production from sugarcane bagasse using immobilized *Candida tropicalis* MTCC 184. Bioresour Technol 268:247–258

132. Torun M (2017) Radiation pretreatment of biomass. Appl Ioniz Radiat Mater Process 2:447–460
133. Usmani Z, Sharma M, Gupta P, Karpichev Y, Gathergood N, Bhat R, Gupta VK (2020) Ionic liquid based pretreatment of lignocellulosic biomass for enhanced bioconversion. Bioresour Technol 304:123003
134. Vaidya AA, Murton KD, Smith DA, Dedual G (2022) A review on organosolv pretreatment of softwood with a focus on enzymatic hydrolysis of cellulose. Biomass Convers Biorefin 12:5427–5442
135. Velmurugan R, Muthukumar K (2012) Ultrasound-assisted alkaline pretreatment of sugarcane bagasse for fermentable sugar production: optimization through response surface methodology. Bioresour Technol 112:293–299
136. Vu ND, Tran HT, Bui ND, Vu CD, Nguyen HV (2017) Lignin and cellulose extraction from Vietnam's rice straw using ultrasound-assisted alkaline treatment method. Int J Polym Sci 1–8
137. Wang F-L, Li S, Sun Y-X, Han H-Y, Zhang B-X, Hu B-Z, Hu B-Z, Gao Y-F, Hu X-M (2017) Ionic liquids as efficient pretreatment solvents for lignocellulosic biomass. RSC Adv 7:47990–47998
138. Wang X, Fang G, Hu C, Du T (2008) Application of ultrasonic waves in activation of microcrystalline cellulose. J Appl Polym Sci 109:2762–2767
139. Wang Z, Qu L, Qian J, He Z, Yi S (2019) Effects of the ultrasound-assisted pretreatments using borax and sodium hydroxide on the physicochemical properties of Chinese fir. Ultrason Sonochem 50:200–207
140. Wu X, Chen L, He W, Qi H, Zhang Y, Zhou Y, Su X, Deng M, Wang K (2020) Characterize the physicochemical structure and enzymatic efficiency of agricultural residues exposed to γ-irradiation pretreatment. Ind Crops Prod 150:112228
141. Xiang Y, Xiang Y, Wang L (2016) Cobalt-60 gamma-ray irradiation pretreatment and sludge protein for enhancing enzymatic saccharification of hybrid poplar sawdust. Bioresour Technol 221:9–14
142. Yan D, Ji Q, Yu X, Li M, Fakayode AO, Yagoub AEA, Chen L, Zhou C (2021) Multimode-ultrasound and microwave assisted natural ternary deep eutectic solvent sequential pretreatments for corn straw biomass deconstruction under mild conditions. Ultrason Sonochem 72:105414
143. Yang CY, Fang TJ (2014) Combination of ultrasonic irradiation with ionic liquid pretreatment for enzymatic hydrolysis of rice straw. Bioresour Technol 164:198–202
144. Yang G, Wang J (2019) Ultrasound combined with dilute acid pretreatment of grass for improvement of fermentative hydrogen production. Bioresour Technol 275:10–18
145. Yang SJ, Yoo HY, Choi HS, Lee JH, Park C, Kim SW (2015) Enhancement of enzymatic digestibility of *Miscanthus* by electron beam irradiation and chemical combined treatments for bioethanol production. Chem Eng J 275:227–234
146. Yao Y, Pan Y, Liu S (2020) Power ultrasound and its applications: a state-of-the-art review. Ultrason Sonochem 62:104722
147. Yin Y, Wang J (2016) Enhancement of enzymatic hydrolysis of wheat straw by gamma irradiation–alkaline pretreatment. Radiat Phys Chem 123:63–67
148. Yuan TQ, Xu F, He J, Sun RC (2010) Structural and physico-chemical characterization of hemicelluloses from ultrasound-assisted extractions of partially delignified fast-growing poplar wood through organic solvent and alkaline solutions. Biotechnol Adv 28:583–593
149. Yuan Y, Jiang B, Chen H, Wu W, Wu S, Jin Y, Xiao H (2021) Recent advances in understanding the effects of lignin structural characteristics on enzymatic hydrolysis. Biotechnol Biofuels 14:205
150. Yunus R, Salleh SF, Abdullah N, Biak DRA (2010) Effect of ultrasonic pre-treatment on low temperature acid hydrolysis of oil palm empty fruit bunch. Bioresour Technol 101:9792–9796
151. Zabed HM, Akter S, Yun J, Zhang G, Awad FN, Qi X, Sahu JN (2019) Recent advances in biological pretreatment of microalgae and lignocellulosic biomass for biofuel production. Renew Sust Energ Rev 105:105–128

152. Zhang J, Zou D, Singh S, Cheng G (2021) Recent developments in ionic liquid pretreatment of lignocellulosic biomass for enhanced bioconversion. Sustain Energy Fuels 5:1655–1667
153. Zhang Q, Zhang P, Pei ZJ, Xu F, Wang D, Vadlani P (2015) Effects of ultrasonic vibration-assisted pelleting on chemical composition and sugar yield of corn stover and sorghum stalk. Renew Energy 76:160–166
154. Zhao X, Zhang L, Liu D (2012) Biomass recalcitrance. Part I: the chemical compositions and physical structures affecting the enzymatic hydrolysis of lignocellulose. Biofuels Bioprod Bioref 6:465–482
155. Zheng B, Yu S, Chen Z, Huo Y-X (2022) A consolidated review of commercial-scale high-value products from lignocellulosic biomass. Front Microbiol 13:933882
156. Zoghlami A, Paes G (2019) Lignocellulosic biomass: understanding recalcitrance and predicting hydrolysis. Front Chem 7:874

Chapter 3
Irradiation-assisted Direct Biomass Conversion Routes for Future Biorefineries: Mechanism, System Analysis, Product Distributions and Pathways

Arghya Datta and Anil Kumar Sarma

Abstract The demand for clean and renewable energy sources has prompted a surge in interest in the utilization of biomass as a feedstock for the manufacturing of fuel and chemicals. A potentially viable strategy involves the utilization of irradiation techniques for the direct conversion of unprocessed biomass into biofuels and high-value compounds. This book chapter presents a comprehensive examination of the existing body of research about biomass conversion methods, such as pyrolysis, saccharification and hydrolysis, utilizing irradiation. The introduction provides an overview of the reasons behind the pursuit of biomass conversion. These reasons include the plentiful and renewable nature of biomass resources, as well as the possibility for greater efficiency when compared to biochemical conversion methods. The subsequent section of the chapter delves into a comprehensive examination of diverse conversion processes that rely on irradiation, encompassing microwave, ultrasonic, gamma, electron beam irradiation, etc. The impacts of irradiation technique on biomass decomposition and product distributions are examined and analyzed. In conclusion, this study provides an overview of the process and the constraints associated with irradiation-assisted biomass conversion. It also highlights the areas that require further research in the future. In general, this book chapter presents a thorough and inclusive examination of the scientific and technological aspects related to the direct conversion of biomass by the utilization of irradiation.

Keywords Biomass irradiation · Gamma radiation · Microwave radiation · Electron beam radiation · Ultrasound · Irradiation techniques

---

A. Datta (✉) · A. K. Sarma
Chemical Conversion Division, Sardar Swaran Singh National Institute of Bioenergy, Wadala Kalan, Kapurthala, Punjab 144601, India
e-mail: arghya.datta.chem@gmail.com

A. K. Sarma
Electrochemical Conversion Division, Sardar Swaran Singh National Institute of Bioenergy, Wadala Kalan, Kapurthala, Punjab 144601, India

© The Author(s), under exclusive license to Springer Nature Switzerland AG 2024
P. Verma and B. Kumar (eds.), *Sustainable Radiation Technologies in Waste-biomass Valorization*, Green Chemistry and Sustainable Technology,
https://doi.org/10.1007/978-3-031-63941-8_3

3.1 Introduction

3.1.1 *Need for Biomass Conversion*

The growing demand for fossil fuels and environmental issues like global warming has shifted the global focus to biomass-based biofuels [74, 77]. In 2023, 29.5% of the world's power came from renewable sources due to lower costs and greener laws [32]. Most renewable energy comes from solar, biomass, wind, geothermal, and hydropower [75]. Renewable energy consumption grew quickly from 4184 TWh in 2009 to 8468 TWh in 2022 with bioenergy consumption at 675 TWh [67]. Bioenergy holds potential as a viable substitute for fossil fuels, contributing to robust economic growth, efficient energy use, national energy stability, and diminished environmental pollution and carbon emissions [21, 73]. Some renewable sources may provide heat but are not suitable for electricity or as a vehicular fuel. Solar energy is better for sparse locations with sufficient sunshine such as tropical villages or areas with vast open spaces, whereas wind is good in coastal areas for energy production. Hydroelectric, geothermal, and ocean thermal energy are all appropriate in different setups, and every renewable source showcases a unique efficiency profile [23]. The optimum choice varies according to the location, intended use, and time of year. However, biofuels are considered to be an integral part of widely used renewable energy sources because biomass is available year-round, allowing for large scale, low-carbon use [68, 78, 87].

Bioenergy accounts for about 10% of worldwide primary energy. Due to the betterment in technology and bioenergy's greater influence over biomass, bioenergy output is expected to rise from 56 to 145 EJ by 2060 [70]. Thus, agricultural residues, algae, and different lignocellulosic biomass are commonly recommended for widespread energy production [25, 45]. Perennial crops including wood, grass/herbal, starch, sugar, and oleaginous seeds can be used to provide energy [1, 26, 27].

3.1.2 *Availability of Biomass*

In 2019–2020, India had a surplus biomass of 242.5 tons with the rice/paddy crop accounting for 20% of India's biomass power potential, or 5.68 MWe out of 28.45 GWe during 2015–18. States with strong rice crop potential for biomass power production include Punjab, Tamil Nadu, Haryana, Andhra Pradesh, Chhattisgarh, Odisha, Madhya Pradesh, Telangana, West Bengal, Bihar, and Uttar Pradesh, accounting for 91.84% of the country's total rice potential (Administrative Staff College of India 2021). India's rice crop has an average surplus factor of 0.21, accounting for 21% of total biomass output. Rice surplus factors vary from 0.08 to 0.5 in selected research states. With 184.55 million tons of biomass produced versus rice, 41.72 million tons are excess, yielding a power potential of 5682.46 Mwe. After Rice (19.97%), Cotton (19.65%), Wheat (15.84%), Maize (6.24%), Potato (5.68%),

Coconut (4.11%), Banana (3.39%), Sugarcane (3.95%), Other Oilseed (2.39%), and Arhar/Tur (2.08%) have the biggest biomass power generation potential [3]. The aforementioned crops accounted for 82.41% of the overall biomass power potential of 28445.52 MWe. The remaining 17.59% of India's estimated biomass power potential came from crops such as Castor, Rapeseed, and Mustard [3].

3.1.3 Biomass Conversion Routes

There are mainly three main processes for converting biomass to energy—chemical, biochemical, and thermochemical process [4]. Chemical Conversion technologies, such as esterification, transesterification, and hydrotreatment, contribute significantly to biodiesel production, offering a feasible substitute for traditional fossil fuels. These processes involve the conversion of vegetable oils or animal fats into biodiesel through chemical reactions. The esterification and transesterification processes are relatively straightforward and widely implemented, making them highly suitable for large-scale biodiesel production [64, 89]. On the other hand, hydrotreatment, which involves using hydrogen to eliminate contaminants from the feedstock, is a more complex and capital-intensive process. Chemical Conversion primarily focuses on esterification, transesterification, and hydrotreatment technologies [34]. These processes find their application in biodiesel production, with varying levels of complexity and ease of implementation. While esterification and transesterification offer high ease of implementation and moderate complexity, hydrotreatment falls on the higher end of both complexity and ease of implementation. The cost associated with these technologies ranges from moderate to high, with biodiesel production being an established market [11, 30, 58].

Biochemical Conversion technologies, including anaerobic digestion and fermentation, contribute to the production of biogas, biomethane, ethanol, and various chemicals [22]. Anaerobic digestion employs microorganisms to break down organic matter and generate a gas mixture primarily composed of CH_4 and CO_2 called biogas [48]. This process is commonly used in industrial-scale biogas plants and wastewater treatment facilities. Fermentation, on the other hand, is utilized to transform sugars into ethanol and various important products. It is widely employed in biofuel production and the manufacturing of industrial chemicals. Both anaerobic digestion and fermentation as key technologies exhibit moderate to high complexity and ease of implementation [60]. Anaerobic digestion is primarily used for biogas and biomethane production, whereas fermentation is employed for the generation of ethanol and other valuable products such as acetic acid and citric acid. The costs associated with these technologies vary, with reactor design and handling microbial consortia being relatively expensive. The biogas market is relatively established, while the biomethane market is emerging [72, 86]. Figure 3.1 shows the typical pathway of a biochemical conversion pathway.

Thermochemical Conversion technologies, such as combustion, pyrolysis, and gasification, offer alternative pathways for energy production [35]. Combustion is

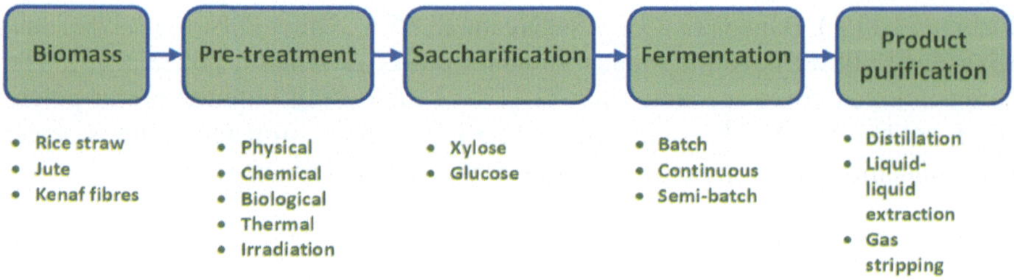

Fig. 3.1 Biochemical conversion of biomass into fuel

a well-established and widely used process for generating heat and electricity by burning biomass. Pyrolysis is a process where biomass breaks down under high temperatures without oxygen, creating biochar, bio-oil, and syngas. Gasification converts biomass into syngas, which can be further utilized for various energy applications. These thermochemical processes provide opportunities for decentralized energy systems and the production of biochar for soil amendment [46, 88]. Combustion is a simple and widely used process for power generation and heat production. Pyrolysis and gasification are moderately complex, with the former yielding biochar and bio-oil, and the latter producing syngas. The costs associated with these technologies range from relatively low to high, depending on scale and feedstock. While combustion has an established market for heat and power, pyrolysis and gasification are still in the early stages of commercialization [50].

3.1.4 Irradiation-Based Conversion

Irradiation-based biomass conversion to energy is one of the most cutting-edge new ways to make clean and sustainable energy. Ionizing radiation, usually in the form of gamma rays or electron beams, is used on biomass, which includes things like food waste and crops grown specifically for energy. Irradiation starts with complicated biochemical events inside the biomass, which break it down into simpler chemicals that can be turned into different types of energy, like biofuels or power. In addition to producing energy, this method also reduces the amount of pathogens and contaminants in biomass, making it a better choice for producing energy overall. Irradiation can also make cellulose and hemicellulose easier to reach, which are important parts of biomass. This makes the conversion process more efficient and complete [12, 52, 81, 85]. Figure 3.2 shows the different pretreatment methods including the irradiation process involved in converting biomass into energy.

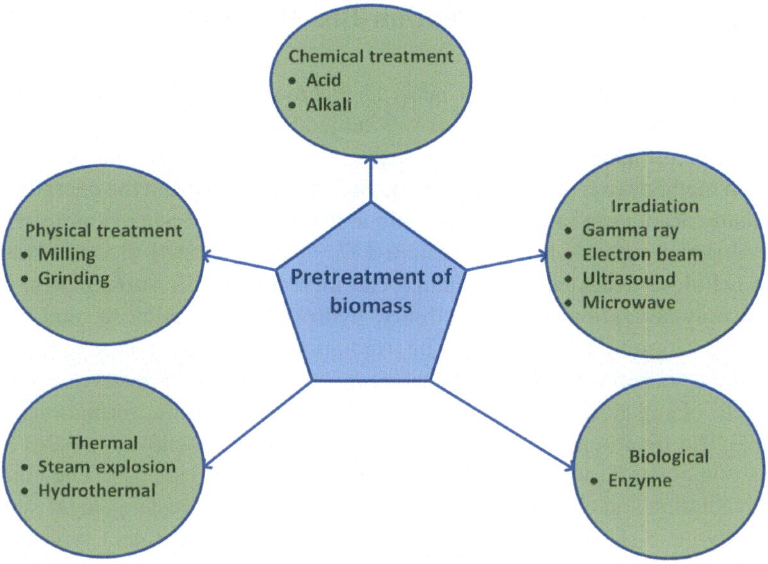

Fig. 3.2 Various methods to pretreat biomass to be converted into fuel

3.1.5 Efficiency Compared to Other Conversion Routes

Several things need to be thought about when comparing how well-irradiation-based biomass conversion works to more standard methods. Conventional methods, like thermochemical or biological conversion, often have problems with not fully utilizing the material or needing extra steps of processing [44, 76]. However, irradiation might make things more efficient generally by getting around these problems. The managed nature of the irradiation process makes it possible to finetune the energy input, which reduces losses and raises conversion rates. Along with that, irradiation can make the material much easier to digest by breaking down complex polymers better than some other ways [7, 40, 41, 80]. Even with these benefits, problems like the need for a lot of energy at the start and the creation of proper irradiation facilities need to be fixed before this method can be used on a large scale and make money [18, 37]. As technology improves and more study is done, irradiation-based biomass conversion could not only compete with but also possibly be more efficient than current methods of turning waste into energy. This shows how powerful it could be as a force for change in the search for sustainable energy options.

3.2 Fundamentals of Irradiation-Based Conversion

Irradiation stands out as a widely adopted technique for the initial treatment of biomass and biofiber. This process involves subjecting these materials to high-energy radiations like ultrasonic waves, microwaves, γ-rays, and electron beams. Different irradiation methods exert distinct effects on both the materials and the overall process mechanisms. Radiation processing technique involves radiolysis reactions using γ-rays produced from cobalt-60 or cesium-137, or concentrated electron beams to degrade cellulose. High-energy radiation can increase specific surface area, reduce cellulose polymerization and crystallinity, hydrolyze hemicellulose, and partially depolymerize lignin [7, 40, 41, 80]. The irradiation pretreatment approach typically hinges on the chosen technology. The assessment of irradiation involves a reduction in the yield of sugars during enzymatic saccharification and the formation of solid remnants. The treatment's efficacy relies on radiation frequency, exposure length, biomass composition, and medium resistance to radiation [18, 37]. The combination of irradiation and chemical pretreatment also affects decreasing sugar synthesis during enzymatic saccharification [84, 94]. Irradiation pretreatment typically falls into four distinct categories to prepare material for the enzymatic saccharification process. There are different types of exposure, such as gamma rays, electron beams, microwaves, and ultrasonication.

3.3 Microwave-Assisted Biomass Conversion

Electromagnetic waves which have 0.3 and 300 GHz frequencies make up microwaves. Most microwave systems used in homes and businesses work with frequencies between 0.9 and 2.45 GHz [24]. Microwave radiation is a type of wave motion that moves in a straight line. This radiation doesn't need a means to pass through; it can go through plastic, glass, and other non-metal objects. Microwaves can change the object in ways other than temperature. Molecular entities interact with the generated electric field induced by microwave energy, leading to the heating of the substance. Microwaves induce alterations in the polarization of entities and ions within the substance without making them hotter. This leads to physical, chemical, and biological processes [49, 65]. A lot of research has been done on how microwave preparation might affect the digestibility of different types of biomass and biofibers, like switchgrass, sweet sorghum bagasse, *Mischanthus*, and corn stover [16, 31, 93, 95].

The size of the biomass particles is decreased by mechanical pretreatment, which makes more surface area for further microwave preparation. When microwaves and chemicals are used together for pretreatment, an acid or alkali is usually used as a trigger. Alkali is used in this process to make the biomass structure bigger and get rid of the lignin part of the biomass [43]. The acid catalyst in this process, on the other hand, changes the hemicellulose and cellulose parts into small sugars like glucose and xylose, which is the major way biofuels are made [43]. Using both microwaves and steam explosions together, high pressure and temperature completely change the makeup of lignocellulosic biomass, making it easier for hydrolytic enzymes to break down cellulose and hemicellulose.

A lot of research has been done on using microwaves to pretreat biomass, like palm biomass [5, 6, 57, 71]. In their study, Akhtar et al. [5] compared the usual method with microwave-assisted chemical preparation of empty fruit bunches (EFB). They found that microwave-assisted chemical treatment removed up to 72% more lignin. Adding chemicals like acids or bases to the microwave pretreatment could make it efficient.

The microwave has been subjected to different kinds of palm waste. In their work, Lai and Idris [47] looked at how microwave pretreatment could change the oil palm trunk (OPT) and frond (OPF) and could break up internally. They were processed at 700 W and 80 °C for one hour in this study. Approximately 41.6% of the cellulose was from OPT, while 64.42% of OPF was freed. They also said that this kind of preparation is better at getting the hemicellulose and cellulose components out of the wood in both OPT and OPF. They found that the best conditions for extracting lignin from OPT were at 100 °C, 1.33 h at 900 W. This resulted in the biggest lignin reduction (22.38%). This study agrees with the previous one that the temperature, reaction time, and microwave power have a big effect on microwave preparation [6].

Researchers have also looked at how microwaves can be used to pretreat other types of biomass, such as kenaf, sago pith, sago bark waste, banana root, *Miscanthus*, and corn stover [36, 82]. A study by Ooi et al. [59] on microwave-assisted alkali pretreatment of kenaf pulp found that 50 °C is the best temperature to change crystalline cellulose to amorphous form and get more sugar out of the sample than when it wasn't treated. In a different study, it was demonstrated that subjecting sago pith to microwave treatment in water can lead to starch swelling and gel formation, making the fiber less clumpy and easier for enzymes to react with Sunarti et al. [82]. A study that looked at introducing *Miscanthus* to microwave at temperatures ranging from 130 to 200 °C found that 180 °C for 20 min is the best temperature for preparation [96]. It was found that temperature is an important part of the microwave process. Pretreatment at high temperatures makes the material more soluble, reduces the time needed for the pretreatment to work, and makes it less resistant to being broken down. On the other hand, the high-temperature cleaning process made a lot of inhibitors that hurt the enzyme-based saccharification and fermentation that followed. Zhang et al. [93] used microwave radiation to be used as the heating source for gasification of corn stover. They observed the ways variations in the supply of air, retention time and reaction temperature influence the quality of the output in the form of heating value and composition. Table 3.1 shows some typical examples of microwave-assisted conversion of different biomass to biofuels and chemicals.

Table 3.1 System analysis for microwave-assisted conversion of different biomass to biofuels and chemicals

Biomass	System analysis		References
Corn stover	Microwave-assisted air gasification	1. Heat source provided by the microwave radiation for gasification 2. The effect of airflow, retention time, and reaction temperature were studied on output concentration, heating value conversion of carbon, etc.	[93]
Empty fruit bunches	Microwave-assisted dilute acid/alkali pretreatment	1. 2.5 M NaOH decreased the 90% of hemicellulose and 10% lignin 2. NaOH and microwave radiation further delignified the biomass by 71% 3. Conventional treatment delignified the biomass only by 36%	[6]
Miscanthus	Microwave-assisted NaOH/H_2SO_4 pretreatment	1. Microwave-assisted H_2SO_4 pretreatment achieved 75.3% sugar yield and 46.7% glucose yield, outperforming conventional methods 2. Microwave conditions enhanced biomass breakdown, increasing sugar release compared to conventional heating, allowing selective glucose and xylose production	[95]
Sago bark waste	Microwave-assisted dilute H_2SO_4 pretreatment	1. Achieved a 60.2% ethanol yield with a concentration of 30.67 g/l 2. The established microwave pretreatment technique resulted in an 80% energy savings compared to previous methods for producing ethanol	[36]
Kenaf pulp	Microwave-assisted NaOH pretreatment	1. At 50 °C yielded significantly higher sugars through enzymatic hydrolysis 2. Produced high glucose yield, avoiding excessive use of concentrated acids	[59]

3.4 Ultrasound-Mediated Conversion

Ultrasonication is another type of irradiation process that is often used to prepare biomass for making energy. An ultrasonic bath or probe-type ultrasonication can be used for this process or be made using piezoelectric or magnetostrictive sensors with frequencies between 20 and 1000 kHz. The waves create a change in the pressure in the medium. There are areas of high pressure (compression) and low pressure (rarefaction) in the pressure wave that moves through the liquid. The cycle's rarefaction can stretch the molecules of the liquid apart, making holes, which are also called bubbles. While the wave traverses the liquid, the bubbles undergo expansion and contraction in response to the wave's movement sequentially pulling additional liquid particles into the bubbles as they increase in size. If the bubbles keep growing and then float to the top, they either stick together because of the forces or break apart when the wave gets smaller. This shrinkage is almost adiabatic, and it can cause temperatures and pressures of up to 1000 atm in some places [61]. The shrinkage causes radicals to form when molecules inside and around the bubbles break apart, and visible light to dissipate when excited molecules lose energy. Additionally, jets emerge from the bubbles, attaining velocities of several hundred kilometers per hour. A lot of different kinds of lignocellulosic biomass and biofiber, like corn straw, kenaf powder, kenaf bast fiber, corn meal, and corn stover [53, 55, 56, 79, 92], have been treated with ultrasound before their biochemical conversion to biofuels. This method finds application in treating tropical waste, including EFB and kenaf fiber. The main conclusion of most of the studies was that ultrasonic pretreatment can help turn waste into energy more efficiently. A study that looked at ultrasound introduction to EFB found that it could help with the acid breakdown when done at low temperature and pressure [90]. The research found that the prepared EFB produced twice as much xylose as the untreated sample after being heated at 100 °C for 40 min. The study that looked at ultrasonic pretreatment of kenaf powder in ionic liquid also found that samples that had been prepared had higher levels of reducing sugar [56]. The amount of hemicellulose in the processed wood changed in a big way in this study. Lu et al. [53] used an ultrasound NATDES/DMSO system to convert corn straw at 140 °C and 40 kHz for 1 h to yield notable platform compounds: 5-HMF (27.54%), furfural (33.64%), pentose (95.63%), and hexose (49.45%). Table 3.2 shows some typical examples of ultrasound-assisted conversion of different biomass to biofuels and chemicals.

3.5 Gamma Irradiation for Biomass Valorization

Gamma rays are a type of high-energy radioactive radiation that can easily pass through most objects. This radiation is made up of very big, high-frequency waves that depend a lot on where they come from. This technology is often used as a tracker in radiotherapy to sterilize food and medical equipment [15]. Recently, a lot of attention has been paid to using this technology, especially to prepare waste and fibers for the production of liquid biofuels. Most of the time, radioactive nuclides like cesium-137 and cobalt-60 are used in this step [52]. This treatment primarily aims to disrupt hydrogen bonds within cellulose, thereby reducing the stability of both intramolecular and intermolecular order. During this process, radiation emitted by the enclosed source will hit the biomass, initiating an impact on them. Gamma radiation's energy is passed to the plant component when the radiation hits it. This causes the atom to lose an electron, which causes the ionization process to happen. When exposed to radiation, the biomass component, mostly cellulose macromolecules, breaks apart, and different types of short- and long-lived radicals are created [29]. During the process, the number of smaller fragments increases gradually with less

Table 3.2 System analysis for ultrasound-assisted conversion of different biomass to biofuels and chemicals

Biomass	System analysis		References
Corn straw	Ultrasound-assisted Metal chloride deep eutectic system/dimethyl sulfoxide treatment	1. Optimally produces platform compounds from corn straw, with yields reaching 27.54% for 5-HMF, 33.64% for furfural, 95.63% for pentose, and 49.45% for hexose at 140 °C for 1 h	[53]
Kenaf bast fibers	Ultrasound-assisted Steam-chemical treatment	1. Treated fibers exhibited higher crystallinity but reduced surface roughness compared to those without steam treatment and untreated fibers 2. Fibers treated without steam displayed a unique nanofiber morphology on microfibers, enhancing surface roughness and suggesting potential for improved composite performance	[79]
Corn meal	Ultrasound-assisted liquefaction	1. Ultrasound pretreatment at 40 kHz, 5 min, and 60 °C enhanced corn meal liquefaction, yielding a 6.82% increase in glucose concentration for bioethanol production 2. The use of ultrasound increased ethanol concentration by 11.15%, reaching 9.67% w/w after 32 h, improving process parameters	[55]
Oil palm empty fruit bunch	Ultrasound-assisted acid hydrolysis	1. Ultrasonic pretreatment significantly enhanced xylose yield in acid hydrolysis at low temperatures, achieving a maximum yield of 58% compared to 22% without pretreatment 2. Acid hydrolysis at higher temperatures (120 and 140 °C) on ultrasonicated OPEFB fiber did not substantially increase xylose formation 3. SEM observations revealed intriguing morphological changes in OPEFB fiber under various acid hydrolysis conditions	[90]
Corn stover	Ultrasound-assisted alkali hydrolysis	1. Ultrasonic waves enhanced saccharification rates in lignocellulose pretreatment. SEM revealed no change in granulated raw material surface, while FTIR shows improved catalytic efficiency by 70% 2. Ultrasonically assisted alkali pretreatment, disrupted hydrogen bonding and decreased crystallinity, improved cellulase catalysis, enhancing lignin degradation and enzyme-based hydrolysis saccharification rates	[92]

chance of forming a polymer. This changes the structure of the biomass, making it easier for the next step, such as the enzymatic saccharification process [9].

Several types of biomass, such as jute fiber, poplar sawdust, wheat straw, and cotton-cellulose [41, 84], have been used to study the possibilities of gamma irradiation technology in the processing of biomass and biofiber. A very small number of studies looked at how gamma irradiation could be used to treat tropical waste and biofiber which was also noted. An investigation where empty fruit bunches (EFB) were subjected to gamma radiation showed that the preparation decreased the amount of lignin in the EFB and increased the amount of cellulose [42]. They used scanning electron microscopy and EDX (SEM–EDX) to observe the significant change in carbon and oxygen levels in the said biomass. Untreated EFB usually has a lot of carbon and not much oxygen. However, the study found that the carbon content went down (9% increase) and the oxygen content went down (16% decrease), which means that the lignin content in the EFB went down. Chen et al. [14] studied the production of hydrogen using dark fermentation and observed a 71.4% increase in hydrogen production at gamma radiation of 20 kGy.

Different doses (10–100 kGy) of gamma-ray have also been used to compare the effects on soft and hardwoods before they are treated [10]. 40 kGy was found to be the best setting for softwood and 90 kGy was needed to properly treat hardwood. Findings indicate that the impact of gamma-ray pretreatment varies among species, meaning that harder woods need a higher dose to damage their cell structures than softwoods. Table 3.3 shows some typical examples of gamma irradiation-assisted conversion of different biomass to biofuels and chemicals.

3.6 Electron Beam Irradiation

Before enzymatic saccharification, biomass is pretreated with electron beam irradiation. This method is employed in welding, drilling, and surface treatment [38]. The most essential accelerator features for commercial application are energy and beam strength which are classed by energy range: low (80–300 keV), medium (300 keV–5 meV), and high (above 5 meV). Irradiation with an electron beam subjects the biomass and fibrous material to a barrage of energetic electrons. Electron beam guns and accelerators accelerate electrons, and changing the irradiation dosage controls electron energy in this pretreatment procedure. It causes heating of the biomass which leads to cellulose breakdown and oxidation of the carbonyl group. The biomass component can also crosslink when subjected to irradiation [54]. Electron beam-exposed biomass has reduced mechanical strength. The hydrogen link between cellulose chains may break, making it less crystalline and more amorphous [28]. Zhang et al. [91] studied using alligator weed which is an invasive species in the region, as an energy source. They observed that at 250 mGy/s electron beam irradiation helped with the reduction of carbon and oil composition and increased gas production. Table 3.4 shows some typical examples of electron beam irradiation-assisted conversion of different biomass to biofuels and chemicals.

Table 3.3 System analysis for gamma irradiation-assisted conversion of different biomass to biofuels and chemicals

Biomass	System analysis		References
Macroalgae *Laminaria japonica*	Gamma irradiation-assisted dark fermentation	1. Gamma irradiation at 20 kGy effectively enhanced fermentative H_2 generation from Laminaria japonica, increasing potential by 71.4% compared to when irradiation was not used 2. Increased amounts of gamma irradiation negatively impacted the generation of hydrogen, reducing small organic compounds and supporting the proliferation of lactic acid bacteria, which compete with microorganisms that produce hydrogen	[14]
Oil palm empty fruit bunch	Gamma irradiation-assisted dilute NaOH pretreatment	1. Significantly increased cellulose and hemicellulose while decreasing lignin content 2. XRD analysis indicated a significant rise in crystallinity index and crystallite size post-pretreatment, while SEM/EDX reveals morphological surface changes dominated by carbon and oxygen	[42]
Khaya senegalensis and *Triplochiton scleroxylon*	Gamma irradiation-assisted hydrolysis	1. High-dose irradiation (40 kGy for softwood and 90 kGy for hardwood) significantly enhanced enzymatic hydrolysis of lignocellulosic materials (*Khaya senegalensis* and *Triplochiton scleroxylon*), increasing 3 times digestibility 2. Softwood exhibited significantly better hydrolysis at the same dosage levels	[10]
Jute	Gamma irradiation effect on mechanical and chemical strength	1. Gamma irradiation significantly reduced tensile strength, elongation, and work done at the break in jute fibers, with the most significant changes observed up to 30 kGy 2. Chemical stability decreased, fluorescence properties diminished, and infrared spectral shape changed with increasing radiation dose, indicating radiochemical alterations in lignocellulose. Thermal stability is lowered in irradiated samples, and the X-ray study shows decreased crystallinity due to gamma irradiation	[41]

3.7 Product Distributions and Pathways

Pretreatment in liquid biofuel production modifies surface shape, structure, and characteristics to promote digestibility during enzymatic saccharification. The chemical composition of biomass is also affected by pretreatment. Lignin and hemicellulose were significantly reduced in EFB after EBI pretreatment.

Primary effects of irradiation pretreatment include changes in biomass and biofiber chemical composition and surface shape. High-crystallinity biomass or fiber may be smooth and include a lot of cellulose. Great-crystallinity biomass provides great tensile strength [62].

Many studies have examined the effects of irradiation pretreatment on tropical biomass such as EFB, kenaf, rubberwood, and bamboo [33, 81, 90]. The study

Table 3.4 System analysis for electron beam irradiation-assisted conversion of different biomass to biofuels and chemicals

Biomass	System analysis		References
Alligator weed	Electron beam irradiation-assisted cracking	1. Reduced oil fraction and coke yields while increasing water and gas yields, with the optimum EBI dose rate at 250 mGy/s	[91]
Bamboo chips	Electron beam irradiation-assisted enzymatic hydrolysis	1. Degraded cellulose and hemicelluloses in bamboo chips, with continuous cellulose degradation and no significant changes in hemicelluloses' molecular weight up to 50 kg	[54]
Rice straw	Electron beam irradiation (EBI) assisted hydrolysis	1. Electron beam irradiation (EBI) at 80 kGy, 0.12 mA, and 1 meV significantly enhanced enzyme digestibility of rice straw, resulting in a glucose yield of 52.1%, compared to 22.6% for untreated straw 2. SEM analysis showed apparent surface damage to pretreated rice straw, and EBI increased the crystalline portion of the straw	[8]

found that biomass preparation affects structure and chemical characteristics. In a 300 kGy irradiation pretreatment investigation of EFB, surface morphology changed significantly [42]. Gamma rays make the firm, intact, rough, and stiff EFB brittle and flaky, according to the study. Rubberwood preparation with irradiation yields similar results. Darji et al. [19] discovered that pretreatment removed most of the fibers in rubberwood.

Alterations in the morphological structure on the surface post-pretreatment could result from irradiation disrupting the intermolecular hydrogen bond, diminishing both intra and intermolecular order in cellulose. High energy and pressure cause cellulose macromolecule scission and fragmentation with poor polymerization [66]. Irradiation also affects biomass pore size. Brunauer–Emmett–Teller (BET) study of kenaf core and cellulose showed considerable pore size increases following irradiation pretreatment [20, 33]. Before biofuel generation, high pore size is crucial for simple access.

Most studies found that biomass chemical structure affects surface morphology. Changes in crystallinity affect surface morphology [17]. XRD analysis is used to assess biomass crystallinity after irradiation. Chen et al. [13] found that the primary diffraction peak for cellulose crystallography is between 22° and 23°, whereas the secondary peak is between 16° and 18°. According to Liu et al. [52], the primary peak (the greatest intensity of the 002 lattice diffraction) is the crystalline region's diffraction intensity, whereas the secondary peak is the amorphous zone's. XRD investigation of irradiated cellulose at 10–100 kGy dosages showed that increasing doses reduced crystallinity index and crystallite size [20]. Another study on OPTT

(Oil Palm Trunk Timber) and OPF (Oil Palm Fronder) irradiation indicated that fast heating during pretreatment transformed the cellulose molecular hydrogen bond, resulting in a significant peak decrease on the main and second peaks [47].

Fourier-transform infrared spectroscopy can assess irradiation pretreatment effects in addition to XRD. This approach is commonly used to measure chemical structural changes following biomass pretreatment [39, 69]. A Fourier-transform infrared (FTIR) study on the irradiated oil palm stem and frond showed that radiation impacted all IR bands [47, 57]. Absorbance variations were seen at 3500–3200, 2840–2690, 1740–1720, 1500–1450, 1300–1001, 1315–1318, and 900–898 cm^{-1}.

In lignin, guaiacyl and syringyl propane units commonly possess one and two methoxy groups each. The existence of guaiacyl propane may restrict the expansion of biomass [63, 83]. Recognition of these structures is possible through analysis of spectra at 1509, 1464, and 1422 cm^{-1} using Fourier-transform infrared (FTIR). The decrease in spectra in this region indicates that the pretreatment removed a significant portion of the lignin present in OPT and OPF within the biomass. Irradiation removes lignin from biomass, allowing enzymes to properly breakdown cellulose and hemicellulose.

After irradiation pretreatment, FTIR measurement revealed considerable changes in biomass absorbance at 1732 and 3300 cm^{-1}, attributable to the vibration of the OH group [2]. According to Liu et al. [51], the shifting and diminution of band 2899 cm^{-1} suggests biomass disturbance from C-H shifting vibration. The study also found that intense radiation might break the cellulose intra and intermolecular H_2 bond. The breakdown of cellulose's carbonyl group was detected at 1603 cm^{-1}. Besides this area, the band shifts between 1164, 1112, and 1058 cm^{-1} due to cellulose C–O–C vibration.

3.8 Future Research and Directions

As demand for clean and renewable energy rises, irradiation-based biomass conversion for biofuels and high-value chemicals becomes more important. However, translating research discoveries into scalable, financially viable applications is difficult. One important component is scaling up irradiation-based conversion technologies. While laboratory-scale tests show promise, industrial-scale implementation requires careful optimization and technical considerations. Process scalability, reactor design, and operational characteristics must be researched to replicate controlled environment efficiency in large-scale operations. Successful commercialization requires economic feasibility, resource availability, and sustainable business concepts. These obstacles must be overcome for the industrial deployment of irradiation-based biomass conversion technology.

Complex biomass conversion processes, especially those requiring irradiation, require a multidisciplinary approach that incorporates scientific knowledge. Irradiation-based biomass conversion requires collaborations across physics, chemistry, biology, and engineering professionals to understand and improve its complex mechanisms. Multidisciplinary collaboration is needed to solve reactor design, process efficiency, and product yield issues. Irradiation-based conversion processes can be improved by using materials science and nanotechnology to build new catalysts and materials. Future research should promote cross-disciplinary collaborations, a holistic understanding of irradiation-based biomass conversion, and innovative solutions to this transformative approach's many challenges.

The existing research recommends emphasizing the need for scaling up processes and adopting multidisciplinary approaches to maximize the effectiveness of using irradiation for transforming biomass into sustainable energy and chemical resources.

3.9 Conclusion

There is a huge scope for converting biomass into biofuels and other value-added products. This will not only lessen the burden of using conventional petroleum-based fuel but also lead to carbon-zero emissions. Biomass residues are available in plenty of amounts to be converted to fuels and other valuable products. However, specific treatments are necessary to break down their complex structure. Among various pretreatment processes, irradiation pretreatment, which involves the use of gamma rays, microwaves, ultrasound, and/or electron beam stands out as an environmentally friendly and effective option. These techniques are advantageous as require fewer chemicals and are more rapid. These techniques facilitate the breakdown of plant material, making it more receptive to the extraction of valuable substances as compared to conventional pre-treatments, thus enhancing biofuel production. Despite its benefits, the energy demand as well as the use of sophisticated machinery associated with irradiation pretreatment is a challenge for large-scale production, and improvements in energy efficiency are needed before its widespread adoption.

Declaration No artificial intelligence tools were utilized in the creation and composition of this chapter.

References

1. Abdul Malek A, Hasanuzzaman M, Rahim NA (2020) Prospects, progress, challenges and policies for clean power generation from biomass resources. Clean Technol Environ Policy 22:1229–1253
2. Abobakr H, Raji M, Essabir H, Bensalah MO, Bouhfid R (2024) Enhancing oriented strand board performance using wheat straw for eco-friendly construction. Constr Build Mater 417:135135
3. Administrative Staff College of India, (Report) (2021) Evaluation study for assessment of biomass power and bagasse cogeneration potential in the country. Ministry of New and Renewable Energy, New Delhi

4. Aishwarya S, Sruthi G, Aditya M, Sivagami K, Chakraborty S (2022) Biomass energy conversion using thermochemical and biochemical technologies. Sustain Clean Energy Prod Technol 93–131
5. Akhtar J, Idris A, Teo CL, Lai LW, Hassan N, Khan M (2014) Comparison of delignification of oil palm empty fruit bunch (EFB) by microwave assisted alkali/acid pretreatment and conventional pretreatment method. Int J Adv Chem Eng Biol Sci 1(2):155–157
6. Akhtar J, Teo CL, Lai LW, Hassan N, Idris A, Aziz RA (2015) Factors affecting delignification of oil palm empty fruit bunch by microwave-assisted dilute acid/alkali pretreatment. BioResources 10(1):588–596
7. Azuma J, Tanaka F, Koshijima T (1984) Enhancement of enzymic susceptibility of lignocellulosic wastes by microwave irradiation. Hakko Kogaku Zasshi (Japan) 62(4)
8. Bak JS, Ko JK, Han YH, Lee BC, Choi I-G, Kim KH (2009) Improved enzymatic hydrolysis yield of rice straw using electron beam irradiation pretreatment. Biores Technol 100(3):1285–1290
9. Basak B, Kumar R, Bharadwaj AS, Kim TH, Kim JR, Jang M, Oh SE, Roh HS, Jeon BH (2023) Advances in physicochemical pretreatment strategies for lignocellulose biomass and their effectiveness in bioconversion for biofuel production. Bioresour Technol 369:128413
10. Betiku E, Adetunji O, Ojumu TV, Solomon BO (2009) A comparative study of the hydrolysis of gamma irradiated lignocelluloses. Braz J Chem Eng 26:251–255
11. Bezergianni S (2013) Catalytic hydroprocessing of liquid biomass for biofuels production. In: Zheng F (ed) Liquid, gaseous and solid biofuels-conversion techniques, pp 299–326. InTech, Rijeka, Croatia
12. Chaturvedi V, Verma P (2013) An overview of key pretreatment processes employed for bioconversion of lignocellulosic biomass into biofuels and value added products. 3 Biotech 3:415–431
13. Chen WH, Tu YJ, Sheen HK (2010) Impact of dilute acid pretreatment on the structure of bagasse for bioethanol production. Int J Energy Res 34(3):265–274
14. Chen Y, Yin Y, Wang J (2022) Biohydrogen production using macroalgal biomass of *Laminaria japonica* pretreated by gamma irradiation as substrate. Fuel 309:122179
15. Chmielewski AG (2023) Radiation technologies: the future is today. Radiat Phys Chem 111233
16. Choudhary R, Umagiliyage AL, Liang Y, Siddaramu T, Haddock J, Markevicius G (2012) Microwave pretreatment for enzymatic saccharification of sweet sorghum bagasse. Biomass Bioenerg 39:218–226
17. Cui T, Li J, Yan Z, Yu M, Li S (2014) The correlation between the enzymatic saccharification and the multidimensional structure of cellulose changed by different pretreatments. Biotechnol Biofuels 7(1):1–10
18. Danu S, Harsojo D, Kardha MS, Marsongko O (2012) Electron beam degradation of oil palm empty fruit bunch. Int J Environ Bioenergy 3(4):168–179
19. Darji D, Alias Y, Mohd Som F, Abd Razak NH (2015) Microwave heating and hydrolysis of rubber wood biomass in ionic liquids. J Chem Technol & Biotechnol 90(11):2050–2056
20. Driscoll M, Stipanovic A, Winter W, Cheng K, Manning M, Spiese J, Galloway RA, Cleland MR (2009) Electron beam irradiation of cellulose. Radiat Phys Chem 78(7–8):539–542
21. Duarah P, Haldar D, Patel AK, Dong C-D, Singhania RR, Purkait MK (2022) A review on global perspectives of sustainable development in bioenergy generation. Biores Technol 348:126791
22. Dutta N, Usman M, Ashraf MA, Luo G, Gamal El-Din M, Zhang S (2023) Methods to convert lignocellulosic waste into biohydrogen, biogas, bioethanol, biodiesel and value-added chemicals: a review. Environ Chem Lett 21(2):803–820
23. Elkhatat A, Al-Muhtaseb SA (2023) Combined "Renewable energy-thermal energy storage (RE–TES)" systems: a review. Energies 16(11):4471
24. Ethaib S, Omar R, Kamal SM, Biak DA (2015) Microwave-assisted pretreatment of lignocellulosic biomass: a review. J Eng Sci Technol 10:97–109
25. Goswami RK, Mehariya S, Verma P, Lavecchia R, Zuorro A (2021) Microalgae-based biorefineries for sustainable resource recovery from wastewater. J Water Process Eng 40:101747. https://doi.org/10.1016/j.jwpe.2020.101747

26. Haldar D, Purkait MK (2020) Micro and nanocrystalline cellulose derivatives of lignocellulosic biomass: a review on synthesis, applications and advancements. Carbohyd Polym 250:116937
27. Haldar D, Purkait MK (2020b) Thermochemical pretreatment enhanced bioconversion elephant grass (Pennisetum purpureum): insight on the production of sugars and lignin. Biomass Convers Biorefinery 1–14
28. Henniges U, Hasani M, Potthast A, Westman G, Rosenau T (2013) Electron beam irradiation of cellulosic materials—opportunities and limitations. Materials 6(5):1584–1598
29. Hon NS (1975) Formation of free radicals in photo-irradiated cellulose. IV. Effect of ferric ions. J Appl Polym Sci 19(10):2789–2797
30. Hossain MZ, Chowdhury MB, Jhawar AK, Xu WZ, Biesinger MC, Charpentier PA (2018) Continuous hydrothermal decarboxylation of fatty acids and their derivatives into liquid hydrocarbons using Mo/Al_2O_3 catalyst. ACS Omega 3(6):7046–7060
31. Hu Z, Wen Z (2008) Enhancing enzymatic digestibility of switchgrass by microwave-assisted alkali pretreatment. Biochem Eng J 38(3):369–378
32. International Energy Agency (2023) Renewables. https://www.iea.org/energy-system/renewables
33. Jeun J-P, Lee B-M, Lee J-Y, Kang P-H, Park J-K (2015) An irradiation-alkaline pretreatment of Kenaf core for improving the sugar yield. Renew Energy 79:51–55
34. Jha S, Nanda S, Acharya B, Dalai AK (2022) A review of thermochemical conversion of waste biomass to biofuels. Energies 15(17):6352
35. Jha S, Okolie JA, Nanda S, Dalai AK (2022) A review of biomass resources and thermochemical conversion technologies. Chem Eng Technol 45(5):791–799
36. Kannan TS, Ahmed A, Farid Nasir A (2013) Energy efficient microwave irradiation of sago bark waste (SBW) for bioethanol production. Adv Mater Res 701:249–253
37. Karthika K, Arun A, Melo J, Mittal K, Kumar M, Rekha P (2013) Hydrolysis of acid and alkali presoaked lignocellulosic biomass exposed to electron beam irradiation. Biores Technol 129:646–649
38. Kashiwagi M, Hoshi Y (2012) Electron beam processing system and its application. SEI Tech Rev 75:47–53
39. Kassim MA, Bhattacharya S (2016) Dilute alkaline pretreatment for reducing sugar production from *Tetraselmis suecica* and *Chlorella* sp. biomass. Process Biochem 51(11):1757–1766
40. Khan A, Labrie J, McKeown J (1986) Effect of electron-beam irradiation pretreatment on the enzymatic hydrolysis of softwood. Biotechnol Bioeng (U S) 28(9)
41. Khan F, Ahmad S, Kronfli E (2006) γ-Radiation induced changes in the physical and chemical properties of lignocellulose. Biomacromol 7(8):2303–2309
42. Kristiani A, Effendi N, Aristiawan Y, Aulia F, Sudiyani Y (2015) Effect of combining chemical and irradiation pretreatment process to characteristic of oil palm's empty fruit bunches as raw material for second generation bioethanol. Energy Procedia 68:195–204
43. Kumar P, Barrett DM, Delwiche MJ, Stroeve P (2009) Methods for pretreatment of lignocellulosic biomass for efficient hydrolysis and biofuel production. Ind Eng Chem Res 48(8):3713–3729
44. Kumar B, Bhardwaj N, Agrawal K, Chaturvedi V, Verma P (2020) Current perspective on pretreatment technologies using lignocellulosic biomass: an emerging biorefinery concept. Fuel Process Technol 199:106244. https://doi.org/10.1016/j.fuproc.2019.106244
45. Kumar B, Verma P (2021) Life cycle assessment: blazing a trail for bioresources management. Energy Convers Manag X, Elsevier 100063. https://doi.org/10.1016/j.ecmx.2020.100063
46. Kumar B, Verma P (2020) Biomass-based biorefineries: an important architype towards a circular economy. Elsevier, Fuel, p 119622
47. Lai LW, Idris A (2013) Disruption of oil palm trunks and fronds by microwave-alkali pretreatment. BioResources 8(2):2792–2804
48. Nair LG, Agrawal K, Verma P (2021a) An insight into the principles of lignocellulosic biomass-based zero-waste biorefineries: a green leap towards imperishable energy-based future. Biotechnol Genet Eng Rev 1–51. https://doi.org/10.1080/02648725.2022.2082223

49. Nair LG, Agrawal K, Verma P (2023) Organosolv pretreatment: an in-depth purview of mechanics of the system. Bioresour Bioprocess, Springer 10(1). https://doi.org/10.1186/s40643-023-00673-0
50. Lanjekar PR, Panwar NL, Agrawal C (2023) A comprehensive review on hydrogen production through thermochemical conversion of biomass for energy security. Bioresour Technol Rep 21:101293
51. Liu Q, Wang S, Zheng Y, Luo Z, Cen K (2008) Mechanism study of wood lignin pyrolysis by using TG–FTIR analysis. J Anal Appl Pyrol 82(1):170–177
52. Liu Y, Chen J, Wu X, Wang K, Su X, Chen L, Zhou H, Xiong X (2015) Insights into the effects of γ-irradiation on the microstructure, thermal stability and irradiation-derived degradation components of microcrystalline cellulose (MCC). RSC Adv 5(43):34353–34363
53. Lu Q, Yan D, Wu P, Chen L, Yagoub AEA, Ji Q, Yu X, Zhou C (2022) Ultrasound-NATDES/DMSO system for corn straw biomass conversion into platform compounds. Renew Energy 190:675–683
54. Ma X, Zheng X, Zhang M, Yang X, Chen L, Huang L, Cao S (2014) Electron beam irradiation of bamboo chips: degradation of cellulose and hemicelluloses. Cellulose 21:3865–3870
55. Nikolić S, Mojović L, Rakin M, Pejin D, Pejin J (2010) Ultrasound-assisted production of bioethanol by simultaneous saccharification and fermentation of corn meal. Food Chem 122(1):216–222
56. Ninomiya K, Kamide K, Takahashi K, Shimizu N (2012) Enhanced enzymatic saccharification of kenaf powder after ultrasonic pretreatment in ionic liquids at room temperature. Biores Technol 103(1):259–265
57. Nomanbhay SM, Hussain R, Palanisamy K (2013) Microwave-assisted alkaline pretreatment and microwave assisted enzymatic saccharification of oil palm empty fruit bunch fiber for enhanced fermentable sugar yield. J Sustain Bioenergy Syst 3(1):29068
58. Norouzi N, Talebi S (2020) An overview on the green petroleum production. Chem Rev Lett 3(1):38–52
59. Ooi BG, Rambo AL, Hurtado MA (2011) Overcoming the recalcitrance for the conversion of Kenaf pulp to glucose via microwave-assisted pre-treatment processes. Int J Mol Sci 12(3):1451–1463
60. Osman AI, Mehta N, Elgarahy AM, Al-Hinai A, Al-Muhtaseb AH, Rooney DW (2021) Conversion of biomass to biofuels and life cycle assessment: a review. Environ Chem Lett 19:4075–4118
61. Pilli S, Bhunia P, Yan S, LeBlanc R, Tyagi R, Surampalli R (2011) Ultrasonic pretreatment of sludge: a review. Ultrason Sonochem 18(1):1–18
62. Poletto M, Ornaghi Junior HL, Zattera AJ (2014) Native cellulose: structure, characterization and thermal properties. Materials 7(9):6105–6119
63. Pu Y, Hu F, Huang F, Davison BH, Ragauskas AJ (2013) Assessing the molecular structure basis for biomass recalcitrance during dilute acid and hydrothermal pretreatments. Biotechnol Biofuels 6(1):1–13
64. Qamar OA, Jamil F, Hussain M, Bae S, Inayat A, Shah NS, Waris A, Akhter P, Kwon EE, Park YK (2023) Advances in synthesis of TiO_2 nanoparticles and their application to biodiesel production: a review. Chem Eng J 460:141734
65. Quitain AT, Sasaki M, Goto M (2013) Microwave-based pretreatment for efficient biomass-to-biofuel conversion. Pretreat Tech Biofuels Biorefineries 117–130
66. Ramos LP (2003) The chemistry involved in the steam treatment of lignocellulosic materials. Quim Nova 26:863–871
67. Ritchie H, Rosado P (2020) Electricity mix. https://ourworldindata.org/electricity-mix
68. Ruiz-Mercado GJ, Segovia-Hernández JG, Castro-Montoya AJ (2018) Transformation towards sustainable bioenergy systems. Clean Technol Environ Policy 20:1385–1385
69. Sabiha-Hanim S, Noor MAM, Rosma A (2015) Fractionation of oil palm frond hemicelluloses by water or alkaline impregnation and steam explosion. Carbohyd Polym 115:533–539

70. Scarlat N, Dallemand J-F (2019) Future role of bioenergy. In: Lago C, Caldés N, Lechón Y (eds) The role of bioenergy in the bioeconomy. Elsevier, Amsterdam, Netherlands, pp 435–547
71. Shahriarinour M, Wahab MNA, Ariff AB, Mustafa S, Mohamad R (2011) Effect of various pretreatments of oil palm empty fruit bunch fibres for subsequent use as substrate on the performance of cellulase production by *Aspergillus terreus*. BioResources 6(1):291–307
72. Sharma I, Rackemann D, Ramirez J, Cronin DJ, Moghaddam L, Beltramini JN, Te'o J, Li K, Shi C, Doherty WO (2022) Exploring the potential for biomethane production by the hybrid anaerobic digestion and hydrothermal gasification process: a review. J Clean Prod 362:132507
73. Singh A, Adsul M, Vaishnav N, Mathur A, Singhania RR (2017) Improved cellulase production by Penicillium janthinellum mutant. Indian J Exp Biol 55:436–440
74. Singh A, Patel AK, Adsul M, Mathur A, Singhania RR (2017) Genetic modification: a tool for enhancing cellulase secretion. Biofuel Res J 4(2):600–610
75. Singh N, Singhania RR, Nigam PS, Dong C-D, Patel AK, Puri M (2022) Global status of lignocellulosic biorefinery: challenges and perspectives. Biores Technol 344:126415
76. Singh R, Krishna BB, Kumar J, Bhaskar T (2016) Opportunities for utilization of non-conventional energy sources for biomass pretreatment. Biores Technol 199:398–407
77. Singhania RR, Patel AK, Pandey A, Ganansounou E (2017) Genetic modification: a tool for enhancing beta-glucosidase production for biofuel application. Biores Technol 245:1352–1361
78. Singhania RR, Ruiz HA, Awasthi MK, Dong C-D, Chen C-W, Patel AK (2021) Challenges in cellulase bioprocess for biofuel applications. Renew Sustain Energy Rev 151:111622
79. Sosiati H, Harsojo (2014) Effect of combined treatment methods on the crystallinity and surface morphology of Kenaf bast fibers. Cellul Chem Technol 48(1–2):33–43
80. Sridar V (1998) Microwave radiation as a catalyst for chemical reactions. Curr Sci 446–450
81. Sun F, Jiang Z, Sun Q, Lu F (2014) Changes in chemical composition and microstructure of bamboo after gamma ray irradiation. BioResources 9(4):5794–5800
82. Sunarti T, Dwiko M, Derosya V, Meryandini A (2012) Effect of microwave treatment on acid and enzymes susceptibilities of sago pith. Procedia Chem 4:301–307
83. Taherzadeh MJ, Karimi K (2008) Pretreatment of lignocellulosic wastes to improve ethanol and biogas production: a review. Int J Mol Sci 9(9):1621–1651
84. Takacs E, Wojnarovits L, Földváry C, Hargittai P, Borsa J, Sajo I (2000) Effect of combined gamma-irradiation and alkali treatment on cotton–cellulose. Radiat Phys Chem 57(3–6):399–403
85. Tang A-M, Liang W-Z (2000) Studies on the structure changes of fast-growing wood fiber treated by ultrasonic wave. Tech Acoust 19(2):78–82
86. Tursi A (2019) A review on biomass: importance, chemistry, classification, and conversion. Biofuel Res J 6(2):962–979
87. Volli V, Gollakota ARK, Purkait MK, Shu C-M (2020) Conversion of waste biomass to bio-oils and upgradation by hydrothermal liquefaction, gasification, and hydrodeoxygenation. Biorefinery Altern Resour Target Green Fuels Platf Chem 285–315
88. Wang Y, Wu JJ (2023) Thermochemical conversion of biomass: potential future prospects. Renew Sustain Energy Rev 187:113754
89. Xie W, Li J (2023) Magnetic solid catalysts for sustainable and cleaner biodiesel production: a comprehensive review. Renew Sustain Energy Rev 171:113017
90. Yunus R, Salleh SF, Abdullah N, Biak DRA (2010) Effect of ultrasonic pre-treatment on low temperature acid hydrolysis of oil palm empty fruit bunch. Biores Technol 101(24):9792–9796
91. Zhang B, Wang J, Aldeen AS, Zhang J, Mwenya S, Wang Y, Xu Z, Zhang H (2022) Enhancing liquid hydrocarbon production from invasive plant via ex-situ catalytic fast pyrolysis coupled with electron-beam irradiation cracking. J Anal Appl Pyrolysis 166:105633
92. Zhang Y, Fu E, Liang J (2008) Effect of ultrasonic waves on the saccharification processes of lignocellulose. Chem Eng & Technol Ind Chem-Plant Equip-Process Eng-Biotechnol 31(10):1510–1515
93. Zhang Y, Fu W, Cui L, Maqsood T, Li B (2023) Experimental microwave-assisted air gasification of biomass for syngas production. Fuel 339:126954

94. Zhu S, Wu Y, Yu Z, Liao J, Zhang Y (2005) Pretreatment by microwave/alkali of rice straw and its enzymic hydrolysis. Process Biochem 40(9):3082–3086
95. Zhu Z, Macquarrie DJ, Simister R, Gomez LD, McQueen-Mason SJ (2015) Microwave assisted chemical pretreatment of *Miscanthus* under different temperature regimes. Sustain Chem Process 3:1–13
96. Zhu Z, Simister R, Bird S, McQueen-Mason SJ, Gomez LD, Macquarrie DJ (2015) Microwave assisted acid and alkali pretreatment of *Miscanthus* biomass for biorefineries. Aims Bioeng 2(4)

Chapter 4
Biomass Vulnerability to Irradiation-Based Treatment for Biorefinery Application: Analytical Techniques for Mechanistic Understanding

Pooja Thakur, Aruna Thakur, and Subhajit Majumder

Abstract Biomass is a promising solution to energy and environmental challenges when used as a renewable resource for producing biofuel and high-value compounds. To enhance the effectiveness of the complex procedures involved in converting biomass into these products, irradiation-based treatments such as gamma irradiation and electron beam irradiation have become viable options. However, the exact mechanisms by which irradiation impacts the susceptibility of biomass remain multifaceted and not yet fully understood. This chapter takes a comprehensive approach to uncover the intricacies of how irradiation changes the structure, composition, and reactivity of biomass. Advanced analytical techniques, including spectroscopy, microscopy, and chromatography, are used to evaluate changes in lignocellulosic components like cellulose, hemicellulose, and lignin. Additionally, the impact of irradiation on biomass microbial communities and enzymatic activities is explored to determine potential synergies or hindrances during subsequent conversion steps. Furthermore, this chapter emphasizes the significance of comprehending the mechanistic intricacies underlying the susceptibility of biomass to irradiation-based treatments. Such insights are crucial for optimizing biofuel and high-value compound production processes, promoting sustainable energy solutions, and advancing the utilization of biomass resources in a transformative manner. By understanding how irradiation affects biomass, we can strike a balance between favorable modifications and potential adverse effects, such as increased energy consumption or reduced biomass stability. This will facilitate the extraction of value-added compounds and

P. Thakur · S. Majumder (✉)
Department of Chemical Engineering, National Institute of Technology (NIT) Hamirpur, Hamirpur, India
e-mail: subhajit@nith.ac.in

A. Thakur
Department of Mechanical and Manufacturing Engineering, National Institute of Advanced Manufacturing Technology (NIAMT), Ranchi, India

© The Author(s), under exclusive license to Springer Nature Switzerland AG 2024
P. Verma and B. Kumar (eds.), *Sustainable Radiation Technologies in Waste-biomass Valorization*, Green Chemistry and Sustainable Technology,
https://doi.org/10.1007/978-3-031-63941-8_4

promote higher biofuel yields, thereby promoting the use of biomass as a renewable energy source and contributing to a more sustainable future.

Keywords Biomass · Irradiation-based treatments · Biofuel · Analytical techniques · Renewable energy source

4.1 Introduction

The extraction of non-renewable fossil fuels poses a significant threat to the environment, including pollution and global warming. Therefore, it is imperative to seek alternative and sustainable approaches to meet our chemical, material, and energy needs. Biomass is the ideal resource for obtaining these commodities. The lignocellulosic biomass, which encompasses agricultural and forestry crops, residues, urban waste, and wood, represents an abundant, renewable resource on Earth [112, 133]. This biomass is characterized by its renewability, carbon neutrality, and environmental sustainability. Structurally, the biomass is composed primarily of cellulose, hemicellulose, and lignin, and its conversion presents opportunities for an extensive range of biofuels, biomaterials, and biochemicals [15, 119, 131]. The use of biomass as a feedstock offers a potentially sustainable alternative to replace crude oil in the production of chemicals and energy. This helps minimize waste and pollution, thereby promoting a cleaner and healthier environment [59]. It is high time that we embrace this alternative and sustainable approach to meet our energy needs.

Biomass materials can be converted into renewable bioenergy through various thermochemical conversion methods, including combustion, torrefaction, pyrolysis, liquefaction, and gasification [17, 79]. The challenge, however, remains in the small-scale production of biofuels such as bioethanol, biohydrogen, and biomethane. These low yields have impeded the exploration of this potential and the resulting economic benefits that could be derived from it. Hence, there is a pressing need to address this challenge and optimize the biofuel production process to maximize the potential of lignocellulosic and microalgal biomass for biofuel production. The pretreatment process aims to break down the structure of the biomass and separate its components, preparing them for further conversion into biofuels [57].

The process of converting biomass into fermentable sugars and biofuels is a critical step in capitalizing on renewable energy sources. Before bioconversion, it is necessary to pretreat the biomass to enhance efficiency and facilitate the production of biofuels. Pretreatment requirements differ depending on the composition and nature of the biomass feedstock. Each type of biomass feedstock necessitates a distinct pretreatment approach to fractionate and assist in component recovery and conversion. It is crucial to identify the suitable pretreatment method based on the structure and composition of the biomass [80]. For example, the highly recalcitrant nature of lignocellulosic biomass demands a varied pretreatment approach related to microalgal biomass. Therefore, selecting the most effective pretreatment method is essential to optimizing the efficiency of the bioconversion process.

The primary objective of the treatment process is to discharge sugar solution for biochemical and biofuel production. Numerous studies have shown that pretreated biomass yields better outcomes compared to untreated ones. While the pretreatment process can be expensive and energy-intensive, especially for lignocellulosic biomass, the method parameters for every single pretreatment vary depending on the category of solvent, procedures, and biomass used. Both microalgal and lignocellulosic biomass undergo four classes of pretreatment, namely physical, chemical, biological, and newly developed methods. The types chosen depend on the biomass's components, as lignocellulosic biomass, which contains lignin, cellulose, and hemicellulose, is more resistant to breakdown. In some cases, pretreatment may be unnecessary for microalgal biomass, particularly for high-value chemicals like alginic acid or agar. Moreover, lignocellulosic biomass, and other discarded materials, such as manure and sewage sludge, have undergone pretreatment, aiming to break down cell walls, rupture cells, and enhance organic solubilization. The ultimate objective is to afford a more usable feed for enzymatic hydrolysis, which enables consequent conversion into biofuel. Various thermochemical conversion techniques can be employed to produce biofuel, including pyrolysis, gasification, hydrothermal liquefaction, torrefaction, and hydrothermal carbonization. Conventional heating has been widely used for the conversion process associated with most of these thermochemical conversion technologies [10, 31, 63, 86, 122, 129].

There are four generations of biofuels, which are classified based on the type of biomass and production techniques used. The first generation includes biofuels resulting from sources such as starch, animal fats, and vegetative oil. Second-generation (2G) biofuels are produced using lignocellulosic biomass, while third-generation (3G) biofuels are obtained from algae [39]. Fourth-generation biofuels, on the other hand, are developed from genetically modified feedstocks. These feedstocks are engineered to enhance the yield of biofuels and to make them more efficient. The use of genetically modified feedstocks is a promising field in biofuel research, as it has the potential to increase the yield of biofuels significantly and make them a sustainable and renewable source of energy [7, 18, 19, 60, 103]. In recent times, there has been a significant focus on the production of 2G and 3G biofuels, with greater attention given to microalgal and lignocellulosic biomass [43, 99]. However, the process of converting these types of biomass into biofuels is more complicated than with other types of biomass such as those containing starch or sugar. This is because the biomass must undergo a pretreatment process before it is transformed into the required biofuel, as shown in Fig. 4.1. Currently, thermochemical methods are used to achieve pretreatment, which is primarily energy-intensive. In conclusion, biofuels offer an alternative to traditional fossil fuels and are produced from organic matter using a range of methods. Despite their potential as a renewable fuel source, the conversion of these types of biomass into biofuels is more complicated and energy intensive.

Biomass pretreatment on a global scale is a promising field, but it requires significant research developments. Previous studies have examined lignocellulose and the formation of inhibitory by-products that hinder the enzymatic biocatalyst, as well as numerous lignocellulose pretreatment processes and economic improvements of

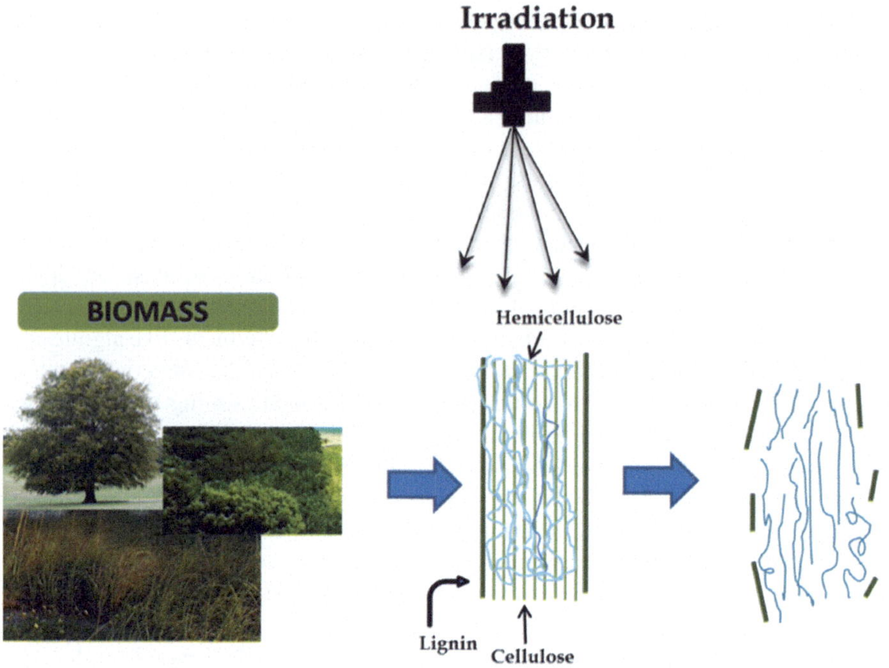

Fig. 4.1 Biomass pretreatment through irradiation

lignin by-products. To keep researchers informed about the latest developments in biomass pretreatment, a comprehensive review of microalgal and lignocellulosic biomass pretreatment is essential. The present chapter provides a comprehensive review of the latest pretreatment developments concerning microalgal and lignocellulosic biomass. The primary objective is to evaluate the feasibility of sustainable production of biofuels and bio-based products. The potential of these resources to contribute to a sustainable future is promising.

4.1.1 Advantages and Challenges of Biomass Composition and Conversion

The pretreatment process is crucial before enzymatic saccharification can take place. It is imperative to find the most efficient pretreatment process as it can significantly reduce production costs, ultimately leading to a lower final product price. The irradiation pretreatment process is currently the most promising approach with several advantages over other pretreatment methods.

1. The irradiation pretreatment is eco-friendly since it requires fewer chemicals for the process.

2. In comparison to other techniques, such as biological pretreatment, which necessitates more than seven days to eliminate lignin material from biomass, the present method requires less than ten minutes.
3. The irradiation process stands out as the most selective method for degrading biomass components, surpassing chemical pretreatment in terms of precision. Unlike the latter, which may degrade cellulose and hemicellulose, irradiation targets the desired biomass component while leaving the others unscathed. This approach offers unmatched efficiency and effectiveness in biomass degradation, making it an indispensable tool in various industries. Its superiority over other methods is undeniable and well-established.
4. Studies have demonstrated that irradiated biomass generates fewer inhibitors that might hinder the enzymatic saccharification and fermentation process. The utilization of water as a soaking buffer during pretreatment has been shown to produce no inhibitors, which significantly improves the performance of the enzymatic saccharification and fermentation process for the production of liquid biofuels.

Although the irradiation pretreatment procedure is simple, it consumes a lot of energy, making it unsuitable for commercial use. Moreover, a specialized reactor is necessary for large-scale biomass pretreatment, which can be costly, energy-intensive, and limiting. As a result, this drawback can increase operational costs [22, 78, 91, 104]. The study of biomass characterization faces several significant challenges that stem from multiple factors, including the intricate nature of biomass, the multiphase bioconversion processes that utilize biomass as feedstock, and the requirement for experimental validation of advanced analytical techniques employed for biomass characterization. Undoubtedly, persistent research efforts will overcome the barriers and lead to the optimization of bioconversion and biorefinery processes. It is important to note that no analytical technique can offer all the necessary facts concurrently with ideal resolution and high sensitivity. Therefore, complementary methods are typically required to comprehensively understand the physical and chemical fundamentals of biomass recalcitrance during bioconversion processing. This approach will inevitably alleviate the challenges associated with experimental validation.

4.2 Irradiation-Based Treatment Techniques

Pretreatment techniques play a crucial role in a range of industries, including biorefineries, bioenergy, and food production, among others. These techniques can be broadly classified into four categories: physical, biological, chemical, and irradiation methods which are mentioned in Fig. 4.2. Each category employs a unique approach to breaking down the complex structure of the cell wall, resulting in a distinct range of yield and products. To gain a comprehensive understanding of these techniques, it is imperative to review the current knowledge of each category. This review will provide the necessary insights to optimize the pretreatment process and enhance the quality and quantity of the end products. Within the realm of various physical techniques, irradiation is emerging as a promising alternative for biomass pretreatment. The process exhibits several features that make it approachable, effective, and energy-efficient, alongside a low energy requirement. Moreover, irradiation is highly selective and easier to handle. The process employs high-energy radiations to break down the chemical bonds of biomass, resulting in a change in its characteristics, such as cellulose crystallinity, depolymerization of lignin, and hydrolysis of hemicelluloses [19]. The ability of irradiation hinges on aspects such as radiation frequency, biomass composition, and time of exposure [99]. Four different radiation methods are generally used in this process: microwaves, gamma rays, electron beams, and ultrasonication. The mechanism of irradiation varies according to the method used.

The utilization of irradiation techniques for biomass pretreatment has been found to significantly enhance biomass digestibility. Irradiation is an indisputably effective pretreatment method that has undergone significant progress from bench scale to pilot scale over the last thirty years, as suggested by recent literature. Ultra sonication

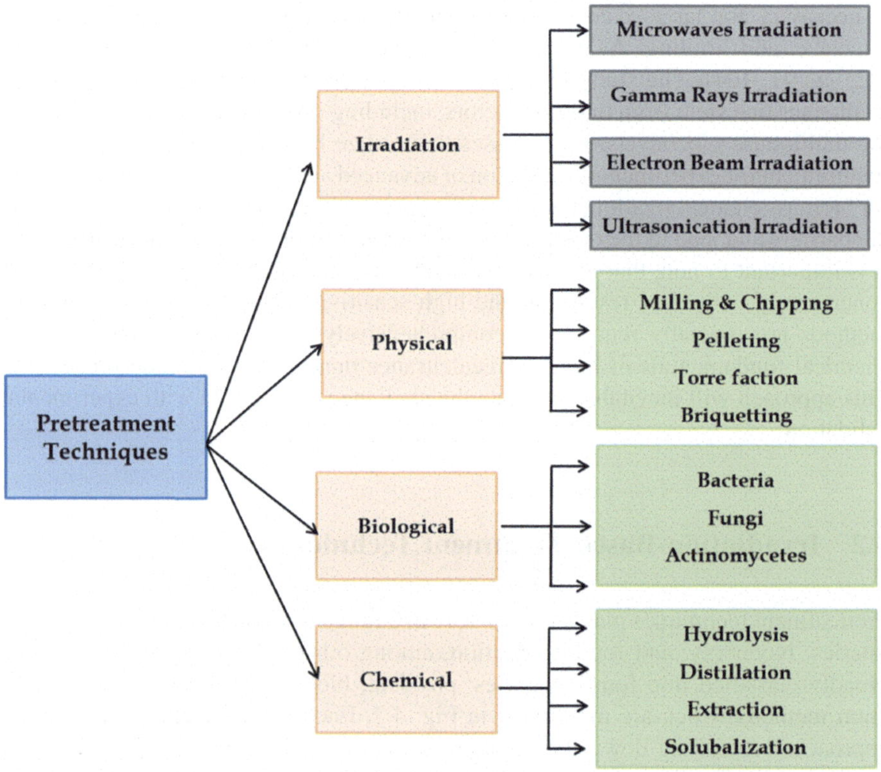

Fig. 4.2 Techniques for biomass pretreatment

generates oxidizing radicals attacking the lignocellulose matrix, which disrupts the chemical networks of lignin, leading to the formation of cavitation bubbles that grow in size and convert unevenly, generating the high temperature and pressure required for pretreatment [103]. A study [43] has unequivocally demonstrated that combining microwave and ultrasonication techniques has a thoughtful impact on the pretreatment process by enhancing hydrolysis and increasing sugar revenue. Although there has been only little information on gamma-ray irradiation, their consequences suggest that it can effectively decrease the lignin content and increase the cellulose content, indicating its potential as a highly promising pretreatment method [18].

4.2.1 Microwave Irradiation

Microwaves are a form of electromagnetic radiation that falls within the wavelength range of 1 mm to 1 m. Microwaves are non-ionizing radiation that selectively transfers energy to different substances, occupying a position in the electromagnetic spectrum between 300 and 300,000 MHz [47]. Over the past three decades, researchers have consistently reported favorable outcomes in lignocellulosic pretreatment using microwave radiation [55, 56, 58]. This technology has now moved from laboratory to pilot scale, proving its effectiveness and potential for large-scale applications [63]. Microwave-supported pretreatment methods for lignocellulosic biomass can be categorized into binary principles: (a) Microwave-supported solvolysis, which depolymerizes biomass to produce value-added chemicals under mild temperatures (<200 °C), and (b) Microwave-supported pyrolysis of lignin without oxygen that converts biomass to bio-oil or bio-gases under high temperatures (>400 °C). Both of these groups of technologies utilize catalysts to achieve their objectives. However, the focus of this discussion is largely on the utilization of microwave-assisted pyrolysis, as this method is increasingly vital in light of the ongoing energy shortage and the emphasis on sustainable development in many countries. Notably, microwave radiation offers several advantages over conventional heating methods. These advantages include greater efficiency, reduced reaction time, and improved selectivity, among others. As such, microwave-assisted pyrolysis represents a promising avenue for research and development in a variety of fields. Notably, it enables rapid heat transfer and has short reaction times. Moreover, it affords selectivity and uniform volumetric heating performance, rendering it an attractive option for those seeking to optimize energy efficiency. Additionally, the degradation or formation of side products is minimal, further highlighting the benefits of microwave radiation. Another advantage of microwave hydrothermal pretreatment is the ability to effectively remove acetyl groups in hemicellulose, which is attributed to the hot spot effect of microwave irradiation [24]. This may be particularly relevant to those seeking to reduce acetyl group levels in hemicellulose for various purposes, including the production of biofuels. Overall, the benefits of microwave radiation make it a compelling option for those seeking to optimize energy efficiency, minimize side products, and selectively target specific components of materials.

In the realm of conventional heating, the transfer of energy takes place from the outer surface of a material toward its core through the process of conduction. This mechanism leads to a potential discrepancy, wherein the outer surface may overheat, while the inner region remains relatively cool. On the other hand, microwaves generate heat by a direct transformation of electromagnetic energy into heat at the molecular level [48]. Microwave heating is a reliable method for achieving uniform heating in various materials [12]. Based on their respective microwave response, materials can be classified into three categories, namely: insulators, absorbers, and conductors. Insulators, such as glass and ceramics, are transparent to microwaves. Conductors, such as metals, reflect microwaves due to their high conductivity. Absorbers or dielectrics, on the other hand, can convert microwaves into heat. Biomass, which is characterized as a low-loss material, requires support from materials that facilitate rapid heating, such as activated carbons, pyrites, graphite, and charcoal.

4.2.2 Ultrasound Irradiation

Ultrasound technology is used in various industries for different purposes [81]. In certain industries, such as chemical processing and medical diagnostics, ultrasound is used as a tool. Typically, ultrasound with a frequency of 20 kHz–1 MHz is utilized in the chemical treating industry, whereas more frequencies are used for remedial and diagnostic purposes [125]. In the context of biomass pretreatment, ultrasound can be used to change the surface morphology and generate oxidizing radicals that chemically attack the lignocellulosic matrix. The use of ultrasound-supported alkali pretreatment can also improve lignin deficiency and enzymatic saccharification degrees by distorting hydrogen bonds among molecules of lignocellulosic and reducing their crystallinity. Nevertheless, it is important to note that ultrasound vibration frequency is insufficient to alter the surface of raw material biomass units [71]. A study was conducted by Subhedar et al. [111] on the utilization of ultrasound-assisted delignification and enzymatic hydrolysis of three types of biomass, namely groundnut shells, pistachio shells, and coconut coir. The study has reported that using ultrasound treatment for the delignification of biomass has resulted in an increase of approximately 80–100% in comparison to the conventional method of using alkali treatments. The ultrasound treatment was carried out with a biomass addition of 0.5%, ultrasound energy of 100 W, and duty series of 80% for 70 min. Under optimized conditions of 0.08% w/v enzyme loading, 3.0% w/v substrate loading, 60W ultrasound power, and 70% duty cycle for 6.5 h, the reducing sugar yields for groundnut shells, pistachio shells, and coconut coir were 21.3 g/L, 18.4 g/L, and 23.9 g/L, respectively, in the case of ultrasound-assisted enzymatic hydrolysis. This study demonstrates the potential of ultrasound technology in enhancing the efficiency of biomass pretreatment and enzymatic hydrolysis in various industries.

The utilization of ultrasound in the extraction of sugar from lignocellulosic waste paper has been shown to have a significant impact. A study conducted by Subhedar et al. [110] demonstrated that the use of ultrasound at a power of 60W and a duty cycle of 70% for 6.5 h, along with a substrate loading of 3.0% (w/v) and a cellulase loading of 0.8% (w/v), resulted in a sugar yield increase by a factor of ~2.4. In addition, ammonia pretreatment of sugarcane bagasse has shown significant potential in cellulose recovery. Ramadoss and Muthukumar [93] conducted a study combining ultrasound with ammonia pretreatment of sugarcane bagasse, which resulted in 95.78% cellulose recovery and 58.14% delignification. The sonication time was 45 min, with 400W energy, 100% amplitude, and 24 kHz frequency, along with a biomass addition of 1 g for each 10 ml of 10% NH_3 at 80v°C. Furthermore, the combination of ammonia with ultrasound has been found to have a synergistic effect on reducing by-product formation, enabling treatment at moderate temperatures, and reducing cellulose crystallinity. This observation is consistent with recent work carried out by Xu et al. [130] on ultrasound-assisted dilute aqueous ammonia (2.0% w/v aqueous ammonia) pretreatment of corn cob, corn stover, and sorghum stalk, using ultrasound at 90W power and 50 kHz frequency. The maximum enzymatic sugar yield was nearly 81% in corn cob, 66% in corn stover, and 57% in sorghum stalk. In addition, Ravindran et al. [95] studied the pretreatment of spent coffee waste by ultrasound-assisted potassium permanganate, which resulted in a 98% cellulose recovery and 46% lignin removal. However, limited literature has been found that provides detailed analyses explaining the underlying mechanism of ultrasound-assisted biomass torrefaction. Therefore, the primary objective of this research is to develop a model for an ultrasound-assisted biomass torrefaction system. This model will provide a better understanding of the torrefaction process and its underlying mechanism by incorporating the effects of ultrasound [113].

4.2.3 Gamma-Ray Irradiation

Gamma-ray radiation, derived from radioisotopes such as Co-60 or Ce-137, has been explored as a lignocellulosic pretreatment methodology. This radiation type can simply penetrate the lignocellulosic organization, triggering modifications of the lignin and leading to breakdowns in the cellulose crystal regions. The formation of free radicals facilitates the breakdown process, which subsequently decays quickly from the amorphous regions after the radiation is terminated. The decay from the crystalline regions at a certain period further causes degradation of the biomass [46]. In a study by Liu et al. [69], the efficacy of γ-irradiation on the bioconversion efficiency of microcrystalline cellulose (MCC) was compared with other pretreatment methods, including ionic liquids, acidic aqueous ionic liquids, 1% HCL, and 1% H_2SO_4. The study involved the irradiation of sugarcane bagasse with gamma radiation from ^{60}Co, with a dose range of 0–2000 kGy. Following this, a comprehensive analysis of the physical, chemical, morphological, and structural properties of the material was conducted. The results of the study reveal that exposure to ionizing radiation led to a substantial 48% increase in soluble sugars. However, it also resulted in a significant decrease in cellulose content from 37.4 to 18.1% at the highest dose of

2000 kGy. These findings provide valuable insights into the potential use of gamma radiation in the processing of sugarcane bagasse [120]. Simultaneously, the reported study shows that irradiating lignocellulosic biomass before enzymatic hydrolysis is highly efficient. It breaks down wood's resistance to hydrolysis, producing glucose without heating or harsh chemicals. A significant portion of cellulose is depolymerized into oligosaccharides and cellobiose, replacing the need for endo 1,4 β-glucanase and cellobiohydrolase hydrolysis [1].

4.2.4 Electron Beam (EB) Irradiation

Ionizing radiation using electron beams (EB) from a linear accelerator is an unequivocally potent pretreatment method that efficiently disrupts the structure of cell wall polymers found in lignocellulosic biomass. This process involves using accelerated beams of electrons to induce cross-link development or chain scission, decrystallization, and/or a reduction in polymerization [40]. The purpose of this process is to generate free radicals that significantly alter the structure of lignin, cellulose, and hemicellulose. A multitude of studies have unequivocally demonstrated that irradiation of sugar maple using EB at prescriptions up to 1000 kGy unequivocally depolymerizes cellulose and hemicellulose structures to varying degrees, resulting in an unequivocal increase in the yield of phenolics [72]. Additionally, research has unequivocally shown that the ideal EB irradiation dosage to handle Korean *Miscanthus sinensis* before enzymatic hydrolysis for fermentable form of sugar manufacture is unequivocally 500 kGy [132]. It is imperative to note that EB is primarily effective in depolymerizing cellulose, but it can be combined with another method for pretreatments, such as alkali, steam explosion, or hydrolyzed hemicellulose and lignin. In conclusion, EB is an unquestionably effective and reliable method for pre-treating lignocellulosic biomass, with an unequivocally wide range of potential applications in various industrial processes (Fig. 4.3).

4.3 Microbial Communities for Biofuel Production

The use of surface-displayed bioactive proteins on engineered microbial cells presents great potential as biocatalysts for biofuels and other yields. Although surface demonstration technologies were originally recognized for phages and bacteria, they have later been useful. In abundant instances, proteins and peptides have been effectively shown to retain their bioactivity. Moreover, the ability to simultaneously display numerous kinds of molecules on an identical cell has recently become possible, significantly expanding the utilization of surface-engineered microorganisms. In comparison to most bacteria, yeasts offer a bigger surface area for display and are easier to extract from a broth after use. Surface mobilization, which was initially established in S. cerevisiae among yeast species, has revolutionized the field

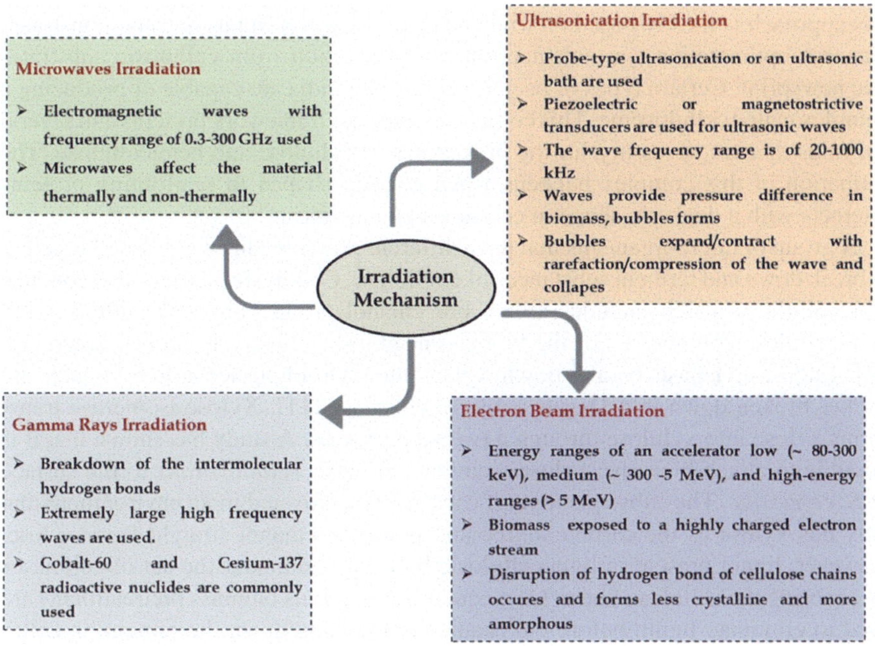

Fig. 4.3 Irradiation mechanism and parameters range

of bioprocessing. This discovery led to the development of similar display systems for other yeasts, including *Pichia pastoris* and *Yarrowia lipolytica*. By displaying enzymes on the surface, yeasts can now break down complex sugar polymers into simple sugars, allowing them to access substrates that were previously inaccessible. This breakthrough has enabled the consolidation of polymer hydrolysis and sugar fermentation into a single operation, resulting in a more efficient and cost-effective process. The potential impact of this consolidated bioprocessing on the conversion of lignocellulosic substrates to ethanol cannot be overstated.

Yeast is only able to directly ferment cellulose when multiple enzymes, including cellobiohydrolases (CBHs), β-glucosidases (BGLs), and endoglucanases (EGs), are co-displayed. The enzymes need to work together to ensure efficient fermentation. EGs turn arbitrarily on the amorphous section of the cellulose chain to create reducing and non-reducing ends. CBHs are required to discharge cellobiose from the bare ends, while BGLs break down the unconstrained oligosaccharides into glucose. This enzymatic synergy is vital for the yeast to directly ferment cellulose. In conclusion, the co-display of CBHs, BGLs, and EGs is essential for the yeast to efficiently break down cellulose into glucose. Any attempt at direct fermentation of cellulose without the simultaneous presence of these enzymes will not be successful. When cells are modified to display several enzymes at once, it's crucial to control the ratios of the different enzyme activities to achieve the desired synergistic effect. Scientists have tried to optimize the ratios of these surface-displayed enzyme activities, although this approach is mostly based on trial and error. However, it has been demonstrated that this optimization can enhance ethanol production from cellulosic substrates like newsprint. Certain types of bacteria called clostridia are capable of producing a complex called cellulosome. This complex comprises framework proteins and several enzymes that are necessary for the breakdown of cellulose and hemicellulose. The formation of this complex happens when cohesin shaped in scaffolding proteins interacts with a dockerin shape in cellulosomal enzymes.

A group of microorganisms that have different proteins can be cultivated together to break down and ferment substances like cellulose. Cellulosic biomass also contains hemicellulose, which includes xylan. For ethanol production, both cellulose and hemicellulose components of the biomass need to be effectively broken down [13, 14]. Endo-β-xylanase breaks down xylan into xylo-oligosaccharides, which are further broken down into D-xylose by β-xylosidase [11]. Xylose isomerase transforms xylose into xylulose through a two-step process. A study has shown that it is possible to attach bioactive xylose isomerase from C. *cellulovorans* to the surface of S. *cerevisiae.* The subsequent yeast can then be cultivated in an environment that only has xylose as the carbon source and produces ethanol straight from xylose. However, lignin present in lignocellulosic biomass can hinder the act of cellulases by covering the cellulose fibers. Consequently, numerous biomass pretreatments are used to eliminate lignin before enzymatically breaking down the remaining cellulosic deposit. The conversion of lignocellulosic biomass to biofuels is a process that relies heavily on microorganisms found in various ecological habitats. As presented in Table 4.1, the majority of cellulolytic microorganisms are taxonomically classified under the phyla *Actinobacteria, Bacteroidetes*, *Firmicutes*, and *Proteobacteria*, which are known to produce essential cellulolytic enzymes for biomass conversion.

4.4 Chemical Modifications and Depolymerization

The use of lignocellulosic biomass is an extremely promising area of study, with significant potential for producing low-cost feedstock for biofuels, natural antioxidant additives, filler materials for plastics, and other high-value green chemicals [134]. This is an area that has attracted considerable attention in recent decades and is one that warrants further exploration and investment [54, 61, 82]. The utilization of biomass as a renewable resource for energy and high-value compound production hinges on effective pretreatment methods [21, 25, 38, 61]. The choice of method often depends on the specific biomass composition, the desired products, and the intended applications [89, 98]. Various tools for inducing chemical modifications and depolymerization in biomass [32, 115, 116, 127] are delineated in the next subsections. Among these, irradiation has emerged as a potent tool for inducing chemical modifications and depolymerization in biomass. This method involves the use of high-energy particles or electromagnetic waves to break chemical bonds within the biomass components, leading to structural changes and enhanced reactivity. Two

Table 4.1 Microorganisms used for biofuel production

Type of biofuels	Produced from	Microbial communities
Ethanol	Xylose	*Zymomonas mobilis* CP4 (pZB5) [22, 126]
	Xylose and glucose	S. *cerevisiae* strain XUSAE57 [52]
	Arabinose	*Zymomonas mobilis* ATCC39676 (pZB206) [22, 126]
	Cellulose	C. *cellulovorans, Clostridium thermocellum* and Thermo *anaerobacterium, saccharolyticum* [3, 36]
	Synthetic medium	*Clostridium autoethanogenum* [64]
	Zymomonas mobilis	*Synechococcuse longatus* PCC 7942 [107]
Butanol	Sucrose and sugarcane	*Clostridium tyrobutyricum* [107]
	Glucose	*Escherichia coli* [5]
Isopentenol	*Bacillus subtilis*	*Escherichia coli* [26]
Isobutanol	*Acinetobacter baylyi*	*Escherichia coli* [26]
	Synthetic medium	S. *cerevisiae* [6]
Hydrogen	*Caldicellulosiruptor bescii*	*Lignocellulose* [16]

common forms of ionizing radiation used for biomass treatment are gamma rays and electron beams. The effects of such rays and alpha particles are discussed below.

4.4.1 Effect of Gamma-Ray Irradiation and Alpha Particles on Biomass

Gamma rays are electromagnetic waves emitted by radioactive isotopes, such as cobalt-60 (^{60}Co) or cesium-137 (^{137}Cs). When biomass is exposed to gamma radiation, high-energy photons penetrate the material, causing ionization and the formation of free radicals. These free radicals initiate a series of chemical reactions, including depolymerization and the breakdown of complex structures like cellulose, hemicellulose, and lignin. Lignin is a naturally occurring polyphenol that is composed of three primary subunits, namely synapse alcohol (syringe, S), coniferyl alcohol (guaiacol, G), and p-coumaryl alcohol (p-hydroxyphenyl, H) [54]. Its complex chemical structure and unique properties make it an important component of plant cell walls, providing structural support and protection against environmental stressors. The identification and characterization of these subunits have significant implications in various fields, including bioengineering, materials science, and biorefinery. Therefore, understanding the composition and properties of lignin is crucial for advancing research and development efforts in these fields. The process of lignin depolymerization yields phenolic monomers, which have shown great potential as a renewable alternative to fossil resources. Specifically, these monomers can be utilized in the production of adhesives, coatings, and other high-value materials. As the world continues to seek sustainable and environmentally friendly solutions, the use of phenolic monomers derived from lignin is becoming an increasingly attractive option. The potential commercial and industrial applications of this technology make it a promising avenue for further research and development [62]. Researchers aimed to investigate the impact of gamma and alpha irradiations on the structural characteristics of lignin by examining the changes in the major absorption bands that are indicative of the lignin structure [67, 126]. The analysis of these changes can provide important insights into the chemical and physical modifications that lignin undergoes under irradiation, and thus can be of great significance in understanding the behavior of lignin-based materials in various applications. The results of the study may have implications for the use of lignin in diverse fields, such as biofuels, bioplastics, and other high-value products, and may contribute to the development of new lignin-based materials with improved performance and stability [126]. Based on the results, it has been confirmed that the broadness of the hydroxyl peak decreases significantly up to a dose of 10 kGy, while the peak intensity slightly decreases at a dose of 20 kGy. This suggests that gamma irradiation on lignin up to a dose of 10 kGy reduces the physical interactions, such as hydrogen bonding, among the lignin molecules. On the other hand, irradiating lignin at a dose of 20 kGy results in a decrease in the number of hydroxyl groups observed in the lignin, as indicated by the reduction in the O–H peak intensity [126]. The formation of phenoxyl radicals through the radiolysis of lignin is a highly productive process that involves the homolytic cleavage of the weak O–H bonds of aromatic units. This reaction results in the elimination of hydrogen atoms, which combine with the aromatic units of lignin, liberating small molecules such as H_2 and H_2O. The excellent electron scavenging activity of hydrogen atoms and aromatic units of lignin makes this reaction highly effective at low doses of up to 20 kGy. These findings have been documented in several studies [88, 94]. There are various advantages associated with gamma-ray irradiation of biomass [54]:

(a) Deep Penetration: Gamma rays can penetrate deep into biomass materials, ensuring uniform treatment.
(b) Controlled Dose: The irradiation dose can be precisely controlled, allowing for tailored modifications based on biomass type and intended applications.
(c) No Residual Radioactivity: The irradiated biomass does not become radioactive, ensuring safety and compliance with environmental regulations.

4.4.2 Effect of Electron Beam Irradiation (EBI) on Biomass

The irradiation pretreatment method, particularly electron beam irradiation (EBI), has been identified as a highly promising approach for biomass pretreatment. This is due to its efficacy in delignifying the biomass, decreasing the crystallinity of cellulose, and increasing the solubilization of sugars [50, 68]. In light of these benefits, electron beam irradiation stands out as a highly attractive pretreatment method for biomass processing applications. The high-speed electrons collide with biomass particles, transferring energy and inducing chemical changes. This method is effective in breaking down molecular structures and initiating depolymerization reactions. Electron beam irradiation induces significant changes in the properties of cellulose, including a decrease in its crystallinity and molecular weight and an increase in its surface area [68]. These structural modifications in the lignocellulosic polymer network facilitate enzymatic saccharification, a process that converts cellulose to glucose and other sugars [50]. This phenomenon has been attributed to the depolymerization of cellulose and the formation of free radicals, which enhance the accessibility of the cellulose to the enzymes. The ability to control these changes through electron beam irradiation makes it a promising technique for the modification of cellulose and the development of new materials with tailored properties. Electron beam irradiation is a highly effective method that enables a rapid dose rate to be achieved within a short period while ensuring optimal energy utilization. With the development of industrial-scale electron beam irradiation facilities, new opportunities have opened up for researchers to explore. The utilization of alkali-mediated EBI treatment of Artemisia ordosica" resulted in its improved hydrolysis. Thus suggesting Artemisia ordosica as suitable feedstock and EBI as an efficient pretreatment strategy for a promising avenue for sustainable energy development Various advantages associated with EBI-induced changes in biomass structure have been reported by Xiang et al. [128]:

(a) Precise Energy Delivery: Electron beams can be precisely controlled for energy delivery, allowing for targeted treatment.
(b) No Residual Radioactivity: Similar to gamma rays, electron beam irradiation does not render the treated biomass radioactive.
(c) Low Heat Generation: Electron beam irradiation generates minimal heat, reducing the risk of thermal degradation in the treated biomass.

Lignocellulosic biomass, a complex matrix of cellulose, hemicellulose, and lignin, holds great potential for sustainable biofuel and high-value compound production. Ionizing radiation, such as gamma rays or electron beams, has emerged as a powerful tool to induce chemical modifications in lignocellulosic biomass, enhancing its conversion efficiency. The intricate processes of chemical modifications and subsequent conversion steps triggered by radiation are as follows.

4.4.2.1 Irradiation-Induced Chemical Modifications

It has been widely acknowledged for several decades that ionizing radiations, such as gamma and EBI, can lead to the degradation of cellulosic materials [87]. When photons from gamma sources or electrons from accelerators interact with cellulose, they deposit their energy in the absorber, resulting in the excitation and/or ionization of the molecules causing scissioning of the main chain [87]. The degradation of cellulosic materials due to ionizing radiations is a well-established phenomenon that has been studied extensively. This process has significant implications for a range of applications, including the preservation of historical documents and the sterilization of medical equipment. As such, a thorough understanding of the mechanisms underlying this phenomenon is of great importance to researchers and practitioners alike.

(a) Cellulose Depolymerization: Cellulose is a linear homopolymer composed of D-anhydroglucopyranose units, commonly referred to as glucose. The monomer units of cellulose are characterized by pyranose rings in the chair conformation with axial hydrogen atoms. The polymer and its fragments exhibit high rigidity, which arises from a high degree of crystallinity and hydrogen bonds, resulting in a significant restriction against conformational transitions [9]. Gamma rays or electron beams induce significant chemical changes in cellulose causing random scission of glycosidic linkages, breaking down cellulose chains into shorter fragments resulting in the formation of carbonyl and carboxy groups at the sites of bond cleavages. A reduced degree of polymerization increases cellulose accessibility for enzymatic hydrolysis in subsequent conversion steps. It has been observed that ionizing radiation exhibits a preference for abstracting hydrogen atoms from the C(1) and C(4) atoms of the pyranose ring. This phenomenon leads to the rupture of the ring and scission of the polymeric chain, resulting in the production of a carbon dioxide molecule and a carbonyl compound in both cellulose and other polysaccharides [28]. Studies have identified over 20 different cellulose macroradicals that form as a result of irradiation [123]. As per the findings of Ershov in 1998, the degradation yield for cellulose is estimated to be between 5 and 7 bonds per 100 eV of absorbed energy from ionizing radiation [28]. Despite its susceptibility to degradation, the crystalline structure of cellulose remains unaffected by irradiation up to several hundred kGy [28]. These findings suggest that while cellulose is sensitive to degradation, its crystalline structure remains stable even under high levels of irradiation. Irradiated cellulose is composed of both water-soluble and water-insoluble fractions. As the dose increases, the amount of soluble cellulosic fractions increases while the degree of polymerization decreases. This finding suggests that the irradiation process affects the chemical structure of cellulose, leading to changes in its solubility and molecular weight [101]. These results have important implications for industries that use cellulose as a raw material, such as the paper and textile industries, as they may need to adjust their processes to account for these changes. Further research is needed to fully understand the effects of irradiation on cellulose and to develop strategies to mitigate any negative impacts.

(b) Hemicellulose Cleavage: Hemicellulose is recognized for its ability to form a hemicellulose-lignin matrix that embeds the cellulose fiber. This matrix is tightly bound to cellulose, which facilitates the transfer of short-lived cellulose radicals to the outer layers of the fiber. This process is crucial for the structural integrity of the fiber and its overall mechanical properties. Also, the interconnection between cellulose, lignin, and hemicellulose is established through hydrogen bonding, which imparts flexibility to the cell wall structure [90]. This structural arrangement is crucial for the mechanical strength and stability of plant cells. The hemicellulose-lignin matrix plays a vital role in the formation of plant cell walls and is a key component in the production of various biomaterials Understanding the properties and behavior of hemicellulose is essential for the development of new and innovative materials that can be used in a wide range of applications. The enzymatic degradation of hemicellulose structure is achieved through the use of biologically produced enzymes [90]. The degradation of hemicellulose into sugar and acetic acid presents the possibility of methane production [100]. However, to produce biofuel, it is imperative to monitor the process to reduce the inhibiting compounds in hemicellulose at higher temperatures [77]. Therefore, process monitoring is essential to ensure the efficient production of biofuel from hemicellulose. Irradiation causes cleavage of glycosidic linkages in hemicellulose, leading to the formation of smaller, more soluble fragments [101]. Increased solubility enhances the accessibility of hemicellulose for further hydrolysis and conversion.

4.4.2.2 Subsequent Conversion Steps

Biomass combustion is not considered the most efficient method for utilizing biomass due to its negative impact on the environment and the inability to recover all the energy stored in the substrates [118]. Instead, the conversion of biomass into solid, liquid, and gaseous forms is regarded as a more efficient and environmentally friendly energy source for various sectors, including heat, power, and transportation fuel [49]. Currently, two primary routes for biomass conversion are employed: thermochemical and biochemical. Thermochemical methods involve using the entire biomass in the presence of a heat source and controllable oxygen atmosphere to transform it into different energy forms [85]. It is worth noting that the conversion of biomass into energy forms through thermochemical methods is a promising approach that can contribute to the sustainable development of the energy sector.

Thermochemical processes have been employed to produce liquid fuels from biomass, utilizing diverse technologies such as direct combustion, torrefaction, hydrothermal liquefaction, pyrolysis, and gasification [85]. These technologies have been developed to address the growing demand for sustainable and renewable energy sources. The production of liquid fuels from biomass through thermochemical processes has the potential to reduce greenhouse gas emissions and mitigate the negative impact of fossil fuels on the environment [96]. The utilization of these technologies requires a thorough understanding of the underlying chemical and physical processes involved, as well as the optimization of process parameters to achieve maximum efficiency and yield. Biomass conversion methods rely on thermal-based processes that can be classified into two categories: low-temperature and high-temperature processes. Low-temperature processes operate at temperatures below 300 °C (e.g., torrefaction and hydrothermal carbonization). High-temperature processes, on the other hand, operate at temperatures above 300 °C and include gasification, combustion, and pyrolysis. These methods are extensively studied and considered promising for the production of biofuels and other value-added products from biomass [92].

(a) Direct Combustion: Biomass combustion is a chemical process that involves the oxidation of carbon and hydrogen. This complex process involves a series of oxidation reactions that result in the formation of carbon dioxide and water during the transformation of carbon and hydrogen. The process is influenced by various factors, including temperature, pressure, and the composition of the biomass. Understanding the chemistry of biomass combustion is essential for developing efficient and sustainable energy systems. Inadequate oxygen quality can result in incomplete combustion, leading to the emission of atmospheric pollutants such as carbon monoxide (CO), nitrogen oxides (NO_x), sulfur dioxide (SO_2), and particulate matter. It is imperative to ensure that the oxygen quality is optimal to prevent the release of these harmful pollutants into the environment. Proper monitoring and control of oxygen levels can help mitigate the negative impact of incomplete combustion on air quality [33]. The process of biomass combustion involves a series of reactions, including drying, pyrolysis, pre-combustion, primary combustion, secondary combustion, and effluent stack gas. The process of combustion is influenced by several parameters, including the size of the biomass, its specific gravity, moisture content, ash percentage, and elemental composition, which encompasses lignin, cellulose, and hemicellulose. It is imperative to note that the anatomical structure of biomass plays a crucial role in the combustion process. The aforementioned parameters have a significant impact on the combustion process of biomass. The size of the biomass, its specific gravity, moisture content, ash percentage, and elemental composition, involving lignin, cellulose, and hemicellulose, all contribute to the combustion process. However, it is important to consider the anatomical structure of the biomass as a crucial aspect that determines the effectiveness of the combustion process. Thus, it is essential to examine and study these parameters in detail to ensure optimal combustion and utilization of biomass as a renewable energy source. The knowledge and understanding of these parameters can facilitate the development of efficient biomass combustion technologies and enhance the sustainability of our energy resources [124]. The combustion of lignin, cellulose, and hemicellulose, which are the primary components of biomass, is a complex process that involves the release of energy in the form of heat and light. To optimize the combustion process, it is essential to understand the various parameters that influence it. These parameters include the magnitude of the biomass, its specific gravity, moisture content, ash percentage, and elemental composition. By carefully controlling these parameters, it is possible to achieve efficient and effective biomass combustion, which can have significant benefits for both businesses and the environment. According to various research studies, the majority of global bioenergy production, approximately 95–97%, is derived from direct biomass combustion [35]. This highlights the significance of biomass combustion as a primary source of bioenergy. The findings of these studies provide valuable insights into the current state of global bioenergy production and can be used to inform future research and development efforts in this field. It is important to note that the accuracy and reliability of these studies are dependent on the quality of data used and the methodology employed. Therefore, further research is required to validate these findings and to explore alternative sources of bioenergy production. Table 4.2 presents the energy content in megajoules per kilogram for various types of biomass. The data presented in these tables is of significant importance to the energy industry, as it provides valuable insights into the energy potential and properties of different types of fuels. The information presented in Table 4.1 can be used to determine the most suitable type of biomass for a particular application, while the data in Table 4.2 can be used to compare the properties of biomass and coal-based fuels and to identify the advantages and disadvantages of each. It is imperative to emphasize that the accuracy and reliability of the data presented in these tables hold paramount significance, given their potential impact on decision-making within the energy industry. As a result, meticulous attention to detail is warranted when reviewing these tables. Any inaccuracies or inconsistencies must be identified and rectified promptly to ensure the integrity of the data and promote informed decision-making. Therefore, it is crucial to ensure that the data is collected, analyzed, and presented rigorously and systematically, using appropriate methods and techniques.

The utilization of biomass in co-combustion with fossil-based fuels is regarded as a cost-effective and intriguing option. This approach has gained significant attention in recent years due to its potential to reduce greenhouse gas emissions and dependence on non-renewable energy sources. The co-combustion of biomass and fossil fuels can be achieved through various methods, including direct co-firing, indirect co-firing, and parallel co-firing. The utilization of biomass in co-combustion can also lead to improved combustion efficiency, reduced emissions, and increased energy security. Therefore, the co-combustion of biomass with fossil fuels is a promising strategy for achieving sustainable energy production and mitigating climate change. However, the

Table 4.2 Representative values of the energy content of selected biomass sources, in KJ/kg [44, 121]

Biomass	Energy content (KJ/kg)
Greenwood	8000
Corn stover	10730
Rice husk	13524
Cotton gin	15500
Corncob	17300
Soybean cake	18300
Groundnut shell	19800
Coconut shell	20000
Hazelnut shell	20470
Wood bark	20570

co-combustion of fossil-based coal with biomass presents several challenges, including increased corrosion, slagging, and fouling due to the high alkali content of the biomass [85]. This approach is further limited by the projected gradual shutdown of fossil-based power plants. To tackle these challenges, it is recommended to shift the focus toward alternative technologies like gasification and pyrolysis, which are expected to become more competitive and widely available shortly. By adopting these technologies, businesses, and industries can mitigate the challenges associated with co-combustion and ensure a sustainable and efficient energy production process.

(b) Liquefaction: Hydrothermal liquefaction is a thermochemical process that decomposes lignocellulosic feedstock into renewable liquid fuel, regardless of whether the feedstock is wet or dry [42]. There are two main subclasses of this process: indirect liquefaction and solvent liquefaction. In indirect liquefaction, the biomass or its liquefied products are first converted into syngas, which is then used to synthesize fuels such as alcohols and alkanes [75]. In contrast, in solvent liquefaction, biomass is directly converted into liquid fuel using proton solvents such as water, alcohols, phenols, sulpholane, and ionic liquid [75]. Solvent liquefaction is preferred over indirect liquefaction due to its remarkable merits, moderate operational conditions, and higher yield of products [37]. The hydrothermal liquefaction process is typically conducted under mild operational parameters, with temperatures ranging from 250 to 500 °C, pressures of 5–35 MPa, and contact periods of 5–60 min. These processes are carried out in the presence of solvents, such as sub/supercritical water, organic solvents, or mixed solvents, which are a combination of water and organic solvents [83]. It is worth noting that the use of solvents is crucial to the success of the hydrothermal liquefaction process. The solvents facilitate the breakdown of the feedstock into smaller molecules, which can then be converted into biofuels or other valuable products. The choice of solvent depends on the specific feedstock and the desired end product. Overall, the hydrothermal liquefaction process is a promising technology for the production of biofuels and other value-added products. Its mild operational parameters and use of solvents make it an attractive alternative to traditional thermochemical conversion. Hydrothermal liquefaction is widely regarded as a more energy-efficient and effective strategy than pyrolysis, primarily due to its superior physicochemical properties. The biocrude produced from the hydrothermal liquefaction process typically exhibits an oxygen content of 10–20 wt.% and a heating value of 30–35 MJ/kg, which is notably higher than those obtained from conventional pyrolysis [42]. This underscores the potential of hydrothermal liquefaction as a promising alternative to conventional pyrolysis, particularly in the context of sustainable and eco-friendly energy production.

(c) Pyrolysis: Pyrolysis is a thermochemical process that is widely employed for the degradation of carbonaceous biomass, including cellulose, hemicellulose, and lignin. This process involves the application of heat in the absence of oxygen, leading to the breakdown of complex organic molecules into simpler compounds. Pyrolysis is considered to be a highly effective method for the conversion of biomass into valuable products such as biofuels, biochar, and other chemicals [2]. Due to its numerous advantages, pyrolysis has gained significant attention in both academic and industrial settings as a promising technology for sustainable energy production and waste management. The thermal decomposition process of biomass is a complex and intricate series of reactions. These reactions involve dehydration, decarboxylation, decarbonylation, hydrogenation, isomerization, aromatization, depolymerization, and charring. The main objective of these reactions is to break down the complex organic compounds present in biomass into simpler compounds. As such, thermal decomposition plays a vital role in the conversion of biomass into useful products such as biofuels and bioproducts. Understanding the mechanisms of these reactions is essential for developing effective and efficient thermal conversion technologies. Therefore, further research is needed to fully comprehend the intricacies of the thermal decomposition process and to develop innovative solutions for sustainable biomass utilization. The process is a crucial aspect of biomass conversion, and understanding the underlying chemical reactions is vital for optimizing the process. The pyrolysis of biomass typically involves a series of steps [23]. Firstly, heat is transferred from its source to the biomass to initiate the reaction. Secondly, the primary vapors undergo an elevated pyrolysis temperature, which contributes to the formation of volatiles and char. The influx of hot vapors to the biomass engenders a heat migration process between the unpyrolyzed fuel and the hot volatiles. This phenomenon, in turn, leads to the condensation of volatiles, which is accompanied by secondary reactions and culminates in the formation of tar. Finally, the primary pyrolytic reaction is accompanied by autocatalytic secondary pyrolysis reactions [23].

It is important to note that the process of pyrolysis is a complex one, and each step is critical to the overall success of the process. Understanding the intricacies of each step is essential for optimizing the yield of desired products and minimizing unwanted by-products. Therefore, further research and development in this area is necessary to fully realize the potential of biomass pyrolysis as a sustainable and environmentally friendly energy source. To optimize liquid yields during pyrolysis, it is essential to employ elevated heating rates, minimum vapor contact periods, and a surrounding temperature of 500 °C [20]. These conditions effectively prevent the thermal or catalytic cracking of primary decomposition products due to the presence of char, as well as the polymerization of these products into char [51]. Furthermore, other types of pyrolysis, such as catalytic and microwave-assisted, carbon dioxide, additives, solar, and hydropyrolysis, can be utilized to enhance product yields.

(d) Gasification: Gasification is a promising technology that has gained significant attention in recent years due to its potential to produce clean energy from a variety of feedstocks, including coal, biomass, and waste materials. It is a thermochemical process that involves the conversion of carbonaceous materials into highly valuable synthesis gases. This process is made possible by introducing a gasifying agent, such as air, oxygen, steam, CO_2, or a combination of these, at a temperature that exceeds 700 °C. The produced gas primarily comprises

CO, H_2, CO_2, and CH_4 [102]. The synthesis gas produced has a wide range of applications in diverse industries such as power generation, chemical production, and transportation fuels. This versatile gas can cater to various industrial needs, ranging from electricity generation to chemical manufacturing and transportation fuel production. Its ability to be employed in different ways makes it a highly valuable resource in the industrial landscape. The gasification process is a complex series of four main steps, which include heating or drying (100–200 °C) to reduce moisture content, pyrolysis, oxidation or partial combustion, and gasification. The initial step involves vaporizing the moisture content (30–60%) of the biomass at approximately 200 °C. The second step, pyrolysis, decomposes various biomass components, including cellulose, hemicellulose, and lignin, into solid residues and volatiles [41]. The third step is oxidation or partial combustion, where the resulting volatiles and char residues are oxidized to CO, CO_2, and H_2O with the aid of a gasifying agent at temperatures above 700 °C. It is essential to note that the gasification process is a critical component of many industrial and academic applications. As such, it is crucial to understand the process's intricacies and the various steps involved. By following the steps outlined above, one can ensure that the gasification process is carried out efficiently and effectively, leading to optimal results. In conclusion, gasification is a highly efficient and versatile process that offers numerous benefits over traditional combustion technologies. Its ability to convert a wide range of feedstocks into valuable synthesis gases makes it an attractive option for businesses and researchers seeking to reduce their carbon footprint and increase their energy efficiency.

4.5 Enhanced Biofuel Yield and Compound Extraction

In the pursuit of sustainable energy and the extraction of high-value compounds from biomass, irradiation has emerged as a transformative tool. The application of ionizing radiation, such as gamma rays or electron beams, induces structural changes in biomass components, unlocking new pathways for enhanced biofuel yield and compound extraction. This section explores the key aspects of how irradiation contributes to the optimization of biofuel production and the extraction of valuable compounds from biomass.

4.5.1 Enhanced Biofuel Yield

Ionizing radiation breaks glycosidic linkages in cellulose, reducing the degree of polymerization. Irradiation induces cleavage of glycosidic linkages in hemicellulose, producing soluble fragments. Lignin undergoes structural modifications, generating reactive intermediates. Modified lignin components can serve as precursors for the synthesis of biofuels and value-added chemicals during subsequent conversion steps.

Irradiation pretreatment significantly improves the efficiency of enzymatic hydrolysis by accelerating the breakdown of cellulose and hemicelluloses. It results in a more accessible substrate for enzymes, leading to enhanced biofuel production.

4.5.2 Compound Extraction

Irradiation-induced lignin modifications yield phenolic compounds. These compounds, known for their antioxidant properties, can be utilized in the pharmaceutical, cosmetic, and food industries. Certain irradiation-induced biomass components possess antioxidant properties that can be used in various applications, including nutraceuticals [54]. Extraction and purification of reactive intermediate molecules unlock opportunities for pharmaceutical and specialty chemical production.

4.6 Analytical Techniques for Mechanistic Understanding

Understanding the intricate mechanisms underlying irradiation-induced changes in biomass is pivotal for optimizing processes and enhancing the efficiency of biomass utilization. Various analytical techniques play a crucial role in deciphering the molecular alterations within biomass components. This section explores the key analytical tools employed for gaining mechanistic insights into how ionizing radiation induces changes in cellulose, hemicellulose, and lignin.

4.6.1 Spectroscopic Techniques

Spectroscopic techniques play a crucial role in understanding the mechanisms underlying irradiation-induced changes in biomass. These techniques provide valuable information about the structural, compositional, and functional alterations that occur in biomass when exposed to various forms of irradiation. Some spectroscopic techniques commonly employed for this purpose are discussed below.

4.6.1.1 Fourier Transform Infrared Spectroscopy (FTIR)

Fourier Transform Infrared Spectroscopy (FTIR) provides insights into molecular vibrations and functional groups in biomass. It can be used to detect changes in cellulose, hemicellulose, and lignin content [83]. Irradiation-induced modifications in chemical bonds can be identified, such as cleavage or formation of bonds using FTIR.

4.6.1.2 Raman Spectroscopy

Raman spectroscopy is useful for studying the crystallinity of cellulose and changes in molecular structures [30]. It can be employed to monitor alterations in biomass composition due to irradiation. Raman spectroscopy has enabled the continuous and accurate quantification of multiple reactant and product concentrations in enzymatic hydrolysis and fermentation reactions [30]. This technological advancement has created new opportunities for the processing of lignocellulosic biomass, as it can now leverage advanced process control (APC) systems to enhance process efficiency. By utilizing APC systems, the processing of lignocellulosic biomass can be optimized to achieve higher yields and reduce costs [30].

4.6.1.3 Nuclear Magnetic Resonance (NMR) Spectroscopy

NMR Spectroscopy provides details about the spatial arrangement of atoms in biomass. It can be applied to investigate changes in cellulose, hemicellulose, and lignin structures due to irradiation. Solid-state NMR methodologies are highly valuable in the study of structural problems in complex biological systems [4]. They are particularly useful in elucidating the mechanisms of biosynthesis and deconstruction for lignocellulosic biomass or individual plant cell wall components, such as cellulose [27]. The ability of solid-state NMR to provide detailed information on the molecular structure and dynamics of these complex systems makes it an essential tool for researchers in the fields of biochemistry, biophysics, and materials science. By leveraging the power of solid-state NMR, researchers can gain a deeper understanding of the fundamental processes that underlie the behavior of biological systems, paving the way for discoveries and innovations in a wide range of fields. Modern solid-state NMR spectroscopy has a diverse range of applications in the study of lignin, hemicellulose, and intact lignocellulose. These applications were initially developed for the investigation of humic acids. The study of lignin using modern solid-state NMR spectroscopy is reported by Orem and Hatcher [84] and Holtman et al. [45], while Komatsu and Kikuchi [53] provide information on the applications of this technique to the study of hemicellulose. The use of modern solid-state NMR spectroscopy in the study of intact lignocellulose has been discussed by Foston et al. [34]. The use of modern solid-state NMR spectroscopy in the study of lignin, hemicellulose, and intact lignocellulose has proven to be an effective tool for the characterization of these complex materials. The technique provides valuable information on the chemical structure and dynamics of these materials, which is essential for understanding their properties and behavior.

4.6.1.4 UV–Visible Spectroscopy

UV–Visible spectroscopy can be utilized to study the presence of chromophores in the biomass and changes in electronic transitions induced by irradiation.

4.6.1.5 Fluorescence Spectroscopy

Fluorescence spectroscopy can provide information on the presence of fluorophores in biomass. Changes in fluorescence intensity and emission spectra can indicate structural alterations.

4.6.1.6 X-Ray Photoelectron Spectroscopy (XPS)

It is useful for studying the elemental composition and chemical state of biomass. It can help identify changes in the surface chemistry induced by irradiation.

4.6.1.7 Electron Paramagnetic Resonance (EPR) Spectroscopy

Electron Paramagnetic Resonance (EPR) spectroscopy, also known as Electron Spin Resonance (ESR) spectroscopy, is a powerful technique that can be employed to understand the mechanisms underlying irradiation-induced changes in biomass [66]. EPR spectroscopy is particularly useful for studying free radicals and paramagnetic species, providing valuable information about the oxidative processes and radical formation in irradiated biomass [66]. By providing direct information about the paramagnetic species involved, EPR spectroscopy complements other analytical techniques and contributes to a comprehensive understanding of the impact of irradiation on biomass [66].

4.6.1.8 Mass Spectrometry

Mass spectrometry can be coupled with other spectroscopic techniques to provide detailed information on the molecular composition of biomass. It can be used to identify and quantify specific molecular species, providing insights into the fragmentation and degradation of biomass components.

4.6.2 Chromatographic Techniques

Chromatographic techniques are valuable tools for understanding the mechanisms underlying irradiation-induced changes in biomass. These techniques allow for the separation and analysis of individual components within complex biomass samples. Here are some chromatographic techniques commonly used for this purpose [65, 74, 134].

4.6.2.1 Gas Chromatography (GC)

Gas Chromatography (GC) can be employed to study changes in the composition of volatile organic compounds (VOCs) released from irradiated biomass. It is particularly useful for analyzing the breakdown products of irradiated biomolecules [108].

4.6.2.2 Liquid Chromatography (LC)

It can be used for the separation and quantification of non-volatile compounds, such as sugars, organic acids, and phenolic compounds. It provides insights into changes in the chemical composition of biomass after irradiation.

4.6.2.3 Ion Chromatography (IC)

IC is useful for analyzing inorganic ions, organic acids, and other polar compounds in biomass. It can be applied to study changes in the ion content of biomass resulting from irradiation.

4.6.2.4 Size Exclusion Chromatography (SEC)

SEC is valuable for studying changes in the molecular weight distribution of polymers in biomass, such as cellulose, hemicellulose, and lignin. It provides insights into irradiation-induced modifications in biomass macromolecular structures.

4.6.2.5 Affinity Chromatography

This technique can be used to study changes in specific biomolecules, such as enzymes or proteins, in biomass after irradiation. It helps in understanding the impact on biologically active components.

4.6.2.6 Supercritical Fluid Chromatography (SFC)

SFC can be applied for the separation of lipids and non-polar compounds in biomass. It is useful for studying changes in lipid composition induced by irradiation.

By combining chromatographic techniques with other analytical methods, such as spectroscopy, researchers can obtain a comprehensive understanding of the complex changes in biomass induced by irradiation. This information is crucial for optimizing processes in various applications, including bioenergy, biorefining, and materials science.

4.6.3 Imaging Techniques

Imaging techniques are crucial for understanding the mechanisms underlying irradiation-induced changes in biomass, as they provide spatial and temporal information about structural, compositional, and functional alterations. Some imaging techniques commonly used for this purpose are discussed as follows.

4.6.3.1 Scanning Electron Microscopy (SEM)

SEM provides detailed surface morphology information. It can be used to visualize structural changes in the biomass at the micro and nanoscale induced by irradiation. SEM enables the observation of cellulose microfibrils and their arrangement in the cell wall. Irradiation-induced changes, such as depolymerization and surface roughening, become apparent under SEM [76]. Changes in the distribution and integrity of hemicellulose can be observed, providing visual evidence of irradiation-induced cleavage and degradation. SEM helps identify regions of hemicellulose loss or alteration. It aids in visualizing lignin distribution within the cell wall. Irradiation-induced modifications, such as lignin fragmentation or redistribution, can be directly observed, contributing to a better understanding of lignin alterations [76]. SEM allows for the examination of pore structures in biomass. Changes in pore size, distribution, and connectivity provide insights into how irradiation affects the accessibility of enzymes or chemicals during subsequent processing steps. It also facilitates the measurement of particle sizes in biomass components. Irradiation-induced fragmentation and depolymerization lead to changes in particle size and distribution, influencing subsequent conversion processes [76]. Integrating SEM with other analytical techniques, such as spectroscopy or elemental analysis, allows for a multifaceted understanding of irradiation-induced changes, linking structural modifications with compositional alterations.

4.6.3.2 Transmission Electron Microscopy (TEM)

It is a sophisticated imaging technique that plays a crucial role in elucidating ultrastructural changes within biomass induced by ionizing radiation. Offering unparalleled resolution at the nanoscale, TEM provides detailed insights into the subcellular architecture, enabling a comprehensive understanding of the molecular transformations occurring in biomass subjected to irradiation [128]. Structural modifications in hemicellulose, induced by ionizing radiation, can be visualized in detail using TEM. This includes the cleavage of glycosidic linkages and changes in hemicellulose distribution. It also provides insights into lignin distribution and the formation of nano complexes within the cell wall. The impact of irradiation on lignin structure, such as fragmentation or reorganization, becomes evident. TEM allows for a detailed analysis of the nanoscale surface topography of irradiated biomass. Surface irregularities, roughness, and modifications in the nanoscale landscape can be visualized. It also enables the observation of cell wall ultrastructure at the nanoscale. The identification of nanoscale defects, such as cracks or dislocations, in the cell wall can be done using TEM [128]. Understanding these defects aids in comprehending the mechanical changes induced by irradiation. In summary, Transmission Electron Microscopy (TEM) stands as a paramount tool for unveiling the nanoscale structural intricacies of biomass subjected to ionizing radiation [128]. By offering high-resolution images and detailed analyses at the nanoscale, TEM contributes significantly to the mechanistic understanding of irradiation-induced changes in biomass, guiding the optimization of processes for sustainable biomass utilization and bio-based product development.

4.6.3.3 Confocal Laser Scanning Microscopy (CLSM)

Confocal Laser Scanning Microscopy (CLSM) is a powerful imaging technique that provides three-dimensional visualization and detailed analysis of the microscale and nanoscale structural changes within biomass. Utilizing focused laser beams, CLSM is particularly valuable for understanding the spatial distribution and organization of cellular components in biomass subjected to ionizing radiation [97]. It enables the real-time observation of changes in the cell wall structure induced by irradiation. This includes alterations in cellulose microfibrils, hemicellulose distribution, and lignin organization at the subcellular level. By using fluorescent probes and dyes, CLSM allows for the selective labeling of specific cellular components [114]. The application of Confocal Laser Scanning Microscopy (CLSM) enables the precise tracking of changes in cellulose, hemicellulose, and lignin within biomass, both during and post-irradiation. CLSM captures images at various focal planes, allowing for the profiling of the biomass surface. This imaging method reveals nanoscale features and alterations on the surface of the biomass resulting from irradiation. CLSM can utilize fluorescent probes to visualize the production and distribution of reactive oxygen species (ROS) within biomass [29]. This is valuable for understanding oxidative reactions triggered by irradiation. In the case of lignocellulosic biomass, CLSM can visualize the changes in microbial colonization and interactions on or within biomass structures following irradiation. By providing high-resolution, in-situ imaging capabilities, CLSM contributes to a detailed mechanistic understanding of how ionizing radiation influences the spatial arrangement, organization, and dynamics of cellular components in biomass structures. This knowledge is pivotal for optimizing irradiation processes in the sustainable utilization of biomass resources.

4.6.3.4 Atomic Force Microscopy (AFM)

Atomic Force Microscopy (AFM) is a powerful imaging technique that operates at the nanoscale, providing detailed topographical and mechanical information about the surfaces of materials. In the context of biomass subjected to ionizing radiation, AFM plays a crucial role in unraveling the nanoscale changes and mechanical properties induced by irradiation. AFM offers high-resolution imaging of biomass surfaces, allowing the observation of nanoscale features, such as cellulose microfibrils, hemicellulose, and lignin structures [70]. It reveals nanoscale morphological features, including surface roughness, cracks, and defects induced by ionizing radiation. This aids in understanding the impact of irradiation on biomass structure at a finer scale. AFM also allows in-situ imaging of live biomass samples during irradiation. This real-time observation captures dynamic changes in surface morphology, enabling a time-resolved understanding of the irradiation process. AFM can measure the thickness of cell walls with high precision. Changes in thickness provide information about the impact of irradiation on the structural integrity of biomass components. AFM generates force-distance curves that characterize mechanical properties such as adhesion, deformation, and rupture forces. Analyzing these curves helps in understanding the effects of irradiation on biomass mechanics.

4.6.3.5 Magnetic Resonance Imaging (MRI)

It is a non-invasive imaging technique widely used in medical diagnostics, but its application has extended to materials science, including the study of plant and biomass structures. In the context of biomass subjected to ionizing radiation, MRI offers unique capabilities for investigating the macroscopic and internal changes induced by irradiation. MRI reveals the internal structure of biomass, offering insights into the arrangement of cellulose, hemicellulose, lignin, and other components. It can potentially be used to map free radicals generated during irradiation [106]. This information is valuable for understanding oxidative processes and their impact on biomass structure. Integrating MRI with other imaging techniques, such as AFM or SEM, enables correlative studies that combine information on different length scales, providing a more comprehensive view of irradiation-induced changes.

4.6.3.6 X-Ray Imaging

X-ray imaging is a versatile and widely used technique for visualizing the internal structure of materials, and it plays a crucial role in studying the effects of ionizing radiation on biomass. X-ray Computed Tomography (CT) provides detailed three-dimensional images of the entire biomass structure, allowing for the visualization of macroscopic changes induced by irradiation. CT scans reveal internal structures, including the arrangement of cellulose, hemicellulose, and lignin. CT scans help in assessing changes in pore size and distribution within biomass [8]. It allows for quantitative density measurements, providing information on the compaction or degradation of biomass components after irradiation [117]. Quantitative analysis of volume fractions helps in understanding the proportion of different biomass components and how they change under irradiation. The use of contrast agents in X-ray imaging enhances the visibility of specific biomass components. This aids in differentiating between cellulose, hemicellulose, lignin, and other constituents. X-ray imaging also helps in detecting cracks and defects in the biomass structure. Understanding these structural changes is essential for assessing the overall integrity of the biomass.

4.6.4 Thermogravimetric Analysis (TGA)

This method is particularly valuable for Sitek et al. [105]. TGA quantifies the mass loss of biomass as a function of temperature. The resulting thermogram provides information about the decomposition kinetics, including the onset temperature, rate of decomposition, and peak temperatures. By applying various heating rates, TGA enables the determination of the activation energy for the thermal degradation of biomass components. This information helps in understanding the energy required for decomposition. TGA can differentiate between the thermal degradation profiles of different biomass components (cellulose, hemicellulose, and lignin). It also measures the residual mass after complete decomposition, providing insights into the remaining components, such as ash content which can indicate the impact of irradiation [109].

4.6.5 Inductively Coupled Plasma Mass Spectrometry (ICP-MS)

Inductively Coupled Plasma Mass Spectrometry (ICP-MS) is a powerful analytical technique that enables the identification and quantification of trace elements and isotopes in a sample. When applied to the study of irradiation-based changes in biomass, ICP-MS provides valuable insights into elemental composition, contamination, and alterations induced by ionizing radiation [73]. ICP-MS is highly sensitive, allowing for the detection of foreign elements that might be introduced during the irradiation process. This aids in assessing potential contamination. It is also employed to detect changes in the oxidation states of elements, providing insights into oxidative processes induced by ionizing radiation in biomass. Integration with other analytical techniques, such as spectroscopy or microscopy, ICP provides a comprehensive understanding of the correlation between elemental changes and structural modifications induced by irradiation [73].

By integrating these analytical techniques, researchers can gain a comprehensive understanding of the mechanisms underlying irradiation-induced changes in biomass, covering aspects of composition, structure, and functionality at various scales.

4.7 Challenges and Future Prospects

Irradiation-based treatment, particularly in the context of biomass, has shown promise for enhancing biofuel and high-value compound conversion. However, several challenges persist, and there are ongoing efforts to address these challenges while exploring future directions for the technology. The following sections delineate the challenges encountered in Irradiation-based treatment techniques and prospects.

4.7.1 Challenges

The effectiveness of irradiation-based treatment depends on various parameters such as dose, energy, and exposure time. Achieving optimal conditions for different types of biomass is challenging and requires careful experimentation. The mechanisms underlying irradiation-induced changes in biomass are complex and not fully understood. A detailed understanding of the chemical and physical changes at the molecular level is crucial for optimizing conversion processes. Irradiation can lead to the formation of inhibitory compounds, such as free radicals or toxic by-products, which may interfere with subsequent bioconversion processes. Scaling up irradiation-based treatments from laboratory to industrial scales poses challenges in terms of equipment design, energy efficiency, and cost-effectiveness. It has also been reported that scaling up irradiation-based treatments from laboratory to industrial scales poses challenges in terms of equipment design, energy efficiency, and cost-effectiveness. Integration of irradiation treatment with subsequent processes, such as enzymatic hydrolysis or fermentation, can be complex. The cost of irradiation facilities and the energy required for the process may currently limit the economic viability of large-scale applications.

4.7.2 Future Prospects

Addressing the challenges and exploring future directions for irradiation-based treatment in biomass conversion requires a multidisciplinary approach, involving collaborations between researchers, industry stakeholders, and policymakers. Various future prospects are discussed as follows:

(a) Precision Irradiation Technologies: The development of precision irradiation technologies may allow for targeted treatment of specific biomass components. This can minimize undesired side effects and enhance the selectivity of the treatment.
(b) Multi-Modal Approaches: Multi-modal approaches may be explored that combine irradiation with other pretreatment methods, such as enzymatic treatments or chemical pretreatments. Synergies between different methods can lead to enhanced overall biomass conversion.
(c) Customized Biomass Varieties: The development of biomass varieties may be investigated that are specifically designed for enhanced conversion through irradiation. This can involve genetic modifications or selective breeding to optimize the biomass for the irradiation-based process.
(d) Biorefinery Integration: Irradiation-based treatment may be integrated into the concept of biorefineries, where multiple value-added products are derived from biomass. This holistic approach can maximize the utilization of biomass resources.
(e) Waste Valorization: The application of irradiation may be explored for the valorization of biomass wastes and by-products. This can contribute to a more circular economy by converting waste streams into valuable bio-based products.
(f) Techno-Economic Assessments: A comprehensive techno-economic assessment needs to be conducted to evaluate the overall feasibility and competitiveness of irradiation-based biomass conversion processes. This will involve a detailed analysis of costs, benefits, and potential market opportunities.

In summary, addressing the challenges and exploring future directions for irradiation-based treatment in biomass conversion requires a multidisciplinary approach, involving collaborations between researchers, industry stakeholders, and policymakers. As technological advancements continue, irradiation has the potential to play a significant role in enhancing the efficiency and sustainability of biofuel and high-value compound production from biomass.

4.8 Conclusion

The exploration of irradiation-based treatments in biomass conversion has unveiled a multifaceted landscape of transformative potential, paving the way for sustainable and efficient utilization of biomass resources. This chapter has delved into the diverse applications of ionizing radiation in inducing structural modifications and chemical transformations within biomass, with a focus on enhancing biofuel production and extracting high-value compounds. In summary of irradiation's impact on biomass, several key reflections emerge. The comprehensive investigations into the mechanisms governing irradiation-induced alterations in biomass have unraveled a nuanced understanding of structural, chemical, and elemental transformations. Techniques such as SEM, TEM, AFM, CT, TGA, and ICP-MS have provided invaluable insights, collectively contributing to a holistic comprehension of the complex processes involved. The chapter has underscored the pivotal role of irradiation in biomass valorization, with a particular emphasis on its efficacy in biofuel production. From enhanced saccharification of cellulose to the depolymerization of lignocellulosic components, irradiation emerges as a transformative tool in streamlining biomass conversion pathways. The irradiation-induced changes in biomass architecture, chemical composition, and reactivity directly impact the efficiency and yield of biofuel production processes.

The integration of irradiation-based treatments into the biorefinery framework emerges as a strategic move toward sustainability. By enhancing the accessibility of biomass components and optimizing conversion processes, irradiation contributes to the development of greener and more efficient biorefinery strategies, aligning with the global imperative for sustainable and renewable resource utilization. While the chapter has highlighted the promising outcomes of irradiation-based treatments, it is essential to acknowledge the existing challenges and opportunities. Factors such as dose optimization, irradiation uniformity, and potential by-product formation necessitate careful consideration. Future research endeavors should address these challenges while exploring innovative avenues for maximizing the benefits of irradiation in biomass conversion. The integration of advanced analytical techniques, ranging from microscopy and spectroscopy to thermal analysis and mass spectrometry, has played a pivotal role in unraveling the intricacies of irradiation-induced changes. This synergy has not only enhanced our mechanistic understanding but also paved the way for correlative studies, providing a more comprehensive view of the transformations occurring at different length scales.

In conclusion, the chapter underscores the transformative role of irradiation in biomass conversion, offering a lens through which we can envision a future where biomass resources are harnessed with unprecedented efficiency and sustainability. As the journey continues, it is with anticipation and optimism that we look toward further discoveries and breakthroughs in irradiation-based treatments, propelling biomass utilization into a new era of innovation and environmental responsibility.

Declaration During the preparation of this book chapter, the authors hereby declare that AI tools have not been used.

References

1. Al Gharib S, Mostafavi M, Belloni J (2023) Radiation-assisted hydrolysis of lignocellulosic biomass. Mechanistic study. Waste Biomass Valorization 14(4):1113–1122
2. Aravind SV, Ahmed G, Kishore N (2023) Pyrolysis of Delonix regia using metal oxide catalysts and solvent effect on fuel fraction of bio-oil. Results Eng 17:100876
3. Argyros DA, Tripathi SA, Barrett TF, Rogers SR, Feinberg LF, Olson DG, Foden JM, Miller BB, Lynd LR, Caiazza NC, Hogsett DA (2011) High ethanol titers from cellulose by using metabolically engineered thermophilic, anaerobic microbes. Appl Environ Microbiol 77(23):8288–8294
4. Asakura T, Ando I (1998) Solid state NMR of polymers. Elsevier
5. Atsumi S, Cann AF, Connor MR, Shen CR, Smith KM, Brynildsen MP, Chou KJ, Hanai T, Liao JC (2008) Metabolic engineering of Escherichia coli for 1-butanol production. Metab Eng 10(6):305–311
6. Avalos JL, Fink GR, Stephanopoulos G (2013) Compartmentalization of metabolic pathways in yeast mitochondria improves the production of branched-chain alcohols. Nat Biotechnol 31(4):335–341
7. Bak JS (2014) Electron beam irradiation enhances the digestibility and fermentation yield of water-soaked lignocellulosic biomass. Biotechnol Rep 4:30–33
8. Barr MR, Jervis R, Zhang Y, Bodey AJ, Rau C, Shearing PR, Brett DJ, Titirici MM, Volpe R (2021) Towards a mechanistic understanding of particle shrinkage during biomass pyrolysis via synchrotron X-ray microtomography and in-situ radiography. Sci Rep 11(1):2656. https://doi.org/10.1038/s41598-020-80228-x
9. Barsbay M, Güven O (2019) Surface modification of cellulose via conventional and controlled radiation-induced grafting. Radiat Phys Chem 160:1–8
10. Belgiorno V, De Feo G, Della Rocca C, Napoli RM (2003) Energy from gasification of solid wastes. Waste Manag 23(1):1–15
11. Bhardwaj N, Kumar B, Verma P (2019) A detailed overview of xylanases: an emerging biomolecule for current and future prospective. Bioresour Bioprocess 6:40. https://doi.org/10.1186/s40643-019-0276-2
12. Bhardwaj N, Kumar B, Verma P (2020a) Microwave-assisted pretreatment using alkali metal salt in combination with orthophosphoric acid for generation of enhanced sugar and bioethanol. Biomass Convers Biorefinery
13. Bhardwaj N, Kumar B, Agrawal K, Verma P (2020) Bioconversion of rice straw by synergistic effect of in-house produced ligno-hemicellulolytic enzymes for enhanced bioethanol production. Bioresour Technol Rep 10:100352
14. Bhardwaj N, Agrawal K, Kumar B, Verma P (2021) Role of enzymes in deconstruction of waste biomass for sustainable generation of value-added products. In: Thatoi H, Mohapatra S, Das SK (eds) Bioprospecting of enzymes in industry, healthcare and sustainable environment, pp 219–250. Springer, Singapore
15. Cao L, Zhang C, Chen H, Tsang DC, Luo G, Zhang S, Chen J (2017) Hydrothermal liquefaction of agricultural and forestry wastes: state-of-the-art review and future prospects. Biores Technol 245:1184–1193
16. Cha M, Chung D, Elkins JG, Guss AM, Westpheling J (2013) Metabolic engineering of Caldicellulosiruptor bescii yields increased hydrogen production from lignocellulosic biomass. Biotechnol Biofuels 6(1):1–8
17. Chen W-H, Lin Y-Y, Liu H-C, Chen T-C, Hung C-H, Chen C-H, Ong HC (2019) A comprehensive analysis of food waste derived liquefaction bio-oil properties for industrial application. Appl Energy 237:283–291
18. Chen W-H, Tu Y-J, Sheen H-K (2011) Disruption of sugarcane bagasse lignocellulosic structure by means of dilute sulfuric acid pretreatment with microwave-assisted heating. Appl Energy 88(8):2726–2734
19. Chen W-H, Ye S-C, Sheen H-K (2012) Hydrothermal carbonization of sugarcane bagasse via wet torrefaction in association with microwave heating. Biores Technol 118:195–203

20. Chintala V (2018) Production, upgradation and utilization of solar assisted pyrolysis fuels from biomass–a technical review. Renew Sustain Energy Rev 90:120–130
21. Chung Y-L, Olsson JV, Li RJ, Frank CW, Waymouth RM, Billington SL, Sattely ES (2013) A renewable lignin–lactide copolymer and application in biobased composites. ACS Sustain Chem & Eng 1(10):1231–1238
22. Colin VL, Rodríguez A, Cristobal HA (2011) The role of synthetic biology in the design of microbial cell factories for biofuel production. BioMed Research International
23. Dabros TM, Stummann MZ, Høj M, Jensen PA, Grunwaldt JD, Gabrielsen J, Mortensen PM, Jensen AD (2018) Transportation fuels from biomass fast pyrolysis, catalytic hydrodeoxygenation, and catalytic fast hydropyrolysis. Prog Energy Combust Sci 68:268–309
24. Dai L, He C, Wang Y, Liu Y, Yu Z, Zhou Y, Fan L, Duan D, Ruan R (2017) Comparative study on microwave and conventional hydrothermal pretreatment of bamboo sawdust: hydrochar properties and its pyrolysis behaviors. Energy Convers Manag 146:1–7
25. Dai L, Liu R, Si C (2018) A novel functional lignin-based filler for pyrolysis and feedstock recycling of poly (L-lactide). Green Chem 20(8):1777–1783
26. Dellomonaco C, Fava F, Gonzalez R (2010) The path to next generation biofuels: successes and challenges in the era of synthetic biology. Microb Cell Fact 9(1):1–15
27. El Hariri El Nokab M, Habib MH, Alassmy YA, Abduljawad MM, Alshamrani KM, Sebakhy KO (2022) Solid state NMR a powerful technique for investigating sustainable/renewable cellulose-based materials. Polymers 14(5):1049. https://www.mdpi.com/2073-4360/14/5/1049
28. Ershov BG (1998) Radiation-chemical degradation of cellulose and other polysaccharides. Russ Chem Rev 67(4):315–334
29. Escada-Rebelo S, Mora FG, Sousa AP, Almeida-Santos T, Paiva A, Ramalho-Santos J (2020) Fluorescent probes for the detection of reactive oxygen species in human spermatozoa. Asian J Androl 22(5):465–471. https://doi.org/10.4103/aja.aja_132_19
30. Ewanick S, Schmitt E, Gustafson R, Bura R (2014) Use of Raman spectroscopy for continuous monitoring and control of lignocellulosic biorefinery processes. Pure Appl Chem 86(5):867–879
31. Fabry F, Rehmet C, Rohani V, Fulcheri L (2013) Waste gasification by thermal plasma: a review. Waste Biomass Valorization 4:421–439
32. Fan L, Lee Y-H, Gharpuray M (2005) The nature of lignocellulosics and their pretreatments for enzymatic hydrolysis. In: Microbial reactions, pp 157–187. Springer
33. Fawzy S, Osman AI, Yang H, Doran J, Rooney DW (2021) Industrial biochar systems for atmospheric carbon removal: a review. Environ Chem Lett 19:3023–3055
34. Foston M, Katahira R, Gjersing E, Davis MF, Ragauskas AJ (2012) Solid-state selective 13C excitation and spin diffusion NMR to resolve spatial dimensions in plant cell walls. J Agric Food Chem 60(6):1419–1427
35. Fouilland T, Grace JR, Ellis N (2010) Recent advances in fluidized bed technology in biomass processes. Biofuels 1(3):409–433
36. Gaida SM, Liedtke A, Jentges AHW, Engels B, Jennewein S (2016) Metabolic engineering of Clostridium cellulolyticum for the production of n-butanol from crystalline cellulose. Microb Cell Fact 15(1):1–11
37. Gollakota A, Kishore N, Gu S (2018) A review on hydrothermal liquefaction of biomass. Renew Sustain Energy Rev 81:1378–1392
38. Gordobil O, Egüés I, Llano-Ponte R, Labidi J (2014) Physicochemical properties of PLA lignin blends. Polym Degrad Stab 108:330–338
39. Goswami RK, Agrawal K, Verma P (2022) Microalgae biomass biorefinery: a sustainable renewable energy feedstock of the future. In: Micro-algae: next-generation feedstock for biorefineries: cultivation and refining processes, pp 1–29. Springer Nature Singapore, Singapore
40. Grabowski CE (2015) The impact of electron beam pre-treatment on the fermentation of wood-based sugars

41. Güleç F, Williams O, Kostas ET, Samson A, Lester E (2022) A comprehensive comparative study on the energy application of chars produced from different biomass feedstocks via hydrothermal conversion, pyrolysis, and torrefaction. Energy Convers Manage 270:116260
42. Guo M, Li C, Facciotto G, Bergante S, Bhatia R, Comolli R, Ferré C, Murphy R (2015) Bioethanol from poplar clone Imola: an environmentally viable alternative to fossil fuel? Biotechnol Biofuels 8:1–21
43. Hassan SS, Williams GA, Jaiswal AK (2018) Emerging technologies for the pretreatment of lignocellulosic biomass. Biores Technol 262:310–318
44. Haykiri-Acma H, Yaman S (2010) Interaction between biomass and different rank coals during co-pyrolysis. Renew Energy 35(1):288–292
45. Holtman KM, Chen N, Chappell MA, Kadla JF, Xu L, Mao J (2010) Chemical structure and heterogeneity differences of two lignins from loblolly pine as investigated by advanced solid-state NMR spectroscopy. J Agric Food Chem 58(18):9882–9892
46. Hong SH, Lee JT, Lee S, Wi SG, Cho EJ, Singh S, Lee SS, Chung BY (2014) Improved enzymatic hydrolysis of wheat straw by combined use of gamma ray and dilute acid for bioethanol production. Radiat Phys Chem 94:231–235
47. Huang Y-F, Chiueh P-T, Kuan W-H, Lo S-L (2016) Microwave pyrolysis of lignocellulosic biomass: heating performance and reaction kinetics. Energy 100:137–144
48. Huang Y-F, Chiueh P-T, Lo S-L (2016) A review on microwave pyrolysis of lignocellulosic biomass. Sustain Environ Res 26(3):103–109
49. Kargbo H, Harris JS, Phan AN (2021) "Drop-in" fuel production from biomass: critical review on techno-economic feasibility and sustainability. Renew Sustain Energy Rev 135:110168
50. Karthika K, Arun A, Melo J, Mittal K, Kumar M, Rekha P (2013) Hydrolysis of acid and alkali presoaked lignocellulosic biomass exposed to electron beam irradiation. Biores Technol 129:646–649
51. Kasmuri NH, Kamarudin S, Abdullah SRS, Hasan H, Som AM (2017) Process system engineering aspect of bio-alcohol fuel production from biomass via pyrolysis: an overview. Renew Sustain Energy Rev 79:914–923
52. Ko JK, Enkh-Amgalan T, Gong G, Um Y, Lee SM (2020) Improved bioconversion of lignocellulosic biomass by Saccharomyces cerevisiae engineered for tolerance to acetic acid. GCB Bioenergy 12(1):90–100
53. Komatsu T, Kikuchi J (2013) Selective signal detection in solid-state NMR using rotor-synchronized dipolar dephasing for the analysis of hemicellulose in lignocellulosic biomass. J Phys Chem Lett 4(14):2279–2283
54. Kumar A, Kumar GA (2023) Modification of lignin properties using alpha particles and gamma-rays for diverse applications. Radiat Phys Chem 202:110562
55. Kumar B, Bhardwaj N, Verma P (2019) Pretreatment of rice straw using microwave assisted $FeCl_3$-H_3PO_4 system for ethanol and oligosaccharides generation. Bioresour Technol Rep 7(2019):100295. https://doi.org/10.1016/j.biteb.2019.100295
56. Kumar B, Verma P (2019) Optimization of microwave-assisted pretreatment of rice straw with $FeCl_3$ in combination with H_3PO_4 for improving enzymatic hydrolysis. Adv Plant Microb Biotechnol 41–48
57. Kumar B, Bhardwaj N, Agrawal K, Chaturvedi V, Verma P (2020) Current perspective on pretreatment technologies using lignocellulosic biomass: an emerging biorefinery concept. Fuel Process Technol 199:106244
58. Kumar B, Bhardwaj N, Verma P (2020) Microwave assisted transition metal salt and orthophosphoric acid pretreatment systems: generation of bioethanol and xylo-oligosaccharides. Renew Energy 158:574–584
59. Kumar B, Verma P (2020) Biomass-based biorefineries: an important architype towards a circular economy. Elsevier, Fuel, p 119622
60. Kumar B, Verma P (2021) Life cycle assessment: blazing a trail for bioresources management. Elsevier, Energy Conversion and Management X, p 100063
61. Letourneau DR, Volmer DA (2023) Mass spectrometry-based methods for the advanced characterization and structural analysis of lignin: a review. Mass Spectrom Rev 42(1):144–188

62. Li C, Zhao X, Wang A, Huber GW, Zhang T (2015) Catalytic transformation of lignin for the production of chemicals and fuels. Chem Rev 115(21):11559–11624
63. Li J, Dai J, Liu G, Zhang H, Gao Z, Fu J, He Y, Huang Y (2016) Biochar from microwave pyrolysis of biomass: a review. Biomass Bioenergy 94:228–244
64. Liew F, Henstra AM, Köpke M, Winzer K, Simpson SD, Minton NP (2017) Metabolic engineering of Clostridium autoethanogenum for selective alcohol production. Metab Eng 40:104–114
65. Lin SY, Dence CW (2012) Methods in lignin chemistry. Springer Science & Business Media
66. List R, Gonzalez-Lopez L, Ashfaq A, Zaouak A, Driscoll M, Al-Sheikhly M (2023) On the mechanism of the ionizing radiation-induced degradation and recycling of cellulose. Polymers 15(23):4483. https://www.mdpi.com/2073-4360/15/23/4483
67. Liu Q, Wang S, Zheng Y, Luo Z, Cen K (2008) Mechanism study of wood lignin pyrolysis by using TG–FTIR analysis. J Anal Appl Pyrol 82(1):170–177
68. Liu Y, Hu T, Wu Z, Zeng G, Huang D, Shen Y, He X, Lai M, He Y (2014) Study on biodegradation process of lignin by FTIR and DSC. Environ Sci Pollut Res 21:14004–14013
69. Liu Y, Zhou H, Wang S, Wang K, Su X (2015) Comparison of γ-irradiation with other pretreatments followed with simultaneous saccharification and fermentation on bioconversion of microcrystalline cellulose for bioethanol production. Biores Technol 182:289–295
70. Lou, Z., Zhang, Y., Li, Y., & Xiao, H. (2023). Study on microscopic physical and chemical properties of biomass materials by AFM. J Mater Res Technol
71. Luo J, Fang Z, Smith RL Jr (2014) Ultrasound-enhanced conversion of biomass to biofuels. Prog Energy Combust Sci 41:56–93
72. Mante OD, Amidon TE, Stipanovic A, Babu SP (2014) Integration of biomass pretreatment with fast pyrolysis: an evaluation of electron beam (EB) irradiation and hot-water extraction (HWE). J Anal Appl Pyrol 110:44–54. https://doi.org/10.1016/j.jaap.2014.08.004
73. Martínez S, Sánchez R, Todolí J-L (2022) Inductively coupled plasma tandem mass spectrometry (ICP-MS/MS) for the analysis of fuels, biofuels and their feedstock using a high temperature total consumption sample introduction system operated under continuous sample aspiration mode. J Anal At Spectrom 37(5):1032–1043
74. Maryandyshev PA, Kangash AI, Pokryshkin SA, Lyubov VK, Trouve G, Brillard A, Brilhac J-F (2021) A study of biofuels by pyrolytic gas chromatography. Solid Fuel Chem 55(3):159–164
75. Mika LT, Cséfalvay E, Németh Á (2018) Catalytic conversion of carbohydrates to initial platform chemicals: chemistry and sustainability. Chem Rev 118(2):505–613
76. Mohd Asyraf K, Khalil HPSA, Noor Aziah S, Mohamad Haafiz Mohamad K, Muhammad Izzuddin S, Aprila, NAS, Rudi D (2016) Irradiation pretreatment of tropical biomass and biofiber for biofuel production. In: Waldemar AM (ed) Radiation effects in materials, pp Ch. 13. IntechOpen, Rijeka
77. Mood SH, Golfeshan AH, Tabatabaei M, Jouzani GS, Najafi GH, Gholami M, Ardjmand M (2013) Lignocellulosic biomass to bioethanol, a comprehensive review with a focus on pretreatment. Renew Sustain Energy Rev 27:77–93
78. Mosier N, Wyman C, Dale B, Elander R, Lee Y, Holtzapple M, Ladisch M (2005) Features of promising technologies for pretreatment of lignocellulosic biomass. Biores Technol 96(6):673–686
79. Nair LG, Agrawal K, Verma P (2022a) An insight into the principles of lignocellulosic biomass-based zero-waste biorefineries: a green leap towards imperishable energy-based future. Biotechnol Genet Eng Rev pp 1–51
80. Nair LG, Agrawal K, Verma P (2022b) An overview of sustainable approaches for bioenergy production from agro-industrial wastes. Energy Nexus 100086
81. Nair LG, Agrawal K, Verma P (2023) Organosolv pretreatment: an in-depth purview of mechanics of the system. Bioresour Bioprocess, Springer 10(1)
82. Okolie JA, Nanda S, Dalai AK, Kozinski JA (2021) Chemistry and specialty industrial applications of lignocellulosic biomass. Waste Biomass Valorization 12:2145–2169

83. Ong HC, Chen W-H, Singh Y, Gan YY, Chen C-Y, Show PL (2020) A state-of-the-art review on thermochemical conversion of biomass for biofuel production: a TG-FTIR approach. Energy Convers Manage 209:112634
84. Orem WH, Hatcher PG (1987) Solid-state 13C NMR studies of dissolved organic matter in pore waters from different depositional environments. Org Geochem 11(2):73–82
85. Osman AI, Mehta N, Elgarahy AM, Al-Hinai A, Al-Muhtaseb AH, Rooney DW (2021) Conversion of biomass to biofuels and life cycle assessment: a review. Environ Chem Lett 19:4075–4118
86. Pan H, Zhifeng Z, Hse CY (2012) Microwave-assisted liquefaction of wood with polyhydric alcohols and its application in preparation of polyurethane (PU) foams. Eur J Wood Wood Prod 70(4):461–470
87. Park CC, Yom SS, Podgorsak MB, Harris E, Price Jr RA, Bevan A, Pouliot J, Konski AA, Wallner PE (2010) American Society for Therapeutic Radiology and Oncology (ASTRO) emerging technology committee report on electronic brachytherapy. Int J Radiat Oncology* Biology* Phy, 76(4):963–972
88. Ponomarev A (2017) Radiolysis of lignin: prospective mechanism of high-temperature decomposition. Radiat Phys Chem 141:160–167
89. Postek MT, Poster DL, Vládar AE, Driscoll MS, LaVerne JA, Tsinas Z, Al-Sheikhly MI (2018) Ionizing radiation processing and its potential in advancing biorefining and nanocellulose composite materials manufacturing. Radiat Phys Chem 143:47–52
90. Preethi KA, Sekar D (2021) Dietary microRNAs: current status and perspective in food science. J Food Biochem 45(7):e13827
91. Quitain AT, Sasaki M, Goto M (2013) Microwave-based pretreatment for efficient biomass-to-biofuel conversion. Pretreat Tech Biofuels Biorefineries 117–130
92. Quraishi M, Wani K, Pandit S, Gupta PK, Rai AK, Lahiri D, Jadhav DA, Ray RR, Jung SP, Thakur VK, Prasad R (2021) Valorisation of CO_2 into value-added products via microbial electrosynthesis (MES) and electro-fermentation technology. Fermentation 7(4):291
93. Ramadoss G, Muthukumar K (2014) Ultrasound assisted ammonia pretreatment of sugarcane bagasse for fermentable sugar production. Biochem Eng J 83:33–41
94. Rao NR, Rao TV, Reddy SR, Rao BS (2015) The effect of gamma irradiation on physical, thermal and antioxidant properties of kraft lignin. J Radiat Res Appl Sci 8(4):621–629
95. Ravindran R, Jaiswal S, Abu-Ghannam N, Jaiswal AK (2017) Evaluation of ultrasound assisted potassium permanganate pre-treatment of spent coffee waste. Biores Technol 224:680–687
96. Rehman A, Rauf A, Ahmad M, Chandio AA, Deyuan Z (2019) The effect of carbon dioxide emission and the consumption of electrical energy, fossil fuel energy, and renewable energy, on economic performance: evidence from Pakistan. Environ Sci Pollut Res 26:21760–21773
97. Reichhardt C, Parsek MR (2019) Confocal laser scanning microscopy for analysis of Pseudomonas aeruginosa biofilm architecture and matrix localization. Front Microbiol 10:677
98. Saha BC (2004) Lignocellulose biodegradation and applications in biotechnology. ACS Publications
99. Saini A, Aggarwal N, Sharma A, Yadav A (2015) Prospects for irradiation in cellulosic ethanol production. Biotechnol Res Int 2015:1–13. (In.)
100. Sánchez O, Sierra R, Alméciga-Díaz CJ (2011) Delignification process of agro-industrial wastes an alternative to obtain fermentable carbohydrates for producing fuel. Altern Fuel 7
101. Sarosi OP, Bischof RH, Potthast A (2020) Tailoring pulp cellulose with electron beam irradiation: effects of lignin and hemicellulose. ACS Sustain Chem & Eng 8(18):7235–7243. https://doi.org/10.1021/acssuschemeng.0c02165
102. Shahabuddin M, Alam MT, Krishna BB, Bhaskar T, Perkins G (2020) A review on the production of renewable aviation fuels from the gasification of biomass and residual wastes. Biores Technol 312:123596
103. Shirkavand E, Baroutian S, Gapes DJ, Young BR (2016) Combination of fungal and physico-chemical processes for lignocellulosic biomass pretreatment–a review. Renew Sustain Energy Rev 54:217–234

104. Sindhu R, Binod P, Pandey A (2016) Biological pretreatment of lignocellulosic biomass–an overview. Biores Technol 199:76–82
105. Sitek T, Pospíšil J, Poláčik J, Chýlek R (2021) Thermogravimetric analysis of solid biomass fuels and corresponding emission of fine particles. Energy 237:121609
106. Sivasubramanian M, Chu CH, Cheng SH, Chen NT, Chen CT, Chuang YC, Yu H, Chen YL, Liao LD, Lo LW (2022) Multimodal magnetic resonance and photoacoustic imaging of tumor-specific enzyme-responsive hybrid nanoparticles for oxygen modulation. Front Bioeng Biotechnol 10:910902
107. Song Z, Chen L, Wang J, Lu Y, Jiang W, Zhang W (2014) A transcriptional regulator Sll0794 regulates tolerance to biofuel ethanol in photosynthetic Synechocystis sp. PCC 6803. Mol & Cell Proteomics 13(12):3519–3532
108. Stas M, Auersvald M, Vozka P (2021) Two-dimensional gas chromatography characterization of pyrolysis bio-oils: a review. Energy Fuels 35(10):8541–8557
109. Su X-J, Zhang C-Y, Li W-J, Wang F, Wang K-Q, Liu Y, Li Q-M (2020) Radiation-induced structural changes of miscanthus biomass. Appl Sci 10(3):1130. https://www.mdpi.com/2076-3417/10/3/1130
110. Subhedar PB, Babu NR, Gogate PR (2015) Intensification of enzymatic hydrolysis of waste newspaper using ultrasound for fermentable sugar production. Ultrason Sonochem 22:326–332
111. Subhedar PB, Ray P, Gogate PR (2018) Intensification of delignification and subsequent hydrolysis for the fermentable sugar production from lignocellulosic biomass using ultrasonic irradiation. Ultrason Sonochem 40:140–150
112. Sun D, Wang B, Wang HM, Li MF, Shi Q, Zheng L, Wang SF, Liu SJ, Sun RC (2018) Structural transformations of Hybrid *Pennisetum* lignin: effect of microwave-assisted hydrothermal pretreatment. ACS Sustain Chem & Eng 7(3):3073–3082
113. Sun M, Zhang M (2023) Multiphysics modeling of ultrasound-assisted biomass torrefaction for fuel pellets production. IISE Trans 55(7):723–730
114. Suzuki T, Matsuzaki T, Hagiwara H, Aoki T, Takata K (2007) Recent advances in fluorescent labeling techniques for fluorescence microscopy. Acta Histochem Cytochem 40(5):131–137. https://doi.org/10.1267/ahc.07023
115. Taherzadeh MJ, Karimi K (2008) Pretreatment of lignocellulosic wastes to improve ethanol and biogas production: a review. Int J Mol Sci 9(9):1621–1651
116. Torget R, Himmel ME, Grohmann K (1991) Dilute sulfuric acid pretreatment of hardwood bark. Biores Technol 35(3):239–246
117. Torgrip RJ, Fernández-Cano V (2017) Rapid X-ray based determination of moisture-, ash content and heating value of three biofuel assortments. Biomass Bioenerg 98:161–171
118. Ullah F, Othman MBH, Javed F, Ahmad Z, Akil HM (2015) Classification, processing and application of hydrogels: a review. Mater Sci Eng, C 57:414–433
119. Van Doren LG, Posmanik R, Bicalho FA, Tester JW, Sills DL (2017) Prospects for energy recovery during hydrothermal and biological processing of waste biomass. Biores Technol 225:67–74
120. Varella LY, Mesquita AZ, de Rezende DB, Mourão IG, Cren ÉC, Lopes NP, Silva WG, Alonso TC (2024) Experimental evaluation of the use of gamma radiation in sugarcane bagasse (Saccharum spp.) as a pretreatment for bioethanol production. Radiat Phys Chem 111611
121. Varol M, Atimtay A, Bay B, Olgun H (2010) Investigation of co-combustion characteristics of low quality lignite coals and biomass with thermogravimetric analysis. Thermochim Acta 510(1–2):195–201
122. Wang M, Huang Y, Chiueh P, Kuan W, Lo S (2012) Microwave-induced torrefaction of rice husk and sugarcane residues. Energy 37(1):177–184
123. Wencka M, Wichlacz K, Kasprzyk H, Lijewski S, Hoffmann SK (2007) Free radicals and their electron spin relaxation in cellobiose. X-band and W-band ESR and electron spin echo studies. Cellulose 14:183–194
124. Williams CL, Rachel ME, Jaya Shankar T (2017) Biomass compositional analysis for conversion to renewable fuels and chemicals. In: Jaya Shankar T (ed) Biomass volume estimation and valorization for energy, pp Ch. 11. IntechOpen, Rijeka

125. Wood RW, Loomis AL (1927) XXXVIII. The physical and biological effects of high-frequency sound-waves of great intensity. Lond, Edinb, Dublin Philos Mag J Sci 4(22):417–436
126. Wu X, Chen L, Chen J, Su X, Liu Y, Wang K, Qin W, Qi H, Deng M (2019) The effect of 60 Co γ-irradiation on the structure and thermostability of alkaline lignin and its irradiation derived degradation products. Waste Biomass Valorization 10:3025–3035
127. Wyman C (1996) Handbook on bioethanol: production and utilization. CRC Press
128. Xiang Y, Xiang Y, Wang L (2017) Electron beam irradiation to enhance enzymatic saccharification of alkali soaked *Artemisia ordosica* used for production of biofuels. J Environ Chem Eng 5(4):4093–4100
129. Xiao W, Han L, Zhao Y (2011) Comparative study of conventional and microwave-assisted liquefaction of corn stover in ethylene glycol. Ind Crops Prod 34(3):1602–1606
130. Xu Q-Q, Zhao M-J, Yu Z-Z, Yin J-Z, Li G-M, Zhen M-Y, Zhang Q-Z (2017) Enhancing enzymatic hydrolysis of corn cob, corn stover and sorghum stalk by dilute aqueous ammonia combined with ultrasonic pretreatment. Ind Crops Prod 109:220–226
131. Yan W-H, Duan P-G, Wang F, Xu Y-P (2016) Composition of the bio-oil from the hydrothermal liquefaction of duckweed and the influence of the extraction solvents. Fuel 185:229–235
132. Yang SJ, Yoo HY, Choi HS, Lee JH, Park C, Kim SW (2015) Enhancement of enzymatic digestibility of *Miscanthus* by electron beam irradiation and chemical combined treatments for bioethanol production. Chem Eng J 275:227–234. https://doi.org/10.1016/j.cej.2015.04.056
133. Zhang Z, Zhao ZK (2010) Microwave-assisted conversion of lignocellulosic biomass into furans in ionic liquid. Biores Technol 101(3):1111–1114
134. Zhu H, Du B, Zhang Z, Wang X, Sun Y, Liu B, Zhou J (2022) Effect of hierarchical HZSM-5 zeolite on the catalytic depolymerization of organosolv lignin to renewable phenols. J Porous Mater 1–13

Chapter 5
Advancements in Microwave-assisted Technologies for the Production of Value-added Products from Lignocellulosic Waste Biomass

Latika Bhatia, Tejaswini Jaiswal, Shruti Singh, and Dilip Kumar Sahu

Abstract The global concerns toward sustainable development have created many paths for the employment of lignocellulosic waste feedstock for producing energy, fuel, chemicals as well as many other value-added substances. Lignocellulosic materials are comprised of the most profuse renewable carbon source. These materials can be either agro-wastes (e.g., rice husk, rice straw, wheat straw, sugarcane bagasse, and corncob) or forest (e.g., aspen and pine wood) residues. The demand for value-added products like furfural, furfuryl alcohol, and formic acid is huge in the global market. The efforts to produce these compounds from lignocellulosic waste biomass are dependent upon various techniques that can alter the structure of this base material. Microwaves signify a competent source of energy to acquire value-added chemicals and fuel from lignocellulosic biomass. Process intensification and energy saving become feasible with microwave-driven green chemical techniques. Volumetric and selective dielectric heating are some unique properties of microwaves that can be effectively exploited for hydrolyzing biopolymers and the further conversion to highly expensive fine chemicals. Microwave heating is a better option in comparison to conventional modes of heating due to its higher reaction rates (speed), throughput, yield, and purity. Uniformity of heat dispersion and less requirement of energy are other merits associated with microwave-assisted processes. Microwave-based pretreatment is a promising approach for pretreating lignocellulosic waste biomass that mediates physical, chemical, or biological reactions. This chapter focuses on some of these characteristics of microwaves and surveys the usage of microwaves as a separate pretreatment method in addition to a co-pretreatment procedure to augment the execution of further thermochemical pretreatments. It also highlights recent advancements in microwave applications for the generation of value-added chemicals, the challenges associated, and future research guidelines to resolve them.

L. Bhatia (✉) · T. Jaiswal · S. Singh · D. K. Sahu
Department of Microbiology & Bioinformatics, Atal Bihari Vajpayee University, Bilaspur, Chhattisgarh, India
e-mail: latikabhatia1@yahoo.co.uk

© The Author(s), under exclusive license to Springer Nature Switzerland AG 2024

P. Verma and B. Kumar (eds.), *Sustainable Radiation Technologies in Waste-biomass Valorization*, Green Chemistry and Sustainable Technology,
https://doi.org/10.1007/978-3-031-63941-8_5

Keywords Biomass pretreatment · Microwave irradiation · Circular economy · Sustainable development · Sustainability

5.1 Introduction

There has been an immense process in the agricultural sector around the globe, as a result of which there has been a boon in the production of agricultural waste and consequently the generation of agro-waste also [1]. There have also emerged the concepts of energy crisis and economic growth worldwide taking ecological concerns along with them. Agro-industrial sectors are now focussing on the approaches that can manage the waste generated through these sectors in an economic and eco-friendly way [2, 3]. Hence came the notion of sustainable management, where several compounds of commercial interest are explored that can be retrieved from these agro-wastes, thereby resolving the problems associated with pollution generated from these wastes along with contributing to its economic competitiveness [4].

These desires gave birth to the philosophy of circular (bio)economy in which useful products are produced in a sustainable manner utilizing residuals, by-products, and waste produced all through the life cycle of industrial processing of numerous resources [5]. In a circular economy, each part of the resource, starting from raw material to waste, is employed to achieve the highest value. It is a future trend that has the potential to resolve environmental challenges raised due to inappropriate handling of wastes and by-products. One of the essential components of the circular economy/circular (bio)economy is biorefinery [6, 7]. Biorefinery is an industrial plant converts several organic residuals (biomass) for producing value-added products like food, energy, chemicals, fertilizers, materials, etc. [8]

A sustainable concept becomes feasible if a circular economy is taken into consideration. In a circular economy, the earth's resources are exploited in such a way that there is a minimum loss of values/services and maximum use of loss of values/services of the materials confined within the products [9]. As a significant constituent, biomass plays a crucial role in the context of circular bioeconomy as it is a source of material and energy [10]. The accomplishment of a sustainable circular bioeconomy becomes possible with the use of waste biomass. This is the reason why biomass waste valorization is becoming popular in the research areas [11].

The usage of waste biomass will not only curtail waste generation but will provide an opportunity to generate novel bioactive compounds, solvents, polymers, and materials [12, 13]. The use of waste biomass also lessens pollution and provides an alternative, sustainable source for producing energy which can be further utilized in producing chemicals [14]. This makes the waste biomass a reliable source of chemical building blocks for the sustenance of future chemical industries [15]. The development of sustainable technology is an outcome of extensive research being done around the globe on the utilization of waste biomass to produce clean and renewable energy. This usage of waste biomass is a sustainable, cost-effective, and eco-friendly approach to generating new chemicals. It allows the development of novel procedures that fully agree with the ideologies of green chemistry [16].

5.2 Renewable Bio-resources

Biomass can be divided into four generations or categories, based on its accessibility and its use in several sectors. Biomass belonging to the first generation is generally agriculture-gained edible crops, comprising wheat, rice, oilseeds, corn, and sugarcane [17]. As many people in various nations are deprived of food and nutrition, there has arisen a concern about food security issues and a debate about food versus fuel [18]. Due to these reasons edible food material cannot be used for producing fuels or chemicals. To resolve these issues and to ensure the sustainable production of various fuels and chemicals, researchers have focused on the waste that is generated out of agricultural practices, agro-industries, and forests [19]. These include wheat straw, rice husk, rice straw, bagasse, fruit peels, and shells of nuts. These lignocellulosic biomass residues belong to the second generation [20]. The use of these biomasses has no competition with edible crops as they are not an essential part of the food chain. In comparison to these lignocellulosic biomasses, the growth rate of algal biomass is very high, with production rates 5–30 times higher than crops [21, 22]. This makes them efficient producers of oil. Moreover, it is possible to cultivate algae in non-cultivable and effluent bodies. Hence, algal biomass is categorized as third-generation biomass and is considered a favorable feedstock more than land biomass. The fourth generation includes GMOs, i.e., Genetically Modified Organisms [23] (Fig. 5.1).

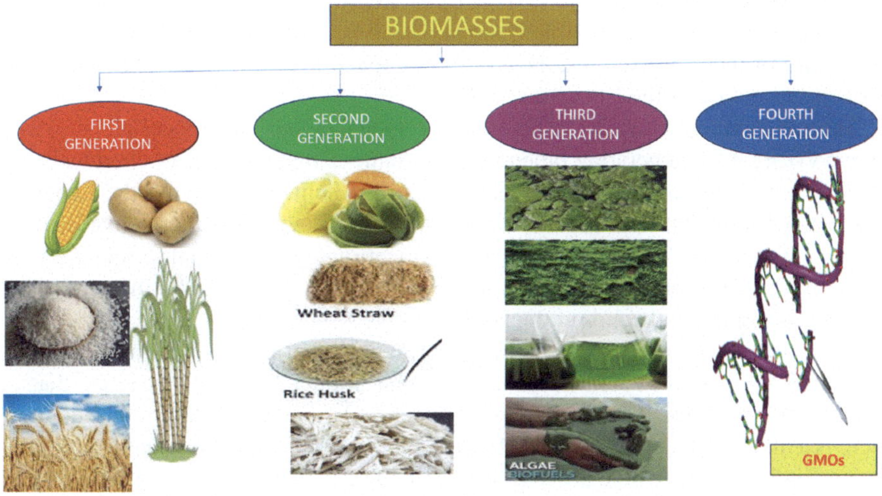

Fig. 5.1 Schematic representation of types of biomasses

Agricultural waste is generated in significant quantities around the globe. Most of the time, it is not utilized appropriately. These wastes are either burnt in the open creating air pollution or are discarded in open lands generating soil and water pollution [24]. In a few places, these agro-residues are used to feed cattle or are used as raw material for the paper industry. These lignocellulosic waste biomasses are the potential feedstocks for the extraction of valuable resources employing sophisticated technologies [25]. Selective bio-materials, chemicals, liquid, and gaseous fuels can be obtained from these bio-resources when they are processed in biorefineries. In comparison to the combustion of coal and petroleum, the combustion of these waste residues generates less SO_2, NOx, and soot. Biomass also supports sustaining carbon neutrality because the combustion of biomass-derived fuel liberates carbon dioxide and CO_2 that can be turned back through plant photosynthesis [11].

5.3 Need for Pretreatment

The pretreatment technologies are pivotal steps for the utilization of biomass, as they make the biomass homogeneous and chemically suitable for further bioconversion into value-added products. Lignocellulosic biomass residues are generally recalcitrant. There is an intricate interaction between lignin, cellulose, and hemicellulose of feedstock [12, 26]. These interactions need to be broken so that the sugars present in cellulose and hemicellulose can become accessible to enzyme treatment or can be converted into monomeric sugars after thermal pretreatment [27]. These monomeric sugars can be further converted into bioethanol and other products. The surface area of the biomass is increased after pretreatment which makes it accessible for further reaction [28]. The selection of methods for pretreatment in a biorefinery system depends upon many factors like sustainability, deployment cost, power requirement, and overall efficiency. Pretreatment methods can be physical, chemical, biological, or a blend of these procedures, have been employed to allow the breakdown of the recalcitrant nature of the LB waste [29].

Thermochemical conversion helps in the management of lignocellulosic waste by valorizing it for generating bioenergy and value-added products or chemicals out of lignocellulosic biomass [30]. Moreover, it also helps in resolving the issues of waste generation. This makes the thermochemical conversion an effective method of pretreatment [31]. In this perspective, gasification, pyrolysis, and torrefaction are recognized and established methods for the valorization of biomass for the generation of gaseous, liquid, and solid products [32].

Nevertheless, the operation of these thermochemical procedures takes place at very high temperatures and needs a persistent supply of thermal energy that is provided by traditional energy sources like fossil fuels [32]. As a consequence, there is a hike in the consumption of fossil fuels, huge emission of greenhouse gases, and encumbering the whole process plant economics. Taking all these problems into account, there should be alternative mechanisms of heating, like microwave heating, which is an energy-efficient and potentially viable solution to the traditional heating mechanism [33].

5.4 Microwave Pretreatment

Microwave pretreatment is an emerging green technology in the process of biorefinery, particularly during its thermal part. It is known for its high reaction speed, homogenous heating, non-touch processing with the least side reactions, and easy control [34, 35]. Its operation depends on the features of the electromagnetic field and the characteristics of the target material. Till recently industrial utilities of microwave methodology were restricted to defrosting food and extracting oil. Current research has explored that this microwave is a promising and feasible green technology in supporting the improved production of biodiesel (biofuel) and other value-added compounds [36].

When microwaves penetrate materials it causes a forceful ionic and dipole rotation conduction. Successively, heat is generated from inside, leading to selective, non-direct contact, and uniform volumetric heating. An outward-flow temperature gradient is generated along with uniform and volumetric heating. This makes microwaving an efficient and quick process [37]. Moisture present in the sample, by its appropriate dielectric property, augments the microwave heating through dipolar polarization, thereby circumventing pretreatment like drying [38].

Microwaves can be used to pretreat biomass waste either like an individual pretreatment or in combination with an additional procedure. These approaches help in the valorization of biomass waste to produce biofuel or value-added products [18]. Microwave can independently pretreat LB if water or a parallel microwave absorber is present as the heat-absorbing media. For the valorization of LBs into value-added bioproducts, it is important to disrupt the recalcitrant structure in the LB. The contact of LB with microwaves permits improved hydrolysis yields [9, 39]. Many researchers have explored the efficiency of microwaves along with other chemical pretreatments like acidic, alkaline, deep eutectic solvent-based, or ionic liquid-based methods. Modifications in the design of microwave vessels have been reported to overcome challenges like hotspots and irregular heating connected with microwave heating [40, 41].

Microwave pretreatment was used by Reddy et al. [42] to yield sugar from corncob. Glycerol was also used during this pretreatment process as it separated alkali and alkaline metals. In an alternative study, when corn stover was pretreated employing microwave-$CaCl_2$, it was found that surface area was increased to 168.93% and about 85.90% hemicellulose was degraded when compared to untreated substrate. As the substrate particle size decreased, its conversion was higher leading to higher yield from corn stover. But the same didn't happen when acid pretreatment was done, as it lowered the yield. Singh et al. [40] reported the latest advancements in biomass pretreatment methods as the use of microwave radiation, for producing biohydrogen

[43]. Microwave-assisted biomass processing has been often explored as an implying intensification application, like torrefaction, gasification, and pyrolysis [38].

5.5 Types of Microwave Treatment

5.5.1 Microwave Torrefaction

Microwave torrefaction takes place at a comparatively low temperature (200–300 °C). During this process, chemically static carbon-dense torrefied char is produced from feedstock in an inert environment [44]. It is easy to transport and store these char pellets. these char pellets are valuable to be treated as a solid fuel [45]. The factors that determine the efficiency of microwave torrefaction are reaction temperature and time, microwave power, microwave cavity mode, and feedstock moisture content [46]. Microwave torrefaction operates in either of two modes, viz., single mode or multimode cavity. However, the multimode cavity is appropriate for processing those feedstocks that have large particle sizes or are huge in quantity [46]. In a single-mode cavity, lower microwave power (500 W) causes a higher heating rate because the microwave radiation is focused directly on the feedstock. Biochar yield is determined by the microwave power during microwave torrefaction. If microwave power is low, it will provide a reduced heating rate, thereby constraining the process of carbonization [29]. If microwave power is high, it enhances the rate of heating but may aggravate secondary reactions. Therefore, optimal microwave power is needed for various forms of feedstock to amplify the production of torrefied char. Along with the heating rate, the microwave power also decides the maximum reaction temperature that can be achieved. A high heating rate and highest temperature can be achieved by increasing the microwave power, leading to the formation of torrefied char (low in yield) and gaseous product (high in yield). If processing time is less, then there is a lower loss of feedstock mass, thereby enhancing the biochar production. The economic efficacy of the microwave torrefaction process can be improved by developing a continuous microwave system and re-utilizing torrefied products (e.g., bio-oil and syngas). This will minimize the external supply of heat [47].

5.5.2 Microwave Gasification

Microwave gasification transforms wastes into syngas by utilizing microwave heating. The chief components of syngas are CO_2, H_2, CH_4, and CO. Electromagnetic energy gets transformed into thermal energy due to direct heating during microwave-assisted gasification. This happens due to the contact of atoms or molecules in the course of heating. The formation of hotspots plays a crucial part in enhancing the gasification rate. The existence of catalyst supports in reducing char and undesirable tar production. Microwave gasification takes place at a temperature of more than 600 °C, which is dependent on the reactor design and the attributes of waste [48]. During gasification, fragmentation, and depolymerization take place and eventually form chiefly syngas and bio-oil. Polymers present in the feedstocks are broken into condensable liquid during the depolymerization reaction, whereas, the covalent links of the polymer are broken to produce incondensable gases during the fragmentation reaction. The syngas (hydrogen + carbon mono-oxide) formation is catalyzed by gasifying agents (for example, carbon dioxide, steam, oxygen, air), during gasification. The produced syngas is commonly used during Fischer–Tropsch reactions for the generation of heat, electricity, and biofuel [49].

5.5.3 Microwave Pyrolysis

Microwave pyrolysis operates around 500–700°C in a limited oxygen or anoxygenic environment for the time of less than one hour. It is a promising technology for conversion of lignocellulosic feedstock into valuable products. Luo et al. [50] found oil palm wastes as a suitable feedstock for producing biochar by microwave pyrolysis as it has a high content of carbon (up to 51 wt%) as feedstocks. As the oxygen content in oil palm wastes is very high (38–41 wt%), microwave pyrolysis can bring down this oxygen content. The proximate and lignocellulosic compositions determine the forms of value-added products from waste. For example, if fixed carbon contents vary from (34–39 wt%), the LB is suitable for biochar production, whereas, liquid and gaseous products are formed if moisture content is high (60 wt%). Various parameters like microwave power, temperature, and flow rate of carrier gas (e.g., nitrogen) determine the quality and yield of products formed during microwave pyrolysis. The heating process gets enhanced by eliciting microwave power, thereby increasing the rate of heating and the final temperature. An increase in the flow rate of carrier gas changes the heating process, as high gas flow rates carry away the generated heat thus dropping the heating rate and overall temperature [51].

5.5.4 Advanced Microwave Pyrolysis

Nitrogen gas is used as the carrier gas in conventional microwave pyrolysis. It helps in creating an inert environment for pyrolysis to take place. The addition of nitrogen in the process increases the cost to the microwave pyrolysis plants if its removal becomes necessary from certain gaseous products (e.g., syngas). This cost can be curtailed by the introduction of vacuum conditions thereby replacing the usage of nitrogen as carrier gas. Vacuum condition during microwave pyrolysis also has one major advantage in that it avoids possible reactions like combustion and re-condensation because gas has a shorter retention time inside the reactor [29]. The vacuum state also generates a negative pressure atmosphere that permits the release of volatile substances at a lower temperature, thereby improving the production and attributes of pyrolytic products [52]. Some researchers have reported that microwave vacuum pyrolysis of LB could be achieved at a lower optimum temperature (336–405 °C) and at a shorter processing time (30 min) in comparison to the usual method that has an optimal temperature of 600 °C and long processing period of about 50 min [20]. However, there is a limitation associated with this technique, where the pyrolytic oil produced during this process has an excessive percentage of polycyclic aromatic hydrocarbon and requires additional modification to produce fine pyrolytic oil [53]. Consequently, pyrolysis through self-purging gas has been proposed as an alternate approach for producing products like bio-oil and gases. This technique uses gaseous products like methane, hydrogen, carbon monoxide, and carbon dioxide produced in pyrolysis as the carrier gas to create a passive (inert) pyrolysis environment. In brief, the gases produced in this process are withdrawn by the reactor as carrier gas [52]. The porosity and carbon content of biochar is improved by the activating agents like water molecules and carbo dioxide which are also used in self-purging gas during pyrolysis [53]. When vacuum pyrolysis and microwave-supported self-purging are employed to activate biochar, they improve the quality and yield of biochar.

5.5.5 Microwave Hydrolysis

In contrast to pyrolysis which deals with limited moisture in feedstock at high temperatures (180–300 °C), microwave hydrolysis takes place at lower temperatures (100–250 °C) in an aqueous medium. The dominant products of microwave hydrolysis are water-soluble monomers like organic acids, glucose, and furans. If the temperature is high (intensive heating) and retention time is long, these products undergo polymerization/condensation forming insoluble products like humins or hydrochar [54]. Microwave hydrolysis needs a polar medium that can absorb and transform wave energy to thermal energy, whereas usual biomass hydrolysis depends on conductive heat transfer. The presence of water in a hydrolytic system has many advantages as it efficiently absorbs microwaves thereby curtailing energy consumption. Usually, the rotation, collision, and orientation of polar species in an electromagnetic field are supposed to decrease the activation energy for reactions. Particular CH_2OH groups present in glucose and cellulose are activated in the presence of microwaves, thereby enhancing their reactions. There are very few findings that discuss microwaves' direct involvement in the decomposition machinery for lignin and hemicellulose [55].

Until recently, there have been limited reports available on the optimization of process factors (e.g., reaction time, particle size, pressure, maximum temperature, and microwave power) for the conversion of waste through microwave advanced pyrolysis procedures. Research done is limited to lab/pilot scale only and has to be further extended for commercial-scale deployment of microwave technology, knowing that this technology has a promising performance in terms of dropping the processing time and enhancing the product's quality and yield. The processing efficiency of the microwave can be optimized by making appropriate changes in microwave frequency, cavity design (for example, size and geometry), and magnetron location [56].

5.6 Value-added Products Obtained from Lignocellulosic Waste Using Different Microwave Technologies

Many fields have explored the advantages of microwave heating in the last decade. This technology supports the organic synthesis, biopolymers as well as extraction of natural products. The cell's water content is selectively heated by microwave which leads to the expansion and breakdown of the cell wall. This facilitates the liberation of the intracellular compounds and their further transformation. This process of MAE helps to enhance the yields of numerous active phytocompounds from lignocellulosic biomass. Microwave technologies support the process of delignification, which performs a significant part in the valorization of LB (Fig. 5.2). The recovery of hemicelluloses and other bioactive compounds (BACs) like antioxidants are augmented by the delignification process. Cellulose accessibility is also improved which allows the further conversion of the left-over solid fraction into value-added products. Bio-oils that are rich in hydrocarbons, phenols, and esters, are produced from lignin, as a consequence of microwave action [57]. Levulinic acid (LevA) is a bio-derived platform chemical that has drawn the attention in past few years. LevA is a versatile chemical building block that makes it a significant compound for the production of several high-end products with acknowledged applications in electronics, coatings, or lubricants. It is also used as a solvent or fuel [38].

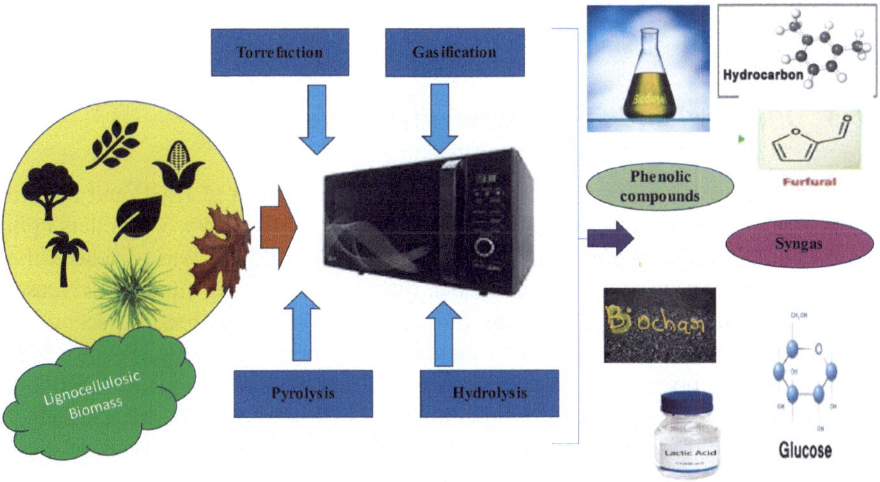

Fig. 5.2 Value-added products obtained after microwave treatment of lignocellulosic biomass

5.6.1 Levoglucosan and Levoglucosenone

Sugars and cellulose are the sources for producing Levoglucosan (LG) Levoglucosenone (LGO), and anhydrosugars. Microwave can efficiently activate the remnant sugars in crude waste softwood hydrolysis lignin (CSHL) at 180 °C (moderate temperature) to provide an enhanced LGO yield (8 wt%), whereas normal pyrolysis produced 6% LGO at a temperature of 310 °C [58]. The cellulose properties are also significant in determining the yield of LG, as weaker hydrogen bonds are present in amorphous cellulose, it yields higher LG than microcrystalline cellulose does. Wet barley straw generated 2.1% of LG, 8.6% of LGO, and 2.6% of furfural during microwave pyrolysis at 180 °C connected with self-induced steam distillation [57].

5.6.2 Hydrocarbons

Lignocellulosic biomass has a high content of oxygen (38–53%), hence it is difficult to generate oils that contain light hydrocarbons during microwave pyrolysis [59]. Catalysts are commonly required to enhance the selectivity of hydrocarbons in bio-oils, like typical pyrolysis. The aliphatic and aromatic hydrocarbons were produced from corn stover during microwave pyrolysis in the presence of SiO_2-modified HZSM-5 (an acid catalyst) [60]. Coke formation causes pore blockage, this process gets suppressed by deposition of SiO2, thereby allowing the deoxygenation reactions to occur in the internal structure of HZSM-5. Hydrocarbons were detected only in the presence of HZSM-5, and not in its absence emphasizing that microwaves alone cannot promote the reactions.

5.6.3 Furfural and 5-hydroxymethyl Furfural

The dehydration of monosaccharides leads to the production of HMF and furfural (Table 5.1). A recent pilot-scale study confirmed that 1.5–3.5 kg wood pellets upon microwave pyrolysis yielded 36–46% bio-oils. In this study. It was also seen that on increasing the power of the microwave, the concentration of furfural decreased (5552 mg g^{-1} at 2000 W to 4203 mg g^{-1} at 3000 W). This is because secondary reactions of furfural are promoted by increasing the power of the microwave [61]. In exposure to an open microwave system, wet barley straw yielded 0.7% HMF and 4.1% furfural [57]. The operating temperature can be lowered by adding acid, which mitigates the evaporation of water in an open system that enables hydrolysis reactions and produces furfural.

C5 and C6 monomers on dehydration produce furfural and HMF, respectively, which can be procured from cellulose, hemicellulose, and starch present in LB. About

Table 5.1 The yield of Levulinic acid with different substrates and catalyst

Substrates	Catalyst	Yield of Levulinic acid	References
Paper towels	10% formic acid	46%	Chen [49]
Potato peel and fungus	H_2SO_4	49%	Lappalainen et al. [77]
Potato peel and fungus	$CrCl_3$	62%,	Lappalainen et al. [77]
Brewery wastes	HCl	160–204 g kg^{-1}	Maceiras et al. [78]
Apple pomace wastes	HCl	49–66 g kg^{-1}	Maceiras et al. [78]
Starch industry waste	HCl	12 g kg^{-1}	Maceiras et al. [78]

49 mol% furfural and 42 mol% HMF a were produced from bamboo concerning hemicellulose and cellulose contents, respectively in the presence of a microwave-assisted biphasic system (acidified aqueous phase and methyl isobutyl ketone) [62]. Furfural and acetic acid were yielded from bamboo 5 wt% to 11.1 wt%, respectively, after two-stage microwave heating in the absence of a catalyst [50]. Current research reported that 1.7 times less energy is required for cellulose conductive heating during microwave heating to accomplish the equivalent temperature [63]. Under the same parameters, cellulose hydrolyzed to glucose and glucose isomerized to fructose 2.3–2.5 times quicker than under conventional heating [62]. The dehydration of C6 sugars gets accelerated in the presence of a microwave. This was suggested in one study in which a significantly higher yield of HMF (23%) was reported from cellulose in the presence of microwave irradiation (versus < 0.1% in conventional heating) [64]. Reactions are accelerated due to localized and microscopic hot spots that are generated on activation of CH_2OH groups as microwave irradiators [63]. It is also important to consider the microwave absorptivity of the solvent media. The use of green solvents like propylene carbonate and c-valerolactone (obtained from carbon dioxide and biomass, respectively) allowed a faster conversion of waste-to-HMF [65].

5.6.4 *Phenolic Compounds*

Pellets of Douglas fir sawdust produced 24.7C% aromatics (C8–C16) during microwave-assisted pyrolysis at 500 °C, using zeolite ZSM-5 and AC powder as a catalyst and absorber respectively [66]. Bio-oils rich in phenols were produced (2.46 mg mL^{-1} yield and 75% selectivity) from H_3PO_4-impregnated corn stover. AC powder was used as an absorber in this process [67]. The phosphoric groups are present on AC and these groups may offer Brønsted acid spots for stimulating the deoxygenation of guaiacols to phenols. Additional deoxygenation of phenols produces aromatic hydrocarbons. With these findings, it is confirmed that the product distribution is controlled by additives in microwave pyrolysis. Thermal degradation of lignin is improved by increasing the power of the microwave, thereby increasing the yield of 3- 3-methylphenol and phenol [61]. The phenols are selectively decomposed in the presence of intensive microwave heating. A high amount of polysaccharide present in LB inhibits the formation of aromatics because this polysaccharide gets converted to char-like materials through carbonization, which blocks lignin activation by microwave [57]. Hydrogenation of aromatics in the presence of RANEYVR nickel catalysts yields cycloalkanes. Cycloalkanes have high selectivity (84.6%) and are used as additives in jet fuel [48].

5.6.5 Syngas and Biochar

Conditions of pyrolysis govern the sharing among syngas, biochar, and bio-oil. Oil production is improved and the yields of solid and gas are reduced by increasing the power, whereas the opposite happens by raising the biomass loading [61]. The intensified contact between carbon and microwave substantially increased the surface area of biochar from 37 $m^2\ g^{-1}$ at 2000 W to 132 $m^2\ g^{-1}$ at 3000 W. Biochar having a high surface area is found suitable for environmental remediation and catalysis. Sawdust pellets produced 3.4% coke, 24.7% aromatics, 34.6% syngas, and 37.4% char on microwave pyrolysis in the existence of a microwave absorber (e.g., AC, the ZSM-5) [66]. Condensation via radical coupling gets promoted in the presence of absorber carbon leading to the increase in yield of char. In the presence of absorbers, pyrolysis reactions occur in a more controllable and homogeneous manner, leading to predictable differences in product distribution. Biomass composition can cause alteration in the heating efficiency under a constant power supply. The rate of diffusion (of volatiles), the rate of heating rate, and the product distribution are affected by the density of the feedstocks. This was demonstrated during microwave pyrolysis of cellulose pellets [68]. The product profile can be changed if chemical pretreatment is adopted. KOH-activated corn grain waste produced 54% hydrogen in the syngas, whereas non-activated corn waste produced 19% H_2 in the syngas [69]. The specific surface area of biochar produced in the former method was 420 > 138 $m^2\ g^{-1}$, making it a possibly more valuable sorbent. Parameters like, e.g., characteristics of microwave absorber, heating rate, feedstock, and feedstock loading should be taken into consideration during microwave pyrolysis experiments to achieve better regulation of selectivity and product yield [70].

5.6.6 Glucose

Cellulose and starch (glucans) present in lignocellulosic biomass are important sources of glucose. Though catalyst-free microwave-supported hydrothermal treatment of LB is frequently executed as a pretreatment or extraction procedure, its usage for producing monosaccharides is not systematically assessed [71]. Cellulose softening temperature is > 180 °C as at this temperature photons of microwave can touch the CH_2OH groups in cellulose as well as activate their rotation for better reactivity. This temperature varies for different LBs. For example, in pubescent remained integrated at 200 °C but hemicellulose was selectively dissolute. It is important to note, that it was the cellulose amorphous phase, where the polar group rotation was predominantly functional, emphasizing the significance of feedstock crystallinity [72].

The power density of microwaves also determines the production of glucose from cellulose. About 11% glucose was produced from cellulose at 800 W L^{-1}, which was substantially more than that produced at 35 W L^{-1}, though the temperature was the same (220 °C). The order of the compound's reactivity to microwave-hydrothermal degradation is alginic acid > hemicellulose > cellulose [47]. Microwave-assisted HTP is an energy-competent approach for hydrolysis of LB with fewer secondary breakdown products.

The addition of salt (sodium chloride) supports the breakdown of cellulose and xylo-oligomers during microwave heating. The addition of salt improves the absorption of the microwave and lowers the power required to achieve the required temperature. The presence of salt increases the adsorption of the proton on the substrate and disrupts the network of hydrogen links. The use of a catalyst lowers the temperature of the microwave needed to produce monosaccharides from cellulose in LB. Hydrolysis becomes feasible at 120–170 °C in the presence of a catalyst. The performance of ultrasound, autoclaving, and microwave coupled with sulphuric acid and nitric acid was investigated on the hydrolysis of fiber waste in terms of glucose yields. Glucose yield was (12–25 mg dL^{-1}) for ultrasound, and autoclaving treatment, whereas it was (48–80 mg dL^{-1}) for microwave treatment [73]. About 87% glucose was obtained from rice waste under microwave heating in the presence of a phosphoric acid-activated wood biochar catalyst.

5.6.7 *Monomeric Phenols*

Phenol production takes place after the removal of lignin from hemicellulose and cellulose. Microwaves can efficiently isolate high-purity lignin from softwood. This method was found superior to the traditional Klason method. About 92% of phenolic compounds, comprising acetovanillone, vanillin, syringol, acetosyringone, and syringaldehyde, were produced from lignin of palm empty fruit bunch, when it was oxidatively depolymerized in the existence of Fe_2O_3 and $Cu(OH)_2$ catalysts, H_2O_2 and sodium hydroxide below microwave heating [51]. Microwave heating at 110°C rapidly decomposed the lignin (7 min) over copper sulfate and hydrogen peroxide [74]. This yield was 2.6 times more than the yield obtained after autoclaving that took place at the same temperature and reaction time. Lignin degrades because microwaves promote the formation of hydroxyl radicals from H_2O_2, catalyzed by $CuSO_4$. This radical attacks lignin. Alternatively, phenolics-rich algae can produce polyphenols under simple microwave-hydrothermal conditions in the absence of a catalyst [75].

5.6.8 Lactic Acid

Lactic acid is obtained from hemicellulose and cellulose present in LBs. Microwaves assist as a method of pretreatment for producing lactic acid via fermentation. About 34% lactic acid and 21% glycolic acid were produced when post-harvest plants of tomato were exposed to microwave heating at 220 °C in the presence of lead(II) nitrate as the catalyst under 3 MPa pressure. It took about 2 min to complete this process, whereas the same result was obtained after 4 h in conventional heating. Microwave-assisted polycondensation of glycolic and lactic acid, produced oligomers that were significant for medical applications. As the polymerization proceeded, the polarity decreased. The target temperature was maintained by higher microwave power. The flow-through microwave reactor supported the constant transformation of milled cellulose [76].

5.6.9 Levulinic Acid and Formic Acid

HMF upon rehydration produces Levulinic acid and formic acid. Microwave heating was found useful for the production of Levulinic acid (17 mol%) from vegetable trash as a cellulose-rich substrate in the presence of Amberlyst 36 (Brønsted acid catalyst), at the temperature of 150 °C for 5 min [64]. The length of fiber determines the rate of conversion. Table 5.1 depicts the yield of Levulinic acid with different substrates and catalysts.

Significantly lower energy is consumed by microwave irradiation to sustain the reaction temperature (90 W) in comparison to that of an oil bath (600 W). This is due to less heat loss in the microwave irradiation [77].

5.6.10 Lipids/fatty Acids and Biodiesel

Lipids like linoleic, oleic, and palmitic acids were effectively extracted from spent hen waste in a dichloroform-methanol at 80 °C, under the microwave [79]. Researchers have compared various pretreatment strategies like thermal-alkaline treatment, freeze-thawing, high-temperature ethanol, and microwave process to compare the effect on the algal oil extraction under their corresponding optimal parameters [68]. The performance of microwaves was the lowest in terms of cell disruption. The extraction efficiency of the microwave was also found lowest, after high-temperature ethanol and thermal-alkaline treatment. However, it is noteworthy that microwave power used was very low in this study (8 W). Intermittent microwave application was found better than continuous irradiation for rupturing the algal cell. About 42% recovery of lipid was extracted from algae under double-stage microwave pretreatment and solvent extraction [80]. When irradiation time was increased in microwave pretreatment, it yielded four times more microalgal oil [81]. Steam reforming of these fatty acids generates hydrogen, whereas biodiesel could be produced on its production esterification/transesterification. Biomass-derived long-chain and short-chain carboxylic acids are potential precursors of biofuel [82]. The time essential to carry out these reactions in the microwave is much less (1 h) in comparison to that in the conventional process (24 h). At elevated temperatures, there is a reduction in the polarity of methanol (transesterification agent) that favors the solubility of lipids, resulting in a better production of algal biodiesel.

5.6.11 Hydrochar

The hydrochar has been synthesized from algae, plant shells, and dairy manure through microwave-assisted hydrothermal treatment at 180–240 °C. The ultimate composition and energy content of hydrochar produced from plant shells were the same, whether it was obtained through autoclaving or from microwave, but the time needed for this production varied significantly. For autoclaving it is 4 h and for microwaving, it is only 20 min [83]. Carbon microspheres are formed in hydrothermal treatment that improves microwave absorption thereby intensifying the hydrochar production from dairy manure [72]. Retention time has a significant influence on the chemical characteristics of hydrochar produced through microwaves. The heating value (HHV) of hydrochar produced by microwave is 15–23 MJ kg^{-1} whereas, it is 17–30 MJ kg^{-1} for conventional hydrochar. This is probably due to the use of higher operating temperatures (up to 300 °C) in the conventional hydrothermal process.

5.7 Fractionation and Pretreatment

Various components in biomass can be effectively valorized to generate a wide variety of value-added chemicals. Fractionation maximizes this process. Efficient and selective liquefaction of hemicellulose and cellulose of woody biomass can be accomplished in microwave acidolysis. The Lignin part remains conserved in the residue. It is possible to selectively separate hemicellulose from other polymers (cellulose and lignin) present in bamboo at 200 °C in microwave-hydrothermal carbonization (HTC), without using any acid [50]. The hemicellulose obtained in this process can function as a raw material for synthesizing film or can be further hydrolyzed to produce monomeric molecules. The lignin and cellulose can be employed for producing aromatics and glucose. It is important to note that the temperature for activation of hemicellulose varies from 180 to 200 °C for various LBs like bamboo, softwood, and wheat straw [50, 84]. The microwave pretreatment possibly allows the process at higher loading rates in anaerobic digesters with less capacity with no compromise in the production of biogas [85]. As microwave-assisted biorefinery is a budding area of research, most of the related research is done at a lab/pilot scale only. The field-scale applications of microwave-assisted biorefinery are a challenge due to constraints like inadequate energy efficiency [85]. Presently, industrial-scale microwave reactors and microwave magnetrons are used for drying purposes in industries like ceramic, food, and wood. Only a few research centers have pilot-scale microwave-assisted biorefineries, for instance, the Biorenewables Development Center and the Green Chemistry Center of Excellence at the University of York, UK. It is assumed that more scale-up research will be done for forthcoming applications at the industrial level.

5.8 Future Perspectives of Microwave-Assisted Pretreatment Technologies

Thus far, the market of bio-based products is dominated by bioprocesses. The usage of costlier enzymes and limitations on the quality of feedstock are however the major constraints during bioprocesses [86]. Carbon-efficient biomass conversion is feasible if biological and thermochemical technologies are integrated, as they have complementary edges. Further studies are needed to explore microwave-based technology that is efficient for utilization in commercial biorefineries. Commercialization has to be integrated with the use of energy that is clean, efficient, and sustainable [74]. Future investigations should focus on how to move the pretreatment process of biomass from the laboratory to marketable and industrial facilities. Moreover, energy integration should be taken into consideration while designing and developing sophisticated processes and techniques. Desirable and comprehensive techno-economic analyses are the key factors determining the success of biorefineries [87]. It is possible to improve and enhance the economic stability of biorefineries, by achieving the values from coproducts, like proteins, lignin, oils, etc. The use of constant procedures for microwave-supported pyrolysis can be extended into bigger central biorefineries in the upcoming time. Upcoming researchers should pay attention to reaction conditions, various forms of microwave absorbers, and upgrading catalyst selectivity to improve the yield during microwave pyrolysis [88]. It is also important to understand the numerical simulation and reaction kinetics of the complete process and pretreating LBs, which brings homogeneity in the process optimization. Future research should also focus on how the reaction process takes place among microwave-absorbing materials and microwave heating, for establishing the optimum process parameters. Important factors that determine the economic viability of microwave treatment are its manufacturing cost, equipment, and human capital [89].

5.9 Conclusion

Biomass is a potential renewable and sustainable carbon resource that can be reused and explored for the production of a variety of biorefinery or value-added products. Biomass waste is a perfect representation of a double-edged sword that is decided by the way it is treated. It could either hurt the environment or produce valuable feedstocks without causing any harm to the environment. There are many benefits associated with the use of microwave-supported processes in contrast to those with established heating. Combined microwave-assisted methods have the advantages of superior heating mechanisms and producing efficient products. It is a quicker, more energy-efficient, and cost-effective treatment process. It is for this reason, that the use of microwave technology for the processing of lignocellulosic wastes has gathered enormous attention over the past decade. To date, microwave processing of lignocellulosic wastes mostly circles pyrolysis, torrefaction, and gasification for producing bio-oil, biochar, and syngas in circular economy and bioenergy applications. However, currently, the use of this technology is restricted to lab/small scale, and massive extensions are required to further drive this technology in the direction of commercial scales. The future recommendations and perspectives that can be drawn based on the above findings will support scientists, researchers, energy executives, and personnel to employ efficiently microwave-supported pretreatment methods for transforming LB into supportable bioenergy and other value-added products.

Declaration NO generative AI tools have been used during the writing process

References

1. Bhatia L, Jha H, Sarkar T, Sarangi PK (2023) Food waste utilization for reducing carbon footprints towards a sustainable and cleaner environment. Int J Environ Res Public Health 20:2318. https://doi.org/10.3390/ijerph20032318
2. Nair LG, Agrawal K, Verma P (2022a) An insight into the principles of lignocellulosic biomass-based zero-waste biorefineries: a green leap towards imperishable energy-based future. Biotechnol Genetic Eng Rev 1–51. https://doi.org/10.1080/02648725.2022.2082223
3. Tapia-Quirós P, Montenegro-Landívar MF, Vecino X, Alvarino T, Cortina JL, Saurina J, Granados M, Reig M (2022) A green approach to phenolic compounds recovery from olive mill and winery wastes. Sci Total Environ 835:155552. https://doi.org/10.1016/j.scitotenv.2022.155552
4. Mikucka W, Zielińska M, Bułkowska K, Witońska I (2022) Valorization of distillery stillage by polyphenol recovery using microwave-assisted, ultrasound-assisted and conventional extractions. J Environ Manage 322:116150. https://doi.org/10.1016/j.jenvman.2022.116150
5. Gonzalez-Rivera J, Campanella B, Ferrari C (2023) From volatiles to solid wastes: towards the full valorization of lavender and rosemary by simultaneous in situ microwaves and ultrasounds irradiation extraction. Ind Crops Prod 194:116362. https://doi.org/10.1016/j.indcrop.2023.116362
6. Nair LG, Agrawal K, Verma P (2022b). An overview of sustainable approaches for bioenergy production from agro-industrial wastes. Energy Nexus 100086. https://doi.org/10.1016/j.nexus.2022.100086

7. Ungureanu N, Vlăduţ V, Biriş S-Ş (2022) Sustainable valorization of waste and by-products from sugarcane processing. Sustainability 14:11089. https://doi.org/10.3390/su141711089
8. Bhatia L, Bachetti R, Ravindra P, Chandel AK (2020) Third generation biorefineries: a sustainable platform for food, clean energy, and nutraceuticals production. Biomass Convers Biorefin.https://doi.org/10.1007/s13399-020-00843-6
9. Kawa-Rygielska J, Pietrzak W, Lennartsson PR (2022) High-efficiency conversion of bread residues to ethanol and edible biomass using filamentous fungi at high solids loading: a biorefinery approach. Appl Sci 12:6405. https://doi.org/10.3390/app12136405
10. Kumar B, Verma P (2020) Biomass-based biorefineries: an important architype towards a circular economy. Fuel Elsevier 288:119622. https://doi.org/10.1016/j.fuel.2020.119622
11. Suriapparaoa DV, Tejasvi R (2022) A review on role of process parameters on pyrolysis of biomass and plastics: present scope and future opportunities in conventional and microwave-assisted pyrolysis technologies. Process Saf Environ Prot 162:435–462
12. Bhardwaj N, Kumar B, Verma P (2022) Microwave-assisted pretreatment using alkali metal salt in combination with orthophosphoric acid for generation of enhanced sugar and bioethanol. Biomass Convers Biorefine 12:1069–1076. https://doi.org/10.1007/s13399-020-00640-1
13. Kumar B, Verma P (2019) Optimization of microwave-assisted pretreatment of rice straw with $FeCl_3$ in combination with H_3PO_4 for improving enzymatic hydrolysis. Adv Plant Microbial Biotechnol 41–48.https://doi.org/10.1007/978-981-13-6321-4_7
14. Bouchez A, Vauchel P, Périno S, Dimitrov K (2023) Multi-criteria optimization including environmental impacts of a microwave-assisted extraction of polyphenols and comparison with an ultrasound-assisted extraction process. Foods 12:1750. https://doi.org/10.3390/foods12091750
15. Gao Y, Remón J, Matharu AS (2021) Microwave-assisted hydrothermal treatments for biomass valorisation: a critical review. Green Chem 23:3502. https://doi.org/10.1039/d1gc00623a
16. Del-Castillo-Llamosas A, Eibes G, Ferreira-Santos P (2023) Microwave-assisted autohydrolysis of avocado seed for the recovery of antioxidant phenolics and glucose. Biores Technol 385:129432. https://doi.org/10.1016/j.biortech.2023.129432
17. Kumar B, Bhardwaj N, Agrawal K, Chaturvedi V, Verma P (2020) Current perspective on pretreatment technologies using lignocellulosic biomass: an emerging biorefinery concept. Fuel Process Technol 199:106244. https://doi.org/10.1016/j.fuproc.2019.106244
18. Bhatia L, Sarangi PK, Singh AK, Prakash A, Shadangi KP (2021) Lignocellulosic waste biomass for biohydrogen production: future challenges and bioeconomical perspectives. Biofuel, bioproduct & biorefining (Biofpr) (Wiley-Blackwell). In Special issue: Biorefineries for waste biomass towards sustainable Bioeconomy (BWTBSB). https://doi.org/10.1002/bbb.2329
19. Pals M, Lauberte L, Ponomarenko J, Lauberts M, Arshanitsa A (2022) Microwave-assisted water extraction of aspen (Populus tremula) and pine (Pinus sylvestris L.) barks as a tool for their valorization. Plants 11:1544. https://doi.org/10.3390/plants11121544
20. Bhatia L (2023) Biofuels—A novel dimension for a sustainable future. Everyman's Science. https://www.everymanscience.com/index.php/emsj/article/view/22
21. Azelee NIW, Adnan SAM, Md Illias R (2019) Assessment of microwave-assisted pretreatments for enhancing pineapple waste delignification. AIP Conf Proc 2155:020003. https://doi.org/10.1063/1.5125507
22. Goswami RK, Mehariya S, Verma P, Lavecchia R, Zuorro A (2021) Microalgae-based biorefineries for sustainable resource recovery from wastewater. J Water Process Eng 40:101747. https://doi.org/10.1016/j.jwpe.2020.101747
23. Kumar S, Konwar J, Purkayastha MD, Kalita S, Dutta MAJ (2023) Current progress in valorization of food processing waste and by-products for pectin extraction. Int J Biol Macromol 239:124332
24. Dutta N, Usman M, Luo G, Zhang S (2022) An insight into valorization of lignocellulosic biomass by optimization with the combination of hydrothermal (HT) and biological techniques: a review. Sustain Chem 3:35–55. https://doi.org/10.3390/suschem3010003

25. Srivastava RK, Sarangi PK, Bhatia L, Singh AK, Shadangi KP (2021) Conversion of methane to methanol: technologies and future challenges. Biomass Convers Biorefine. https://doi.org/10.1007/s13399-021-01872-5
26. Bhatia L, Johri S (2015) Biovalorization potential of peels of *Ananas cosmosus* (L) Merr. For ethanol production by *Pichia stipitis* NCIM 3498 & *Pachysolen tannophilus* MTCC 1077. Indian J Exp Biol 53:819–827
27. Bhatia L, Paliwal S (2010) Banana peel waste as substrate for ethanol production. Int J Biotechnol Bioeng Res 1(2):213–221
28. Bhardwaj N, Kumar B, Agrawal K, Verma P (2020) Bioconversion of rice straw by the synergistic effect of in-house produced ligno-hemicellulolytic enzymes for enhanced bioethanol production. Bioresour Technol Rep 10:100352. https://doi.org/10.1016/j.biteb.2019.100352
29. Agung M, Rahim I, Amir F (2023) Microwave torrefaction technology in biochar production: a review. AIP Conf Proc 2630:020011. https://doi.org/10.1063/5.0125902
30. Kang K, Zhang T, Sun G, Zhu M, Li K, Li D (2021) Valorization of tree leaves waste using microwave-assisted hydrothermal carbonization process. GCB Bioenerg.https://doi.org/10.1111/gcbb.12882
31. Okoliea JA, Epelleb EI, Tabat ME, Orivri U, Amenaghawon AN, Okoye PU, Burcu G (2022) Waste biomass valorization for the production of biofuels and value added products: a comprehensive review of thermochemical, biological and integrated processes. Process Saf Environ Prot 159:323–344
32. Liu J, Hou Q, Ju M, Ji P, Sun Q, Li W (2020) Biomass pyrolysis technology by catalytic fast pyrolysis, catalytic Co-pyrolysis and microwave-assisted pyrolysis: a review. Catalysts 10:742. https://doi.org/10.3390/catal10070742
33. Hoang AT, Ni˘zeOng HC, Mofijur M, Ahmed SF, Ashok B, Bui VTV, Chau MQ (2021) Insight into the recent advances of microwave pretreatment technologies for the conversion of lignocellulosic biomass into sustainable biofuel. Chemosphere 281:130878
34. Kumar B, Bhardwaj N, Verma P (2020) Microwave assisted transition metal salt and orthophosphoric acid pretreatment systems: Generation of bioethanol and xylo-oligosaccharides. Renew Energ 158:574–584. https://doi.org/10.1016/j.renene.2020.05.006
35. Mikulski D, Kłosowski G (2022) Delignification efficiency of various types of biomass using microwave-assisted hydrotropic pretreatment. Sci Rep 12:4561. lhttps://doi.org/10.1038/s41598-022-08717-9
36. Zhu Y, Luan Y, Zhao Y, Liu J, Duan Z, Ruan R (2023) Current technologies and uses for fruit and vegetable wastes in a sustainable system: a review. Foods 12:1949. https://doi.org/10.3390/foods12101949
37. Foong SY, Liew RK, Lam SS (2020) Valorization of biomass waste to engineered activated biochar by microwave pyrolysis: progress, challenges, and future directions. Chem Eng J 389:124401. https://doi.org/10.1016/j.cej.2020.124401
38. Salgado-Ramos M, Tabasso S, Calcio Gaudino E, Moreno A, Mariatti F, Cravotto G (2022) An innovative, green cascade protocol for grape stalk valorization with process intensification technologies. Appl Sci 12:7417. https://doi.org/10.3390/app12157417
39. Nair LG, Agrawal K, Verma P (2023) Organosolv pretreatment: an in-depth purview of mechanics of the system. Bioresour Bioprocess 10:50(1–29). https://doi.org/10.1186/s40643-023-00673-0. Springer
40. Singh E, Mishra R, Kumar A, Shukla SK, Lo S-L, Kumar S (2022) Circular economy-based environmental management using biochar: driving towards sustainability. Process Saf Environ Prot 163:585–600
41. Singh J, Bhonsle AK, Atray N (2022) Life-cycle analysis of a hydrocarbon biorefinery. In: Maity SK, Gayen K, Bhowmick TK (eds) Hydrocarbon biorefinery. Elsevier, pp 387–408
42. Reddy BR, Sridevi V, Swami Undi GSNVKSN (2022) Synthesis of renewable carbon biorefinery products from susceptor enhanced microwave-assisted pyrolysis of agro-residual waste: a review. Process Saf Environ Prot 164:354–372. https://doi.org/10.1016/j.psep.2022.06.027
43. Foong SY, Chan YH, Lam SS (2023) Microwave processing of oil palm wastes for bioenergy production and circular economy: recent advancements, challenges, and future prospects. Biores Technol 369:128478

44. Siritheerasas P, Waiyanate P, Sekiguchi H, Kodama S (2017) Microwave torrefaction of moist municipal solid waste (MSW) pellets with the addition of char from agricultural residues. In: 11th pure and applied chemistry international conference, PACCON 2017, vol 757 KEM, pp 156–160
45. Das PVPC, Mathimani T, Pugazhendhi A (2021) A comprehensive review on the factors affecting thermochemical conversion efficiency of algal biomass to energy. Sci Total Environ 766:144213
46. Yek PNY, Chen X, Peng W, Liew RK, Cheng CK, Sonne C, Sii HS, Lam SS (2021) Microwave co-torrefaction of waste oil and biomass pellets for simultaneous recovery of waste and co-firing fuel. Renew Sust Energ Rev 152:111699
47. Remon J, Santomauro F, Chuck CJ, Matharu AS, Clark JH (2018) Production of fermentable species by microwave-assisted hydrothermal treatment of biomass carbohydrates: reactivity and fermentability assessments. Green Chem 20(19):4507–4520. https://doi.org/10.1039/C8GC02182A
48. Lee SB, Jeong GT (2015) Catalytic conversion of chitosan to 5-hydroxymethylfurfural under low temperature hydrothermal process. Appl Biochem Biotechnol 176(4):1151–1161. https://doi.org/10.1007/s12010-015-1636-9
49. Chen SS, Wang L, Yu IK, Tsang DC, Hunt AJ, Jerome F, Zhang S, Ok YS, Poon CS (2018) Valorization of lignocellulosic fibers of paper waste into levulinic acid using solid and aqueous Brønsted acid. Biores Technol 247:387–394. https://doi.org/10.1016/j.biortech.2017.09.110
50. Luo Y, Fan J, Budarin VL, Hu C, Clark JH (2017) Microwave-assisted hydrothermal selective dissolution and utilisation of hemicellulose in Phyllostachys heterocycla cv. Green Chem 19(20):4889–4899. https://doi.org/10.1039/C7GC02300F
51. Panyadee R, Posoknistakul P, Jonglertjunya W, Kim-Lohsoontorn P, Laosiripojana N, Matsagar BM, Wu KC-W, Sakdaronnarong C (2018) Sequential fractionation of palm empty fruit bunch and microwave-assisted depolymerization of lignin for producing monophenolic compounds. ACS Sustain Chem Eng 6(12):16896–16906. https://doi.org/10.1021/acssuschemeng.8b04246
52. Kong SH, Lam SS, Yek PNY, Liew RK, Ma NL, Osman MS, Wong CC (2019) Self-purging microwave pyrolysis: an innovative approach to convert oil palm shell into carbon-rich biochar for methylene blue adsorption. J Chem Technol Biotechnol 94(5):1397–1405
53. Lam SS, Tsang YF, Yek PNY, Liew RK, Osman MS, Peng W, Lee WH, Park YK (2019) Co-processing of oil palm waste and waste oil via microwave co- torrefaction: a waste reduction approach for producing solid fuel product with improved properties. Process Saf Environ Prot 128:30–35
54. Ferreira-Anta T, Flórez-Fernández N, Torres MD, Mazón J, Dominguez H (2023) Microwave-assisted hydrothermal processing of *Rugulopteryx okamurae*. Mar Drugs 21:319. https://doi.org/10.3390/md21060319
55. Tabasso S, Grillo G, Cravotto G (2016) Microwave-assisted-valerolactone production for biomass lignin extraction: a cascade protocol. Molecules 21:413. https://doi.org/10.3390/molecules21040413
56. Ethaib S, Omar R, Kamal SMM, Biak DRA, Zubaidi SL (2020) Microwave-assisted pyrolysis of biomass waste: a mini review. Processes 8:1190. https://doi.org/10.3390/pr8091190
57. Bouxin FP, Clark JH, Fan J, Budarin V (2019) Combining steam distillation with microwave-assisted pyrolysis to maximise direct production of levoglucosenone from agricultural wastes. Green Chem 21(6):1282–1291. https://doi.org/10.1039/C8GC02994F
58. De Bruyn M, Fan J, Budarin V, Macquarrie D, Gomez L, Simister R, Farmer T, Raverty W, McQueen-Mason S, Clark J (2016) A new perspective in bio-refining: Levoglucosenone and cleaner lignin from waste biorefinery hydrolysis lignin by selective conversion of residual saccharides. Energy Environ Sci 9(8):2571–2574. https://doi.org/10.1039/C6EE01352J
59. Huang Y-F, Chiueh P-T, Kuan W-H, Lo S-L (2018) Product distribution and heating performance of lignocellulosic biomass pyrolysis using microwave heating. Energ Proc 152:910–915. https://doi.org/10.1016/j.egypro.2018.09.092
60. Zhang X, Lei H, Zhu L, Wu J, Chen S (2015) From lignocellulosic biomass to renewable cycloalkanes for jet fuels. Green Chem 17(10):4736–4747. https://doi.org/10.1039/C5GC01583A

61. Nhuchhen D, Afzal M, Dreise T, Salema A (2018) Characteristics of biochar and bio-oil produced from wood pellets pyrolysis using a bench scale fixed bed, microwave reactor. Biomass Bioenerg 119:293–303. https://doi.org/10.1016/j.biombioe.2018.09.035
62. Sweygers N, Alewaters N, Dewil R, Appels L (2018) Microwave effects in the dilute acid hydrolysis of cellulose to 5-hydroxymethylfurfural. Sci Rep 8(1):7719. https://doi.org/10.1038/s41598-018-26107-y
63. Sweygers N, Harrer J, Dewil R, Appels L (2018) A microwave-assisted process for the in-situ production of 5-hydroxymethylfurfural and furfural from lignocellulosic polysaccharides in a biphasic reaction system. J Clean Prod 187:1014–1024. https://doi.org/10.1016/j.jclepro.2018.03.204
64. Chen SS, Yu IK, Tsang DC, Yip AC, Khan E, Wang L, Ok YS, Poon CS (2017) Valorization of cellulosic food waste into levulinic acid catalyzed by heterogeneous Brønsted acids: temperature and solvent effects. Chem Eng J 327:328–335. https://doi.org/10.1016/j.cej.2017.06.108
65. Yu IK, Tsang DC, Yip AC, Hunt AJ, Sherwood J, Shang J, Song H, Ok YS, Poon CS (2018) Propylene carbonate and c-valerolactone as green solvents enhance Sn (IV)-catalysed hydroxymethylfurfural (HMF) production from bread waste. Green Chem 20(9):2064–2074. https://doi.org/10.1039/C8GC00358K
66. Zhang ZB, Lu Q, Ye XN, Wang TP, Wang XH, Dong CQ (2015) Selective production of levoglucosenone from catalytic fast pyrolysis of biomass mechanically mixed with solid phosphoric acid catalysts. Bioenerg Res 8(3):1263–1274. https://doi.org/10.1007/s12155-015-9581-6
67. Yang Z, Lei H, Zhang Y, Qian K, Villota E, Qian M, Yadavalli G, Sun H (2018) Production of renewable alkyl-phenols from catalytic pyrolysis of Douglas fir sawdust over biomass-derived activated carbons. Appl Energ 220:426–436. https://doi.org/10.1016/j.apenergy.2018.03.107
68. Hao X, Peng H, Xu P, He M, Dou B (2019) Production of H2 by steam reforming in schizochytrium algae oil of cell disruption and extraction via ultrasound method. Int J Hydrogen Energy 44(30):15779–15786. https://doi.org/10.1016/j.ijhydene.2018.09.220
69. Grycova B, Pryszcz A, Lestinsky P, Chamradova K (2018) Influence of potassium hydroxide and method of carbonization treatment in garden and corn waste microwave pyrolysis. Biomass Bioenerg 118:40–45. https://doi.org/10.1016/j.biombioe.2018.07.022
70. Dong S, Zhang X, Hu P, Zhang W, Han J, Hu P (2019) Biomass-derived carbon and polypyrrole addition on SiC whiskers for enhancement of electromagnetic wave absorption. Chem Eng J 359:882–893. https://doi.org/10.1016/j.cej.2018.11.101
71. Santhoshkumar A, Ramanathan A (2019) Energy and life cycle assessment of solar assisted microwave pyrolysis of waste biomass. In: National conference on recent advances in fuel cells and solar energy IOP conference series: earth and environmental science, vol 312, pp 012017. https://doi.org/10.1088/1755-1315/312/1/012017
72. Gao Y, Liu Y, Zhu G, Xu J, Xu H, Yuan Q, Zhu Y, Sarma J, Wang Y, Wang J, Ji L (2018) Microwave-assisted hydrothermal carbonization of dairy manure: chemical and structural properties of the products. Energy 165:662–672.https://doi.org/10.1016/j.energy.2018.09.185
73. Abu-Saied M, Taha TH, Elnaggar EM, Amer RA, Mansy A, Elkady GM (2018) Green production of bio-ethanol from cellulosic fiber waste and its separation using polyacrylonitrile-co-poly methyl acrylate membrane. Cellulose 25(11):6621–6644. https://doi.org/10.1007/s10570-018-2038-2
74. Dessie W, Luo X, He F, Liao Y, Duns GJ, Qin Z (2023) Lignin valorization: a crucial step towards full utilization of biomass, zero waste and circular bioeconomy. Biocatal Agric Biotechnol 51:102777. https://doi.org/10.1016/j.bcab.2023.102777
75. Yuan Y, Zhang J, Fan J, Clark J, Shen P, Li Y, Zhang C (2018) Microwave assisted extraction of phenolic compounds from four economic brown macroalgae species and evaluation of their antioxidant activities and inhibitory effects on a-amylase, a-glucosidase, pancreatic lipase and tyrosinase. Food Res Int 113:288–297. https://doi.org/10.1016/j.foodres.2018.07.021
76. Gavila L, Constantı M, Medina F (2015) D-Lactic acid production from cellulose: dilute acid treatment of cellulose assisted by microwave followed by microbial fermentation. Cellulose 22(5):3089–3098. https://doi.org/10.1007/s10570-015-0720-1

77. Lappalainen K, Vogeler N, Kärkkäinen J, Dong Y, Niemelä M, Rusanen A, Ruotsalainen AL, Wäli P, Markkola A, Lassi U (2018) Microwave-assisted conversion of novel biomass materials into levulinic acid. Biomass Convers Biorefine 8(4):965–970. https://doi.org/10.1007/s13399-018-0334-6
78. Maiti S, Gallastegui G, Suresh G, Pachapur VL, Brar SK, Le Bihan Y, Drogui P, Buelna G, Verma M, Galvez-Cloutier R (2018) Microwave-assisted one-pot conversion of agro-industrial wastes into levulinic acid: an alternate approach. Biores Technol 265:471–479. https://doi.org/10.1016/j.biortech.2018.06.012
79. Safder M, Temelli F, Ullah A (2019) Extraction, optimization, and characterization of lipids from spent hens: an unexploited sustainable bioresource. J Clean Prod 206:622–630. https://doi.org/10.1016/j.jclepro.2018.09.214
80. Onumaegbu C, Alaswad A, Rodriguez C, Olabi A (2019) Modelling and optimization of wet microalgae scenedesmus quadricauda lipid extraction using microwave pretreatment method and response surface methodology. Renew Energ 132:1323–1331. https://doi.org/10.1016/j.renene.2018.09.008
81. Cancela A, Perez L, Febrero A, Sanchez A, Salgueiro J, Ortiz L (2019) Exploitation of *Nannochloropsis gaditana* biomass for biodiesel and pellet production. Renew Energ 133:725–730. https://doi.org/10.1016/j.renene.2018.10.075
82. Umrigar VR, Chakraborty M, Parikh P (2018) Catalytic activity of zeolite Hb for the preparation of fuels' additives: its product distribution and scale up calculation for the biofuel formation in a microwave assisted batch reactor. J Environ Chem Eng 6(6):6816–6827. https://doi.org/10.1016/j.jece.2018.10.059
83. Maceiras R, Alfonsín V, Seguí L, González JF (2021) Microwave assisted alkaline pretreatment of algae waste in the production of cellulosic bioethanol. Energies 14:5891. https://doi.org/10.3390/en14185891
84. Gil A, Siles J, Martın M, Chica A, Estevez-Pastor F, Toro-Baptista E (2018) Effect of microwave pretreatment on semi-continuous anaerobic digestion of sewage sludge. Renew Energ 115:917–925. https://doi.org/10.1016/j.renene.2017.07.112
85. Kor-Bicakci G, Ubay-Cokgor E, Eskicioglu C (2019) Effect of dewatered sludge microwave pretreatment temperature and duration on net energy generation and biosolids quality from anaerobic digestion. Energy 168:782–795. https://doi.org/10.1016/j.energy.2018.11.103
86. Usmani Z, Sharma M, Gupta VK (2023) Converting biowaste streams into energy–leveraging microwave assisted valorization technologies for enhanced conversion. J Energy Inst 107:101161
87. Yangin-Gomec C, Sárvári Horváth I, Martín C (2023) Energy. production from biomass valorization. Energies 16:4300. https://doi.org/10.3390/en16114300
88. Holliday MC, Parsons DR, Zein SH (2022) Microwave-assisted hydrothermal carbonisation of waste biomass: the effect of process conditions on hydrochar properties. Processes 10:1756. https://doi.org/10.3390/pr10091756
89. Arpia AA, Chena WH, Lame SS (2021) Sustainable biofuel and bioenergy production from biomass waste residues using microwave-assisted heating: a comprehensive review. Chem Eng J 403:126233

Chapter 6
Microwave-assisted Technologies for Municipal Solid Waste Conversion to Value-added Chemicals and Fuels

Navneet Kumar Mishra, Lei Wu, Jun Zhou, and B. Rajasekhar Reddy

Abstract Microwave-assisted Municipal Solid food waste conversion represents an innovative and promising approach to address both waste management and resource utilization sustainably. By harnessing the power of microwave irradiation, this method offers an effective and efficient means of converting Municipal food waste into valuable products, contributing to environmental protection and circular economy initiatives. Municipal Food waste presents a significant global issue, with its production anticipated to rise substantially in the coming years. The management of post-consumption Municipal food waste is explored as an avenue for bio-fuel and biochar production. Pyrolysis-gasification processes are identified as promising routes for producing high-value products like bio-fuel, biochar/ash, and water with minimal environmental impact. Issues related to proper post-consumption Municipal food waste technological processing mechanisms at both industrial and community levels, along with limited research on sustainability and life cycle assessments are addressed in this chapter. Recent studies have demonstrated the viability of microwave-assisted hydrothermal conversion for food waste, showcasing its potential to produce reducing sugars and versatile chemical blocks. The optimization of the technologies to strike a balance between resources, energy consumption, and environmental considerations remains a challenge. Through diverse microwave-assisted technologies, including anaerobic digestion, extraction of bio-fuel and biochar production, and integrated pyrolysis-gasification processes, the potential for resource recovery, environmental improvement, and sustainable agriculture is substantial.

Keywords Microwave · Food waste · Circular economy · Bio-fuels · Hydrothermal conversion

N. K. Mishra · B. Rajasekhar Reddy (✉)
Department of Fuel, Minerals and Metallurgical Engineering, IIT (ISM) Dhanbad, Dhanbad 826004, India
e-mail: rajasekhar@iitism.ac.in

L. Wu · J. Zhou
School of Chemistry and Chemical Engineering, Xi'an University of Architecture and Technology, Xi'an 710055, China

© The Author(s), under exclusive license to Springer Nature Switzerland AG 2024
P. Verma and B. Kumar (eds.), *Sustainable Radiation Technologies in Waste-biomass Valorization*, Green Chemistry and Sustainable Technology,
https://doi.org/10.1007/978-3-031-63941-8_6

6.1 Introduction

6.1.1 Background

Unwanted or unusable material is encompassed by the term "waste." The Glossary of Environment Statistics by the United Nations Statistics Division defines waste as materials that do not constitute primary products (i.e., goods created for the market), have no further utility according to the generator's production, transformation, or consumption objectives, and are intended for disposal [1]. The swift rise in urbanization, economic growth, population, and improved life facilities is rapidly escalating waste generation, leading to various socio-economic and environmental challenges [2, 3]. In 1960, the global population was 3.1 billion, and by 2010, it had nearly doubled, reaching almost 7 billion. According to scientific projections, it is anticipated to reach 9.3 billion by the year 2050 [4]. The growing population significantly contributes to the substantial generation of municipal solid waste. In 2016, the annual production of municipal solid waste reached 2.01 billion tons. Projections suggest an annual increase of 3.40 billion tons by the year 2050 for this quantity [4]. Every year, a total of 1.3 billion tons (Bt) of municipal solid waste (MSW) is generated, with a daily rate of production being 1.2 kg per person. According to a 2012 report from the World Bank, the significant quantity of municipal solid waste (MSW) is expected to grow to 2200 million tons annually by the year 2025 [5]. Effectively managing this considerable MSW volume poses a critical challenge to environmental sustainability.

6.1.2 Indian Municipal Solid Waste Dilemma

In the current scenario, India holds the second position in the worldwide list of the most densely populated nations [6]. The substantial population in the country contributes to a total municipal solid waste generation of 1.60 million tons per day (MTPD). Out of this, 152749.5 Tons per day (TPD) is effectively gathered with 95.4% of collection efficiency. A significant portion, constituting 50%, equivalent to 79956.3 Tons per day (TPD), undergoes treatment, whereas 18.4% amounting to 29427.2 Tons per day (TPD), is used in landfill sites. Notably, 31.7% of the total generated waste, equivalent to 50655.4 Tons per day (TPD), is not tracked within the existing waste management framework. Figure 6.1 provides a breakdown of Municipal solid waste management details across different states [6].

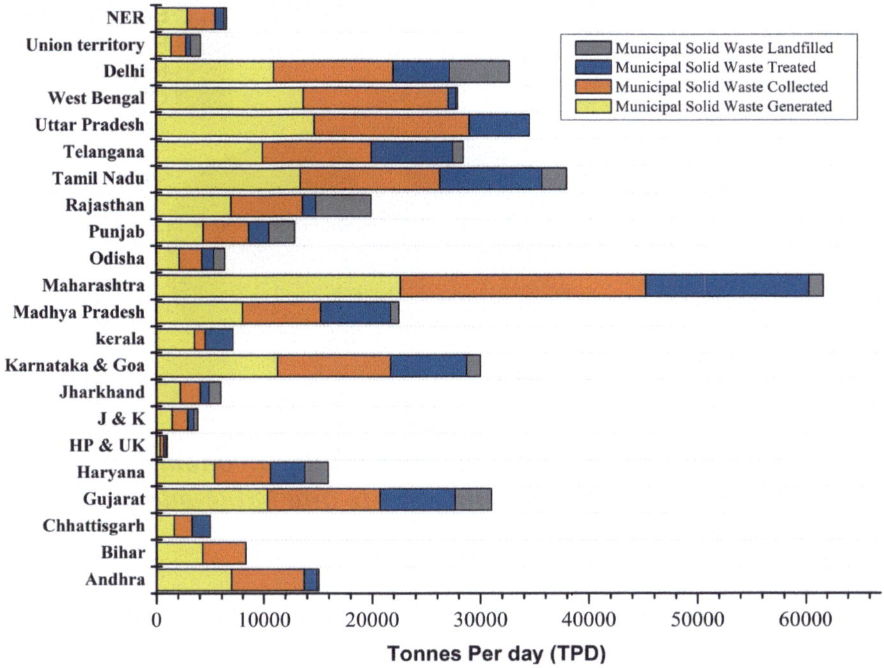

Fig. 6.1 Statistical data of MSW generation across different Indian states

6.1.3 Characteristics of Municipal Solid Waste—Indian and Global Scenario

In the context of India, MSW is categorized into biodegradable materials, including Edible waste and organic waste, and non-degradable materials, encompassing glass, cellulosic material, plastic, rubber, metallic elements, and other similar substances [7]. Figure 6.2 depicts the typical composition distribution of MSW in India. The characteristics and composition of MSW in India differ significantly from those in other developed countries. Notably, Indian MSW prominently features a substantial fraction of food/biodegradable waste, accounting for the largest share at 51%, in contrast to paper, which constitutes only 7%. Indian MSW exhibits higher moisture content, approximately 50%, in contrast to the moisture levels observed in the US and European countries, which typically range between 20 and 30% [8]. The calorific value of MSW in India is reported to be less than half, ranging from 800 to 1000 kcal/kg, in comparison to the calorific values observed in developed countries, which typically fall within the range of 2000–4000 kcal/kg [9]. Conversely, the chemical makeup of Indian MSW encompasses nitrogen at 0.64%, P at $(0.67 \pm 0.15)\%$, K at 0.68%, and a C/N ratio of 26 ± 5 [10].

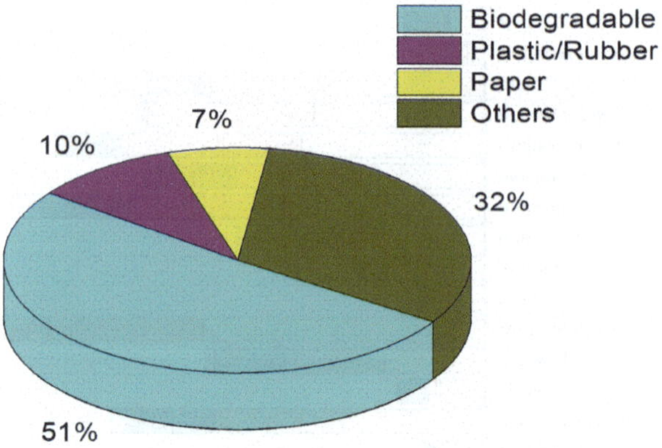

Fig. 6.2 The usual makeup of MSW in India

6.1.4 Advancing Sustainability: The Circular Economy Paradigm for Transforming Waste into Value-added Product

The global need for energy is continuously on the rise, with a projected reach of seventeen BT of oil equivalent (Btoe) by 2035. At the same time, CO_2 emissions are expected to increase, falling within the range of 29–43 gigatonnes per year, as reported by the International Energy Agency (IEA) [11]. Industries should focus on enhancing efficiency, optimizing operations, and embracing innovative waste utilization practices, including the promotion of a circular economy. Biodegradable components (waste food) in municipal solid waste, found worldwide, offer significant potential as raw materials for producing valuable products like fuels and chemicals. Innovation in processes is crucial for promoting the circular economy, attracting significant attention across diverse sectors in numerous countries, spanning industry, commerce, and services.

6.1.5 Motivation for Processing Municipal Food Waste

The bio-refinery concept, which involves recycling food wastes to generate valuable commercial goods and energy, is gaining momentum in the scientific community as a sustainable and environmentally friendly option [12] as shown in Fig. 6.3. Technological processes like anaerobic digestion, carbonization, gasification, pyrolysis, etc., are employed to undergo the transformative conversion of waste into biorefinery products, including bio-fuels, biochar, bio-fertilizers, and secondary chemicals. Selective extraction of valuable components from organic materials entails a comprehensive method employing biochemical, chemical, or thermal/physical techniques [13]. Following extraction, these components are then modified to produce enhanced-value food products or additives.

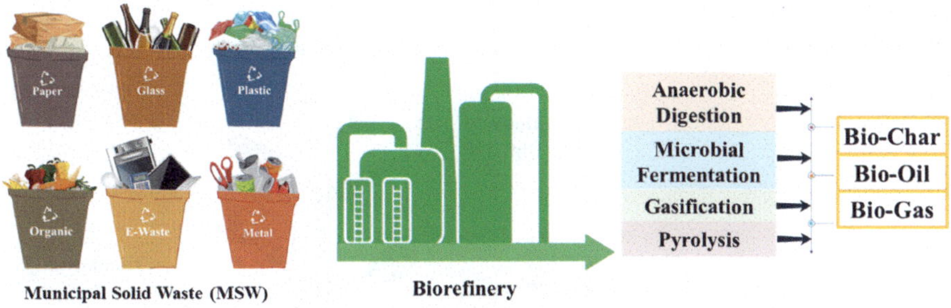

Fig. 6.3 Biorefinery (Waste to valuable products)

Different techniques have been previously used for extracting bio-fuels, biochemicals, and biochars. These methods encompass solvent-based extraction, supercritical carbon dioxide extraction, pyrolysis, gasification, catalytic upgradation, pressurized liquid extraction, and microwave/ultrasound-assisted extraction. These processes come with drawbacks such as huge operating costs, minimum yield, increased energy consumption, and the use of inorganic solvents [14, 15]. In this scenario, the application of microwaves has exhibited considerable promise in terms of reducing costs, enhancing energy efficiency, and saving time.

Microwave heating, functioning as an indirect heating method, enhances efficiency by expediting energy transfer. This approach reduces equipment size and operational units, enables precise heat control, and minimizes thermal gradient [16, 17]. Previously, it has found widespread application in the drying of diverse fruits and vegetables. Nevertheless, not all materials possess the requisite high absorption capacity for efficient microwave pyrolysis and gasification. To enhance the effectiveness of this method, numerous researchers have explored the addition of various microwave absorbers or catalysts, such as metal oxides, to the raw materials salts, activated carbon, and acids [18–20]. This section of the book furnishes details on extracting diverse compounds from food waste utilizing microwave-based techniques.

6.2 Overview of Municipal Food Waste Management: Disposal, Recycling, and Valorization Applications

In this section, we provide a succinct summary of the primary applications for handling municipal food waste, categorized into disposal, recycling, and valorization methods as depicted in Fig. 6.4.

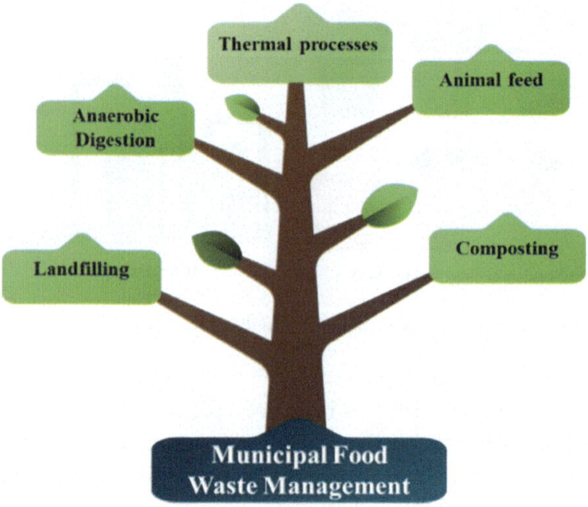

Fig. 6.4 Municipal food waste management's methods

6.2.1 *Landfill Disposal for Municipal Food Waste*

The conventional and time-honored approach to Municipal food waste disposal involves collecting and depositing it in landfill sites. Although landfilling is a straightforward and cost-effective method, it comes with several drawbacks. These include heightened transportation costs, a substantial land footprint that becomes more problematic as land values increase, and notable environmental impacts such as leachate, gaseous emissions, and dust [21].

6.2.2 *Municipal Waste Transformation Through Composting*

Composting, a straightforward and established process, involves the breakdown of complex biomass, particularly food waste used as feedstock, into nutrient-rich organic material. This resulting compost serves as an effective soil amendment [22]. While composting is an effective method, it presents certain drawbacks, including the need for a relatively huge land area because of extended reaction times, the production of a lower value-added product, transportation costs associated with delivering the product to designated locations, and the management of a moist end product [23].

6.2.3 *Anaerobic Digestion: Extracting Energy from Municipal Food Waste*

In recent years, anaerobic digestion has garnered significant attention as the most investigated method for processing organic waste, biomass, and food waste. This heightened interest is driven by its potential to yield substantial amounts of fuel gas, namely, hydrogen and methane, with applications in diverse energy contexts [24, 25]. Anaerobic digestion plants offer benefits, including the creation of value-added methane gas, a compact land footprint, and a well-established processing method. Anaerobic digestion plants face challenges, including potential emissions containing sulfur-organic compounds and methane, leading to odor and greenhouse warming. Furthermore, handling residual slurry and liquid by-products, even when repurposed as liquid fertilizer, involves expenses related to odor control and transportation/storage. Efficiently separating digestible food waste at its origin presents an extra operational challenge [26].

6.2.4 Municipal Food Waste as Animal Feed

For centuries, waste food and leftovers have been utilized as feed for animals, but in recent times, this practice has undergone increased commercialization [27]. Using food waste for animal feed offers advantages, including its contribution to the food chain system and its minimal environmental impact [28]. Drawbacks may arise, such as the necessity for separating food from municipal solid waste (MSW) if it hasn't been segregated at the source. Ensuring quality becomes essential to prevent the inclusion of bones and seafood shells in both the initial feed material and the end product. Additionally, there is a need for swift processing, as any delay may lead to the deterioration of food waste, rendering it inedible unless costly preservatives or enzymes are employed [29].

6.2.5 Thermal Treatment of Municipal Waste Food

Primary thermal processes are gasification, incineration, and pyrolysis. One of these techniques involves the use of pyrolysis technology, or hybrid alternatives like catalytic pyrolysis and hydrothermal carbonization [30]. This method stands out for its convenience, flexibility, and capacity to effectively process a diverse array of biomass and Municipal waste materials [31]. Given the pivotal role of pyrolysis in advancing sustainability, extensive research has been undertaken in recent years. This chapter of the book focuses on the application of microwave pyrolysis technology to convert residual lignocellulosic biomass in association with waste materials into diverse bio-fuels. The involved thermochemical conversion processes target the production of distinct fractions, including bio-oil in liquid form, bio-gas in gaseous form, and biochar in solid form.

6.3 Pyrolysis

Pyrolysis constitutes a thermal degradation process of substances occurring in the absence of O_2 and other oxidants [32]. It is influenced by a range of factors, including temperature, heating rate, duration of pyrolysis, raw material composition, particle size, catalysts, and the rate of the carrier gas flow. Additionally, pyrolysis is primarily carried out using two heating techniques: conventional heating and microwave heating [33]. Considering the rate of heating, conventional heating pyrolysis can be further subdivided into slow (0.1–1.0 °C/s), moderate (10–200 °C/s), and flash (>1000 °C/s) pyrolysis as shown in Fig. 6.5 [34].

In conventional heating, heat is transferred from the surface of the furnace or source to the particles, whereas microwave heating involves the conversion of energy directly within the particles. Microwave heating presents advantages such as reduced reaction times, increased rates of heating, selectivity, energy efficiency, and decreased emissions of pollutants as compared with conventional heating [36, 37]. Microwave pyrolysis (MP) is a process where electromagnetic energy is converted into heat within feedstocks without physical contact. Recently, there has been a growing interest among researchers in exploring different applications of microwave heating pyrolysis. Examples include microwave vacuum pyrolysis [38], fixed-bed microwave pyrolysis [39], and microwave flash pyrolysis. The field has witnessed increased combination and innovation recently, driven by the significant potential of microwave heating's internal heating and low energy consumption to enhance both the pyrolysis process and the resulting products.

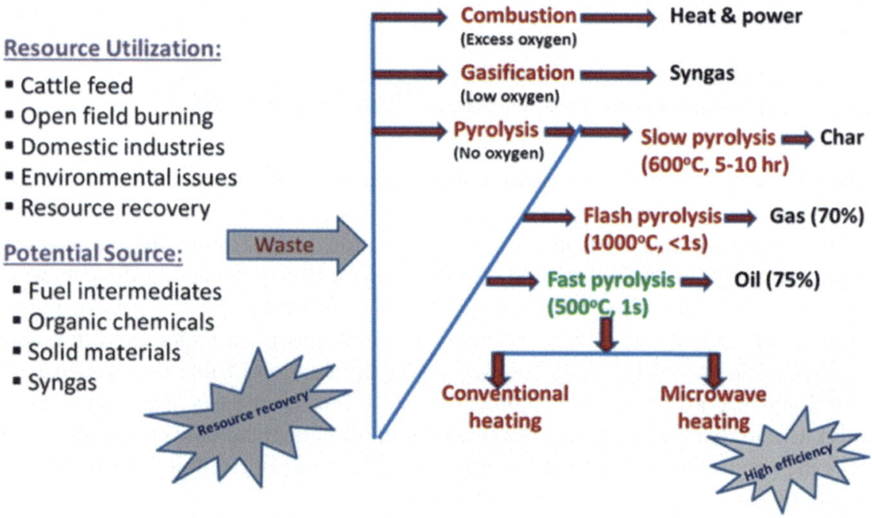

Fig. 6.5 Unleashing technologies to transform waste into valuables [35]

6.4 Microwave Properties

Microwave heating (MH) represents an energy conversion process occurring within feedstocks, distinguishing itself from Traditional heating through non-contact methods. Key aspects of microwave heating involve internal heating with a temperature gradient reversal, the presence of localized high-temperature spots, and an elevated level of selectivity. Microwave radiation consists of two perpendicular fields in space: an oscillating electric field and an oscillating magnetic field, both operating within the frequency range of 300 MHz. In technical terms, materials capable of absorbing microwave radiation are referred to as dielectrics. Consequently, the process of microwave heating is also known as dielectric heating. A dipole refers to a set of equal and opposite charges positioned at a specific distance from each other. Although naturally present in dielectric materials, dipoles can also be generated when subjected to microwave radiation [40].

When subjected to microwave radiation, heat is primarily generated through the following mechanisms [41]:

1. Polarization Loss: This phenomenon takes place when the polarization in substances cannot adjust rapidly to alterations in the external electric field. In theory, polar molecules with electrical dipole moments align and rotate under the influence of the electromagnetic field, leading to oscillation. Following collisions and friction between molecules, some energy is converted into heat.
2. Conduction Loss: In conductive materials, the alignment of free electrons is prompted by the electric field, resulting in the generation of currents within the materials and causing the dissipation of electrical heat.
3. Hysteresis Loss and Eddy Currents Loss: These losses arise from the magnetic properties of materials. Hysteresis loss occurs due to the lagging of magnetization behind the changing magnetic field [42], while eddy current loss results from circulating currents induced by the changing magnetic field [43]. Both contribute to the conversion of energy into heat.

In practical applications, microwave heating tends to be nonuniform due to the uneven distribution of dielectric materials and electromagnetic fields, resulting in the formation of hot spots. Uniform mixing of susceptors with feedstock can overcome this issue to a certain extent. These localized areas exhibit higher temperatures than the surrounding regions. Research suggests that these hot spots can exert a notable impact on both the yield and properties of pyrolysis products [36].

Materials are classified into three categories based on their interaction with microwaves:

- Insulators: These materials, such as glass wool and Teflon®, do not interact with microwaves, allowing them to pass through without absorption or reflection.
- Conductors: Including bulk metals, conductors reflect microwaves rather than absorbing them.
- Absorbers or Susceptors: Materials like carbon, char, and H_2O fall into this category, effectively absorbing microwaves and converting them into heat.

6.5 Microwave Pyrolysis Outputs from Municipal Food Waste

6.5.1 *Solid-fraction*

The residual carbon material known as biochar, produced using Microwave Pyrolysis, is predominantly made up of elemental carbon obtained from the thermal breakdown of organic constituents. It consists of unaltered organic compounds, including solid additives, and may also incorporate carbon nanoparticles formed through secondary reactions in the gas phase (Table 6.1). The significance of biochar should not be disregarded, as it has the potential to actively participate as a reactant in both heterogeneous and catalytic heterogeneous reactions [44]. Some of the literature was focused on producing biochar using microwave pyrolysis. Considering food waste, Kadlimatti et al. [45] achieved a maximum biochar yield of 60.03 wt% under the optimum pyrolysis conditions, with a fixed carbon content of 48.71 wt%. Additionally, the biochar produced had a higher heating value of 33.35 MJ/kg, indicating its potential as a valuable energy source [45].

Grycova et al. [46] found that waste coffee can be effectively utilized for the production of activated carbon sorbents with significantly improved sorption properties. The activated sorbents had a BET surface area of 1794 m^2/g. The comparison of adsorptive characteristics of the activated carbons reported in the literature provides insights into the performance of the sorbents prepared from waste coffee in relation to other materials [46]. Nyambura et al. [47] demonstrate that the use of susceptors during the co-pyrolysis of kitchen food waste and rice straw has a significant impact on the properties of the resulting biochar. The biochar produced under specific susceptor conditions exhibited enhanced thermal, surface, and fuel properties, making it suitable for high combustion efficiency fuel applications. The results propose that utilizing this biochar in microwave pyrolysis-based integrated with heat and power systems is a viable option. Bio-Char produced as BET surface area of 40.1133 m^2/g with a higher heating value of 20.550 MJ/kg and having a fuel ratio of ≤ 5.267. These data points highlight the favorable properties of the biochar produced under specific susceptor conditions, indicating its potential as a higher combustion efficiency fuel [47].

6.5.2 *Liquid Fraction*

The oil produced through Microwave Pyrolysis is an intricate blend of various organic compounds, potentially accompanied by inorganic species. When dealing with biomass, particularly food waste, the liquid or oil fraction, known as bio-oils, is discovered to be rich in oxygen, chemically intricate, unstable, and exhibits lower miscibility in traditional fuels (Table 6.2). Hence, there is a necessity to enhance the liquid products by reducing the oxygen content and eliminating residual components.

Table 6.1 Microwave pyrolysis of MSW: a focus on solid fraction literature

Source material	Purposes	Pyrolysis processing conditions	Conclusion	Reference
Food waste	Optimization methodologies using the design of the experiment	Temperature: 350–484 °C Microwave power: 450 W, 2450 MHz Duration: 22–38 min Carrier gas: Nitrogen, 33–67 mL/min	At optimal pyrolysis conditions (400 °C temp, 30 min residence time, and 50 mL/min under N_2 atmosphere), a microwave power of 450 W produced 60.03 wt% biochar with a fixed carbon (FC) content of 48.71 wt%	Kadlimatti et al. [45]
Waste coffee	Adsorbent preparation	KOH activation Final temperature: 800 °C Microwave power: 400 W, 2450 MHz Duration: 20 min	Microwave pyrolysis yields biochar with an impressive surface area of 1044.0 m^2/g	Grycova et al. [46]
Kitchen food waste and Rice straw Co-pyrolysis	Preparation of biochar with and without susceptor and its Condition optimization	Temperature: 600 °C Microwave power: 1500 W, 2450 MHz Duration: 15–25 min, N_2 Gas flow rate: 50 mL/min	Biochar, produced at a 1:1 ratio under $ZnCl_2$, showed lower ash content, higher heating value (HHV) of 20.550 MJ/kg, energy yield of 55.944%, and fuel ratio of less than 5.267, demonstrating good fuel properties	Nyambura et al. [47]
Kitchen waste (KW)	Converting of kitchen waste into different value-added products	140–700 W KW: Feed: Susceptor ratio: 20:0 to 20:20 (g/g), 200–600 °C	420 W is the optimal microwave power for high energy efficiency (55%) and KW conversion. Increasing the KW: susceptor ratio decreases furan derivative selectivity, increases simple phenolics, and yields substantial CO and H_2 gases	Suriapparao and Vinu [48]

(continued)

Table 6.1 (continued)

Source material	Purposes	Pyrolysis processing conditions	Conclusion	Reference
Palm kernel	NH_3 conversion	Microwave power: 700 W Atmosphere: Vacuum	MP produced a higher biochar surface area of 419 m^2/g	Su et al. [49]
Orange peel waste	Biochar production	Microwave power: 1000 W Temperature range: 100–700 °C Duration: 15 min Gas: 5 L/min pure CO_2 or steam	Activation with steam resulted in a reduction of biochar yield of 31 wt% but a substantial increment in BET surface area from 95.6 to 305.1 m^2/g Top of Form	Sial et al. [50]

Anis et al. [51] investigated the use of microwave-assisted pyrolysis as well as distillation processes to convert waste cooking oil followed by fresh cooking oil into bio-oil as well as liquid bio-fuels. The bio-oil obtained from the process contained a prominent proportion of aliphatic hydrocarbons, including alkanes and alkenes, as well as small amounts of aromatic compounds. The study also provided the percentage yields of these fuel fractions, indicating that the average yields were approximately 10 wt% for gasoline fraction, 44 wt% for jet fuel fraction, 43 wt% for green-diesel fraction, and about 3 wt% for heavy fuel fraction. These findings demonstrate the successful separation of different fuel fractions from the bio-oil acquired through the distillation of waste cooking oil, highlighting the potential for producing valuable liquid bio-fuels from this process [51].

Liu et al. [52] explore the effect of organic components in SS on bio-oil yield and composition, as well as the potential reaction mechanisms involved in the process. The microwave pyrolysis process achieved a product yield of $43.10 \pm 2.23\%$ bio-oil, $48.07 \pm 1.26\%$ biochar, and $8.83 \pm 1.65\%$ bio-gas. Gas chromatography–mass spectrometry analysis identified 145 different products as components of the bio-oil derived from microwave pyrolysis. The bio-oil composition included nitrogen-containing compounds, anhydrosugars, phenols, and aromatic structures. Proteins and lipids in sewage sludge were the principal contributors to bio-oil production, with the decomposition of amino acids leading to higher yields of N_2-containing compounds in the bio-oil. Findings contribute to an in-depth understanding of the microwave pyrolysis process, bio-oil composition, and the role of organic components in sewage sludge in bio-oil production [52].

Table 6.2 Microwave pyrolysis of MSW: a focus on liquid fraction literature

Source material	Pyrolysis processing conditions	Conclusion	Reference
Waste cooking oil, fresh cooking oil	Temperature range: 400–550 °C, 900 W Susceptor: charcoal Reaction time: 60 min, N_2 flow 100 mL/min	The highest bio-oil yields for waste cooking oil and fresh cooking oil were achieved at a processing temperature of 500 °C	Anis et al. [51]
Sewage sludge (SS)	Microwave oven: Max 1400 W, 2.45 GHz Inert gas: N_2 injected at a constant flow rate of 300 mL/min, Temperature setting: 900 °C	The residues obtained from the microwave pyrolysis of SS show a high percentage of bio-oil (43.10 ± 2.23%) and a low proportion of bio-gas (8.83 ± 1.65%)	Liu et al. [52]
Coconut shell waste	Feedstock: Coconut shells of various sizes (1–3 mm) Microwave power levels: 300–600 W, 2450 MHz Pyrolysis times: 5–30 min	Achieving an optimum bio-oil yield of 34.6% requires a microwave power of 593.6 W, a raw material size of 2.9 mm, and a heating duration of 28.5 min. The most abundant compounds include phenol, dimethoxy phenol, guaiacol, hydroxyanisole, and methoxyphenol	Anggraini et al. [53]
Pine Wood Sawdust	Processing time: 20 min Activated carbon (AC): 10 wt% Microwave power: 100–800 W, 2.45 GHz Nitrogen: 800 mL/min, Temperature: 505 ± 2 °C	The peak oil yield (24 wt%) was achieved at 505 ± 2 °C, using a sample heated at 800 W with 10 wt% of activated carbon (AC)	Khelfa et al. [54]
Waste cooking oil	Temperature range: 350–550 °C, HZSM-5 with metal oxide, Ex-situ mixed, ex-situ separate, in-situ separate catalysis Microwave absorber: SiC beads	The peak bio oil yield was achieved at 350 °C, with a decrease in yield observed as the temperature increased. Notably, the ex-situ mixed pyrolysis of CaO and HZSM-5 resulted in a reduction in PAH concentrations from 35.47 to 9.18%	Wu et al. [55]
Plastic waste and waste edible oil	Temperature range: 400–550 °C Microwave absorber: Coconut husk activated carbon (25 g, 50 g, 100 g, 125 g, 150 g) Weight Ratio WP: UCO = 50 g: 50 g	A peak bio-oil yield of 84 wt% was achieved in 28 min, with a heating rate of 29 °C/min, at a temperature of 581 °C, using 50 g of activated carbon	Lam et al. [38]

Anggraini et al. [53] focused on Enhancing the production efficiency of pyrolytic oil from coconut shells through the application of microwave-assisted pyrolysis. He identified the best parameters for producing liquid smoke from coconut shells, attaining a notable yield of 34.6% through the use of microwave power of 593.6 W, a raw material size of 2.9 mm, and a heating duration lasting 28.5 min. The liquid smoke contained various chemical components, and the composition varied with the applied power during the pyrolysis process. Some of the identified compounds included phenol, guaiacol, methoxyphenol, hydroxyanisole, 1,2-cyclopentanedione, 2-Hydroxy-3-methyl, 2-methylphenol, 1,2-benzenediol, 1,4-dimethoxy-2-methylbenzene, benzaldehyde, and 2-propanone. The analysis of chemical composition offered valuable insights into the complex structure of liquid smoke derived from coconut shells and how different levels of microwave power affect its composition [53].

6.5.3 Gas Fraction

The gas generated during the pyrolysis process primarily consists of combustible gases like H_2, CO, C_2H_2, CH_4, C_2H_4, C_2H_6, and similar. Additionally, there may be the presence of other gases such as CO_2 and pollutants (SO_2, NO_x), albeit in reduced concentrations (Table 6.3). Gas resulting from pyrolysis, characterized by substantial levels of H_2 and CO, can be employed in syngas applications. It is recognized that syngas (comprising $H_2 + CO$) with varying H_2/CO molar ratios are well-suited for diverse applications.

Lin et al. [56] demonstrated that using a CO_2 reaction atmosphere in microwave pyrolysis of food waste effectively improved the bio-gas yield, with the syngas proportion being significantly enhanced compared to conventional pyrolysis. This finding highlights the potential of CO_2 as a reaction medium for enhancing syngas production from food waste pyrolysis. The syngas yield was found to increase significantly from 46.73 L/kg in a nitrogen environment to 67.35 L/kg in a pure carbon dioxide environment during the microwave pyrolysis of waste food [56].

Domínguez et al. [57] discuss the traditional and microwave-induced pyrolysis of coffee hulls to generate a fuel gas rich in hydrogen. Microwave-assisted pyrolysis produces much higher gas yields than conventional pyrolysis over the complete range of temperatures. The H_2/CO ratio decreases as temperature increases in microwave-assisted pyrolysis, while it increases in conventional heating. The existence of char somewhat promotes secondary reactions of oil components and gas formation, and this impact is more evident in microwave-assisted treatment compared to conventional treatment [57].

Liu et al. [58] explore the effects of temperature on hydrogen-rich bio-gas production and the mechanisms underlying microwave pyrolysis. In resistive heating-assisted pyrolysis, the yield of H_2 increased from 0.12 mmol/g at 600 °C to 1.69 mmol/g at 900 °C. The yield of CO_2 increased from 0.22 mmol/g at 600 °C to

Table 6.3 Microwave pyrolysis of MSW: a focus on gas fraction literature

Source material	Pyrolysis processing conditions	Conclusion	Reference
Food waste	1400 W, 2.45 GHz N_2 or CO_2 30 mL/min	Under CO_2, the yield was 67.9 wt% (equivalent to 67.35 L/kg syngas), while under N_2, it was 55.63 wt% (46.73 L/kg syngas) at 1400 W microwave power. The experiment demonstrated that higher microwave power resulted in greater bio-gas production, particularly in the presence of CO_2	Lin et al. [56]
Coffee hull	Microwave power: 130, 270, 420 W THP (temperature holding period): 500, 800, 1000 °C N_2 flow rate: 60 mL/min	Under Microwave Pyrolysis, a syngas yield of 68.72 wt% was observed. Microwave Pyrolysis demonstrated higher syngas production but lower CO_2 and CO generation than resistive heating pyrolysis	Domínguez et al. [57]
Sewage sludge	1400 W, 2.45 GHz Nitrogen flow rate: 300 mL/min Pyrolysis temperatures: 600–900 °C	In Microwave Pyrolysis, H_2 varied from 1.84 mmol/g to 3.67 mmol/g. A high H/C ratio under microwave pyrolysis suggests that more amounts of hydrogen atoms directly changed to tar instead of converting into bio-gas, attributable to side reactions	Liu et al. [58]
Sludge	Maximum microwave power: 1600 W, 2.45 GHz Pyrolysis temperature: 1100 °C Reaction time: 20 min Carrier gas: Nitrogen (N_2)	Achieving a bio-gas yield of 50.15 wt% was observed. Effective control of moisture content enhances the generation of H_2 by promoting the steam reforming reaction	Lin et al. [59]
Food waste	Microwave power: 300–600 W Nitrogen gas flow: 300 mL/min Catalysts: CaO, MgO, CuO, Fe_2O_3, MnO_2, $MgCl_2$, $CuCl_2$, NaCl	The bio-gas yield was 39 wt% at 600 W without catalysts and increased to 42 wt% at 400 W for $CuCl_2$. Higher microwave power consistently led to an increase in bio-gas yield. Notably, MgO, MnO_2, and $CuCl_2$ were found to significantly enhance the rate of heating and final temperature	Liu et al. [60]

1.98 mmol/g at 900 °C. The yield of CO decreased from 0.08 mmol/g at 600 °C to 0.01 mmol/g at 900 °C. The yield of CH_4 increased from 0.02 mmol/g at 600 °C to 0.11 mmol/g at 900 °C. While in MP, the production of H_2 rrose from 0.05 mmol/g at 600 °C to 0.68 mmol/g at 800 °C, but decreased to 0.27 mmol/g at 900 °C. The yield of CO_2 increased from 0.16 mmol/g at 600 °C to 1.57 mmol/g at 900 °C. The yield of CO decreased from 0.05 mmol/g at 600 °C to 0.01 mmol/g at 900 °C. The yield of CH_4 increased from 0.01 mmol/g at 600 °C to 0.05 mmol/g at 900 °C. Overall, the study found that conventional pyrolysis resulted in higher H_2 yields at higher temperatures, while microwave pyrolysis showed a more gradual increase in H_2 yield with temperature, but with lower overall yields compared to conventional pyrolysis [58].

6.6 Conclusions

Municipal solid waste (MSW) is a source of valuable organic components such as food, plastic, and biomass residues. Effective utilization of the MSW through various thermochemical techniques paves the path for MSW recycling. Microwave heating is an electromagnetic radiation heating technology, getting significant attention from the research community due to its rapid, volumetric, efficient heating mechanism for converting solid feedstocks to valuable biochar, bio-oil, and gaseous products. However, microwave heating is highly influenced by the dielectric properties of the materials, and the MSW components are generally poor microwave absorbers. So, it is important to add a susceptor that can effectively convert microwave energy to heat energy for effective thermal conversion of MSW. The susceptors are broadly classified as carbonaceous, non-carbonaceous, and metallic. The widely used microwave absorbers are as follows: activated carbon, Biochar, and graphite are the carbonaceous susceptors, SiC is the non-carbonaceous susceptor, and Fe, $MgCO_3$, and other metal oxides are the metallic susceptors. Among the different thermochemical conversion methods, pyrolysis is a promising technology due to its versatility of the feedstock handling, product output, and operational flexibility. Apart from the type of susceptor, feed-to-susceptor ratio, microwave power, feed-susceptor contact pattern, amount of input mass, and type of microwave delivery highly influence the product distribution and quality. Therefore, careful optimization of the above parameters is crucial for the production of targeted compounds. Moreover, future research should be focused toward the development of high-capacity continuous microwave pyrolysis reactors for the MSW conversion.

Declaration We declare that we have not used any AI tools or technologies in preparing this book chapter contents.

References

1. Abdel-Shafy HI, Mansour MSM (2018) Solid waste issue: sources, composition, disposal, recycling, and valorization. Egypt J Pet 27(4):1275–1290. https://doi.org/10.1016/j.ejpe.2018.07.003
2. Malinauskaite J, Jouhara H, Czajczyńska D, Stanchev P, Katsou E, Rostkowski P et al (2017) Municipal solid waste management and waste-to-energy in the context of a circular economy and energy recycling in Europe. Energy 141:2013–2044. https://doi.org/10.1016/j.energy.2017.11.128
3. Nair LG, Agrawal K, Verma P (2022a) An insight into the principles of lignocellulosic biomass-based zero-waste biorefineries: a green leap towards imperishable energy-based future. Biotechnol Genetic Eng Rev 1–51
4. Kaza S, Yao L, Bhada-Tata P, Van Woerden F (2018) What a waste 2.0: a global snapshot of solid waste management to 2050. World Bank Publications
5. Bhada-Tata P, Hoornweg DA (2012) What a waste?: a global review of solid waste management
6. CPCB (Central Pollution Control Board) (2022) Status report on municipal solid waste management. https://cpcb.nic.in/status-of-implementation-of-solid-waste-rules/
7. Cheng H, Hu Y (2010) Municipal solid waste (MSW) as a renewable source of energy: current and future practices in China. Biores Technol 101(11):3816–3824. https://doi.org/10.1016/j.biortech.2010.01.040
8. Patel U, Ahluwalia IJ (2018) Solid waste management in India an assessment of resource recovery and environmental impact
9. Sharholy M, Ahmad K, Mahmood G, Trivedi RC (2008) Municipal solid waste management in Indian cities—a review. Waste Manage 28(2):459–467. https://doi.org/10.1016/j.wasman.2007.02.008
10. Joshi R, Ahmed S (2016) Status and challenges of municipal solid waste management in India: a review. Cogent Environ Sci 2(1):1139434
11. Chu S, Majumdar A (2012) Opportunities and challenges for a sustainable energy future. Nature 488(7411):294–303. https://doi.org/10.1038/nature11475
12. Kumar B, Verma P (2020) Biomass-based biorefineries: an important architype towards a circular economy. Fuel 119622. Elsevier
13. Zarifi MH, Sadabadi H, Hejazi SH, Daneshmand M, Sanati-Nezhad A (2018) Noncontact and nonintrusive microwave-microfluidic flow sensor for energy and biomedical engineering. Sci Rep 8(1):139
14. Carvalho GR, Monteiro RL, Laurindo JB, Augusto PED (2021) Microwave and microwave-vacuum drying as alternatives to convective drying in barley malt processing. Innov Food Sci Emerg Technol 73:102770
15. Nair LG, Agrawal K, Verma P (2022b) An overview of sustainable approaches for bioenergy production from agro-industrial wastes. Energy Nexus 100086
16. Hu Z, Ma X, Chen C (2012) A study on experimental characteristic of microwave-assisted pyrolysis of microalgae. Biores Technol 107:487–493. https://doi.org/10.1016/j.biortech.2011.12.095
17. Kumar B, Bhardwaj N, Verma P (2020b) Microwave assisted transition metal salt and orthophosphoric acid pretreatment systems: Generation of bioethanol and xylo-oligosaccharides. Renew Energ 158:574–584
18. Bu Q, Lei H, Ren S, Wang L, Holladay J, Zhang Q et al (2011) Phenol and phenolics from lignocellulosic biomass by catalytic microwave pyrolysis. Biores Technol 102(13):7004–7007. https://doi.org/10.1016/j.biortech.2011.04.025
19. Li L, Ma X, Xu Q, Hu Z (2013) Influence of microwave power, metal oxides and metal salts on the pyrolysis of algae. Biores Technol 142:469–474. https://doi.org/10.1016/j.biortech.2013.05.080
20. Wan Y, Chen P, Zhang B, Yang C, Liu Y, Lin X, Ruan R (2009) Microwave-assisted pyrolysis of biomass: catalysts to improve product selectivity. J Anal Appl Pyrol 86(1):161–167. https://doi.org/10.1016/j.jaap.2009.05.006

21. Oldfield TL, White E, Holden NM (2016) An environmental analysis of options for utilising wasted food and food residue. J Environ Manage 183:826–835
22. Chiew YL, Spångberg J, Baky A, Hansson P-A, Jönsson H (2015) Environmental impact of recycling digested food waste as a fertilizer in agriculture—a case study. Resour Conserv Recycl 95:1–14
23. Wei Y, Li J, Shi D, Liu G, Zhao Y, Shimaoka T (2017) Environmental challenges impeding the composting of biodegradable municipal solid waste: a critical review. Resour Conserv Recycl 122:51–65
24. Mir MA, Hussain A, Verma C (2016) Design considerations and operational performance of anaerobic digester: a review. Cogent Eng 3(1):1181696
25. Molino A, Nanna F, Ding Y, Bikson B, Braccio G (2013) Biomethane production by anaerobic digestion of organic waste. Fuel 103:1003–1009
26. MØller J, Boldrin A, Christensen TH (2009) Anaerobic digestion and digestate use: accounting of greenhouse gases and global warming contribution. Waste Manage Res 27(8):813–824
27. Dou Z, Toth JD, Westendorf ML (2018) Food waste for livestock feeding: feasibility, safety, and sustainability implications. Glob Food Sec 17:154–161
28. Salemdeeb R, Zu Ermgassen EKHJ, Kim MH, Balmford A, Al-Tabbaa A (2017) Environmental and health impacts of using food waste as animal feed: a comparative analysis of food waste management options. J Clean Prod 140:871–880
29. Jinno C, He Y, Morash D, McNamara E, Zicari S, King A et al (2018) Enzymatic digestion turns food waste into feed for growing pigs. Anim Feed Sci Technol 242:48–58
30. Saqib NU, Baroutian S, Sarmah AK (2018) Physicochemical, structural and combustion characterization of food waste hydrochar obtained by hydrothermal carbonization. Biores Technol 266:357–363
31. Rago YP, Surroop D, Mohee R (2018) Assessing the potential of biofuel (biochar) production from food wastes through thermal treatment. Biores Technol 248:258–264
32. Kumar B, Bhardwaj N, Agrawal K, Chaturvedi V, Verma P (2020a) Current perspective on pretreatment technologies using lignocellulosic biomass: an emerging biorefinery concept. Fuel Process Technol 199:106244
33. Zhu Y-J, Chen F (2014) Microwave-assisted preparation of inorganic nanostructures in liquid phase. Chem Rev 114(12):6462–6555
34. Li J, Dai J, Liu G, Zhang H, Gao Z, Fu J et al (2016) Biochar from microwave pyrolysis of biomass: a review. Biomass Bioenerg 94:228–244. https://doi.org/10.1016/j.biombioe.2016.09.010
35. Suriapparao DV, Tanneru HK, Reddy BR (2022) A review on the role of susceptors in the recovery of valuable renewable carbon products from microwave-assisted pyrolysis of lignocellulosic and algal biomasses: Prospects and challenges. Environ Res 215:114378. https://doi.org/10.1016/J.ENVRES.2022.114378
36. Huang YF, Chiueh PT, Lo SL (2016) A review on microwave pyrolysis of lignocellulosic biomass. Sustain Environ Res 26(3):103–109. https://doi.org/10.1016/j.serj.2016.04.012
37. Zhang S, Dong Q, Zhang L, Xiong Y (2015) High quality syngas production from microwave pyrolysis of rice husk with char-supported metallic catalysts. Biores Technol 191:17–23. https://doi.org/10.1016/J.BIORTECH.2015.04.114
38. Lam SS, Wan Mahari WA, Ok YS, Peng W, Chong CT, Ma NL et al (2019) Microwave vacuum pyrolysis of waste plastic and used cooking oil for simultaneous waste reduction and sustainable energy conversion: Recovery of cleaner liquid fuel and techno-economic analysis. Renew Sustain Energ Rev 115(June):109359. https://doi.org/10.1016/j.rser.2019.109359
39. Zhao X, Wang M, Liu H, Zhao C, Ma C, Song Z (2013) Effect of temperature and additives on the yields of products and microwave pyrolysis behaviors of wheat straw. J Anal Appl Pyrol 100:49–55
40. Kelly RM, Rowson NA (1995) Microwave reduction of oxidised ilmenite concentrates. Miner Eng 8(11):1427–1438
41. Jones DA, Lelyveld TP, Mavrofidis SD, Kingman SW, Miles NJ (2002) Microwave heating applications in environmental engineering—a review. Resour Conserv Recycl 34(2):75–90.https://doi.org/10.1016/S0921-3449(01)00088-X

42. Bartoli M, Frediani M, Briens C, Berruti F, Rosi L (2019) An overview of temperature issues in microwave-assisted pyrolysis. Processes 7(10):1–14. https://doi.org/10.3390/pr7100658
43. Morgan SP Jr (1949) Effect of surface roughness on eddy current losses at microwave frequencies. J Appl Phys 20(4):352–362
44. Menéndez JA, Domínguez A, Fernández Y, Pis JJ, Menéndez JA, Doménguez A, et al (2007) Evidence of self-gasification during the microwave-induced pyrolysis of coffee hulls. Energ Fuels 21(1):373–378.https://doi.org/10.1021/ef060331i
45. Kadlimatti HM, Mohan BR, Saidutta MB (2019) Microwave-assisted pyrolysis of food waste: optimization of fixed carbon content using response surface methodology. Biofuels.
46. Grycova B, Pryszcz A, Lestinsky P, Chamradova K (2017) Preparation and characterization of sorbents from food waste. Green Process Synth 6(3):287–293
47. Nyambura SM, Li C, Li H, Xu J, Wang J, Zhu X et al (2023) Microwave co-pyrolysis of kitchen food waste and rice straw: effects of susceptor on thermal, surface, and fuel properties of biochar. Fuel 352:129093
48. Suriapparao DV, Vinu R (2021) Recovery of renewable carbon resources from the household kitchen waste via char induced microwave pyrolysis. Renew Energ 179:370–378. https://doi.org/10.1016/j.renene.2021.07.044
49. Su MH, Azwar E, Yang Y, Sonne C, Yek PNY, Liew RK et al (2020) Simultaneous removal of toxic ammonia and lettuce cultivation in aquaponic system using microwave pyrolysis biochar. J Hazard Mater 396:122610
50. Sial TA, Lan Z, Khan MN, Zhao Y, Kumbhar F, Liu J et al (2019) Evaluation of orange peel waste and its biochar on greenhouse gas emissions and soil biochemical properties within a loess soil. Waste Manage 87:125–134
51. Anis S, Alhakim R, Khoiron AM, Kusumastuti A (2021) Microwave-assisted pyrolysis and distillation of cooking oils for liquid bio-fuel production. J Anal Appl Pyrol 154:105014
52. Liu Y, Liu Y, Chen R, Wang H, Gao H, Wang Y, Wang J (2024) Microwave pyrolysis of sewage sludge for bio-oil production: effects of organic components and mechanisms. Fuel Process Technol 253:108023
53. Anggraini SPA, Suprapto S, Juliastuti SR, Mahfud M (2024) Optimization of pyrolytic oil production from coconut shells by microwave-assisted pyrolysis using activated carbon as a microwave absorber. Int J Renew Energ Dev 13(1):145–157
54. Khelfa A, Rodrigues FA, Koubaa M, Vorobiev E (2020) Microwave-assisted pyrolysis of pine wood sawdust mixed with activated carbon for bio-oil and bio-char production. Processes 8(11):1437
55. Wu Q, Wang Y, Peng Y, Ke L, Yang Q, Jiang L et al (2020) Microwave-assisted pyrolysis of waste cooking oil for hydrocarbon bio-oil over metal oxides and HZSM-5 catalysts. Energy Convers Manage 220:113124
56. Lin J, Ma R, Luo J, Sun S, Cui C, Fang L, Huang H (2020) Microwave pyrolysis of food waste for high-quality syngas production: positive effects of a CO_2 reaction atmosphere and insights into the intrinsic reaction mechanisms. Energ Convers Manage 206(3688):112490. https://doi.org/10.1016/j.enconman.2020.112490
57. Domínguez A, Menéndez JA, Fernández Y, Pis JJ, Nabais JMV, Carrott PJM, Carrott MMLR (2007) Conventional and microwave induced pyrolysis of coffee hulls for the production of a hydrogen rich fuel gas. J Anal Appl Pyrol 79(1–2 special issue):128–135. https://doi.org/10.1016/j.jaap.2006.08.003
58. Liu Y, Chen T, Gao B, Meng R, Zhou P, Chen G et al (2021) Comparison between hydrogen-rich biogas production from conventional pyrolysis and microwave pyrolysis of sewage sludge: is microwave pyrolysis always better in the whole temperature range? Int J Hydrogen Energy 46(45):23322–23333
59. Lin J, Liu S, Han Z, Ma R, Cui C, Sun S (2023) Scaled-up microwave pyrolysis of sludge for hydrogen-rich biogas and life cycle assessment: Parameters synergistic optimization, carbon footprint analysis and technology upgrade. Chem Eng J 452:139551
60. Liu H, Ma X, Li L, Hu Z, Guo P, Jiang Y (2014) The catalytic pyrolysis of food waste by microwave heating. Biores Technol 166:45–50. https://doi.org/10.1016/j.biortech.2014.05.020

Chapter 7
Microwave-assisted Technologies for Microalgae Biomass Valorization

Hiba N. Abu Tayeh, Yoram Gerchman, and Hassan Azaizeh

Abstract Microalgae are gaining much attention as an alternative feedstock for biofuel production. Microalgae have many advantages, including wide geographic distribution, high biomass productivity, and high carbon dioxide (CO_2) capture efficiency. Nevertheless, the high cost of oil production from microalgae limits the commercial application of oil and the production of bioactive compounds. Downstream processing including harvesting, dewatering, and extraction reports cost production of 70–80%, suggesting these technologies must be improved. Up to now, several attempts have been made to pretreat microalgae biomass via physical, chemical, mechanical, and biological methods. Among these different pretreatment methods have advantages but also have limitations. It is difficult to disrupt the microalgae cells due to polymers, such as algaenan and sporopollenin, in their cell wall. Conventional methods utilized for cell disrupting and extracting are expensive and are often with low productivity. In recent years, microwave-assisted thermal conversion of biomass has attracted significant attention as it could improve process efficiency and product quality. Microwave (MW) heating has been proven to be an effective, alternative heating method for dewatering and extracting biomass. This chapter deals with a few of these aspects and reviews the recent developments in the MW-assisted thermal treatment of microalgae biomass to chemicals and fuels, both as an individual pretreatment method and as a co-treatment method to improve the performance of other thermo-chemical pretreatments. Finally, it raises some of the

H. N. Abu Tayeh (✉)
Institute of Agricultural Engineering, Agricultural Research Organization (ARO), Volcani Institute, 68 Hamacabim Road, 7505101 Rishon LeZion, Israel
e-mail: hibaa@volcani.agri.gov.il

Y. Gerchman
Oranim Academic College, 36006 Tivon, Israel

Faculty of Natural Sciences, University of Haifa, 3600600 Tivon, Israel

H. Azaizeh
Department of Environmental Science, Water Science and Biotechnology, Tel Hai College, 12208 Upper Galilee, Israel

© The Author(s), under exclusive license to Springer Nature Switzerland AG 2024
P. Verma and B. Kumar (eds.), *Sustainable Radiation Technologies in Waste-biomass Valorization*, Green Chemistry and Sustainable Technology,
https://doi.org/10.1007/978-3-031-63941-8_7

challenges related to the use of MW and future research aspects and directions that are discussed in the chapter.

Keywords Microalgae biomass valorization · Pyrolysis · Hydrogen · Microwave irradiation · Renewable energy · Bio-crude · Bio-oil · Bio-char · Hydrothermal liquefaction

7.1 Introduction

Recently, there has been a notable change in interest surrounding microwave-assisted technologies for the valorization of microalgae biomass. Microalgae, unicellular photosynthetic organisms, have the unique ability to produce a diverse array of bioactive compounds of high value, including proteins, carbohydrates, pigments, lipids, essential fatty acids, vitamins, and antioxidants [1]. Due to these valuable properties, microalgae have wide applications in many industries such as cosmetics, food, biofuels, feed, and health products [2, 3]. Furthermore, microalgae have proven used in bioremediation processes, treatment of wastewater, CO_2 mitigation, and the production of biofuels [4–6].

Given the diverse range of microalgal species, *Chlorella* stands out as promising for economical applications, in addition to *Dunaliella*, *Botryococcus*, *Chlamydomonas*, *Arthrospira*, and others [7]. The extensive growth of *Chlorella* throughout Asia, the United States, and Europe is attributed to its wide acceptance for human use [8], consolidating its dominant position in the global microalgae market along with the widely known *Arthrospira* (*Spirulina*) [9].

The effectiveness of extraction of high-value products and biofuel production depends on the efficiency of extracting biochemicals. An optimal extraction should exhibit increased selectivity for extracting specific biochemicals while minimizing the extraction of unwanted ones. Different techniques have been used to extract biochemical compounds from microalgae, where the initial step involves the disruption of the microalgal cell wall. Then the choice of the extraction method mainly depends on the stiffness of the cell wall and the components targeted for interest [10]. These methods can be grouped into classes—chemical, thermal/thermo-chemical, mechanical, electromagnetic, biological, or current methods, based on the underlying process (Fig. 7.1). Each method and each category possess its own set of advantages and disadvantages as well (Table 7.1).

Conventional solvent extraction techniques, commonly utilized, are characterized by labor-intensive processes, the use of substantial amounts of hazardous organic solvents, potential exposure of bioactive compounds to excessive oxygen, light, and heat, and the likelihood of inducing changes in stereochemistry [30]. In the specific case of extraction of microalgae lipids with solvent, the methanol-chloroform (1:1) yields the maximum amounts compared to various solvents for four species of microalgae [31]. Nevertheless, the deployment and removal of chloroform and methanol on an industrial scale can significantly impact the environment and pose

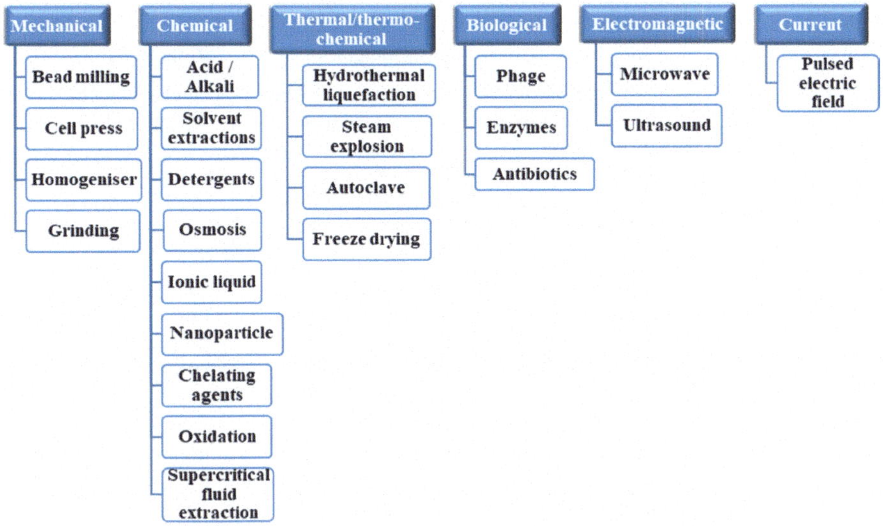

Fig. 7.1 Microalgae cell disruption methods [10]

health risks to humans. Conventional approaches, like grinding (with or without liquid nitrogen freezing), are known for their high efficiency but are expensive and unfeasible for large-scale uses.

Microwave-assisted extraction (MAE) is considered a non-conventional method and is a key area of advancement in bioactive compound extraction from microalgae. This technique utilizes microwave energy to break down the microalgae cell walls, facilitating the release of intracellular compounds. This has proven to be a more efficient and rapid method compared to traditional extraction techniques, leading to higher yields of bioactive components.

Microwave (MW) heating usage in chemical processes started at the end of the seventies in analytical laboratories. The use of MW energy was first described for extraction in 1986 [32]. MW energy is generated as high-frequency waves (300 MHz–300 GHz) of 0.001–1 m wavelength [10]. When MW is used at a frequency close to 2.45 GHz, polar compounds and water absorb energy causing dielectric heating [33]. The absorbance of the MW induces vibration of water and other polar molecules common in wet biomass, leading to temperature rises in intracellular liquids, evaporation of water and internal pressure and rupture of cells [34, 35]. In addition, MW disrupts hydrogen bonds, initiating the movement of ions and enabling increased penetration of solvent into the sample [36, 37]. Characterization of water by the high dielectric constant ensures a more efficient transfer of thermal energy to cell walls during MW heating [38]. Unlike convective heating, MW heating is not limited by thermal conduction or convection, thus allowing for a faster temperature rise [39], with maximum temperature depending on heat loss rate and power

Table 7.1 Pros and cons of microalgae cell disruption methods

Method	Advantages	Disadvantages	References
Bead milling	Effective cell wall disruption, wet samples can be extracted, solvent-free	Low efficiency where cells are rigid, an additional step required to remove beads, maintenance costs can be high, as is energy consumption	Patel et al. [11], Theegala [12]
High-pressure homogenizer	Suitable at room temperature, effective for extraction of neutral lipids, solvent-free, easy, and simple	High energy consumption, not suitable for high molecular weight extraction, high maintenance cost	Patel et al. [11], Karim et al. [13], Menegazzo and Fonseca [14]
Cell press	Simple and easy to use, no solvent required, minimal contamination, industry standard for oil recovery from seeds	Cell disruption is ineffective, high-energy consumption, slow method	Kapoore et al. [10], Patel et al. [11], Theegala [12], Menegazzo and Fonseca [14]
Microwaves	Effective disruption of the cell wall, high recovery amounts of components, relatively low energy requirement, fast heating and thus short reaction time, reduced solvent usage	Heat generated, high maintenance cost	Karim et al. [13], Menegazzo and Fonseca [14], Esquivel-Hernández et al. [15], Vasistha et al. [16]
Alkali/Acid	Low energy requirement	Disposal of acid/alkali after extraction needed, degradation of carotenoid	Gong et al. [17]
Nanoparticles	Non-toxic	Expensive, requirement to remove nanoparticles, technology in infancy	Seo et al. [18]
Extraction with supercritical fluid (e.g., liquid CO_2)	Polarity of solvent is a tunable, fast process, environmentally friendly, and generally recognized as a safe solvent	Costly equipment, reduced extraction of polar compounds, not suitable for some microalgae species	Menegazzo and Fonseca [14], Esquivel-Hernández et al. [15], Bharte and Desai [19]
Grinding	Easy and fast, efficient at a laboratory scale	Time-consuming, degradation of some of the bioactive	Kapoore et al. [10], Karim et al. [13]

(continued)

Table 7.1 (continued)

Method	Advantages	Disadvantages	References
Enzymatic hydrolysis	Effective cell wall breakdown, mild treatment, low energy requirements, no harmful solvent applied, no thermal-degradation of compounds	High cost of enzymes, long reaction time, enzymes must be disposed of after use, large-scale application is limited	Karim et al. [13], Chen et al. [20], Onumaegbu et al. [21]
Autoclave	Low cost, extraction is efficient, solvent-free	High energy requirement, not suitable for pigments as it degrades and releases non-target compounds	Patel et al. [11], Bharte and Desai [19], Onumaegbu et al. [21]
ionic liquids	Effective extraction, mild treatment, reusability	The cost of chemicals can be high, solvent toxicity, and complex process	Vasistha et al. [16], Pan et al. [22], Radošević et al. [23]
Pulse electric field	High selectivity, mild treatment, suitable for the extraction of thermolabile compounds, low energy consumption, environmentally friendly	High energy requirements, high cost, not feasible at large scale	Vasistha et al. [16], Barba et al. [24], Han et al. [25], Raso et al. [26]
Ultrasound	Simple, cost-effective, useful for developing a high-yield and low-time extraction process	High energy requirements, difficulties in upscale	Karim et al. [13], Mena-García et al. [27], Deshmukh et al. [28], Sati et al. [29]

applied [33, 39]. Two primary categories of MWs are in use, open and closed vessels. Closed vessel techniques regulate pressure and temperature and avoid loss of water, drying of sample, and loss of MW efficiency, but require sophisticated high-pressure bearing, MW-compatible vessels. MAE has been documented for recovering essential compounds, including proteins, carbohydrates, lipids, vitamins, and photosynthetic pigments, from diverse algal groups [10, 40–44]. Before employing alternative extraction methods, MW pretreatments of biomass have been found to significantly improve extraction efficiency due to their capacity to render cellular boundaries more permeable to solvents [45].

7.2 Microwave-Assisted Technology, Instrument Design, and Performance

7.2.1 *Microwave-Assisted Extraction of Biochemical Compounds from Microalgae*

Commercialization of the production of oil and bioactive compounds from microalgae is limited by the high production cost [10, 40]. Traditional methods used so far for disruption and extraction of the cell are costly and of low efficiency. Microwave-assisted Extraction offers the extraction of different components such as carbohydrates, lipids, proteins, pigments, and vitamins (Table 7.2). The MAE can result in higher yields and decrease the processing times and products with high purity [10].

7.2.2 *Microwave-Assisted Extraction of Lipids for Biodiesel*

Lipid extraction from algae usually requires breaking the cell walls. Studies show that MAE is a technology with viable potential for lipids extraction from microalgae biomass. Several scientists have highlighted MAE with promising results, especially when evaluating its performance compared to other extraction techniques (Fig. 7.2). The research conducted by Qv et al. [46] demonstrated that optimizing the MAE process resulted in the extraction of 57.02% of the lipid content of *Dunaliella tertiolecta*, better than the results of 45.94% obtained with optimized ultrasonic-assisted extraction. In a separate study conducted by Wahidin et al. [47], using a mixture of methanol/hexane (1:2, v/v) as the extraction solvent, a higher amount of lipid from *Nannochloropsis sp.* was extracted with MW irradiation, yielding 38.31%, while 23.01% was obtained in water bath-assisted extraction. Dai et al. [48] also described an improved efficiency of MAE for lipid extraction in microalgae compared to ultrasonic and conventional heating extractions. Their work investigated different organic solvents, and the mixture of n-heptane and isopropanol yielded the highest lipid extraction efficiency, performing better than the single solvents of n-heptane and isopropanol. The authors suggested that this could be due to the synergistic advantages of the solvent mixture, where n-heptane exhibited high lipid solubility, and isopropanol increased the energy absorption of MW. To investigate the application of a low-toxicity solvent, Iqbal and Theegala [49] investigated the potential of methyl soyate as a co-solvent where ethanol was used to extract the lipids from *Nannochloropsis sp.* Methyl soyate, a biodiesel produced from marketed soybean oil, was considered. The results using this co-solvent were similar to those obtained with a chloroform/ethanol mixture and Soxhlet extraction. The authors attributed this process to the ability of methyl esters to absorb efficiently the MW energy, produce hot solvents, and thus increase the dissolution of neutral lipids. Also, another study

Table 7.2 Microwave-assisted extraction for selected microalgae biomass, operating conditions, and yield

References	Algae species	Solvent	Operating conditions	Extracted added-value compounds
Qv et al. [46]	*Dunaliella tertiolecta*	Chloroform/ Methanol (2:1, v/v)	490 W, 160 s, liquid/solid ratio 100 mL/g	57.02% of the lipid content
Wahidin et al. [47]	*Nannochloropsis sp.*	Methanol/ Chloroform	700 W, 5 min, 65	38.31% lipid content
Dai et al. [48]	Dry microalgae powder	n-Heptane/ Iso-propanol (2:1)	1000 W, 40 min	30 wt.%
Iqbal and Theegala [49]	*Nannochloropsis sp.*	40% biodiesel in ethanol	1200 W, 15 min	56.6±7.9% lipid of dry wt
Pan et al. [50]	C. sorokiniana N. salina G. sulphuraria	Ionic solvents [BMIM][HSO4]	800 W, 120 °C, 60 min, liquid/solid ratio 5g/g	0.23 g lipid/g dry algae 0.10 g lipid/g dry algae 0.19 g lipid/dry algae
Onumaegbu et al. [51]	*Scenedesmus quadricauda*	Methanol and sulfuric acid mixture	600 W, 8 min+3.5h	49%
Esquivel-Hernández et al. [52]	*A. platensis*	Methanol/ethyl acetate/light petroleum (1:1:1 v/v)	400 W, 50 °C, 1 bar	629±0.13 μg carotenoids/g
Zhao et al. [53]	*H. pluvialis*	Ethanol and ethyl acetate (2:1, v/v)	141W, 83s, and a solvent of 9.8 mL	594±3.02 μg astaxanthin from 100 mg of the dried algal sample
Xiao et al. [54]	*L. japonica, U. pinnatifida, S. fusiforme*	Ethanol	300W, 60 °C, 10 min, liquid/ solid ratio 15:1 mL/g	5.13 mg, 109.3 mg, 2.12 mg, fucoxanthin all from 100g
Esquivel-Hernández et al. [15]	*A. platensis*	Non-polar solvents (limonene and ethyl acetate)	400 W, 60 °C, 15 min, 1 bar, 0.14 ratio of dry weight biomass/solvent (w/v)	C-phycocyanin: 2.28±0.10 μg/g, A-phycocyanin: 4.11±0.03 μg/g α-tocopherol: 37.86±0.78 μg/ g, β-carotene: 123.64±1.45 μg/ g
Juin et al. [55]	*Porphyridium purpureum*		10 s, 40 °C 10 s, 100 °C 10 min, 100 °C	Phycoerythrin: 73.7±2.3 μg/mg Phycocyanin: 3.48 μg/mg Allophycocyanin: 35.1±1.1 μg/mg

performed by Pan et al. [50] showed that MAE was better than conventional heating and ultrasonic-assisted extractions when ionic solvents were used as an extractant. The MAE obtained nearly 20 times more lipids for *C. sorokiniana*, four times more for *N. salina*, and 12 times more for *Galdieria sulphuraria* compared to conventional oil bath heating. These findings indicated that MW irradiation enhances the lipid extraction efficiency of the ionic liquid, probably due to the MW-absorbing properties of the triazolium intermediates in the ionic liquid. The ionic liquid [C4mim][HSO4] seemed to be an effective extractant in combination with MW uses when compared to a chloroform/methanol mixture. Research conducted by Krishnan et al. [56] reported similar encouraging results. They achieved lipid yields of 19.2% using an aqueous solution of 1-octyl-3-methylimidazolium acetate ([Omim][OAc]) with a concentration of 2.5%, whereas the conventional IL less system only produced a yield of 10.9%.

More recently, Onumaegbu et al. [51] performed a work in which they modeled and improved the extraction of lipids from wet *Scenedesmus quadricauda* microalgae by using MW pretreatment. The experimental approach employed a Box-Behnken

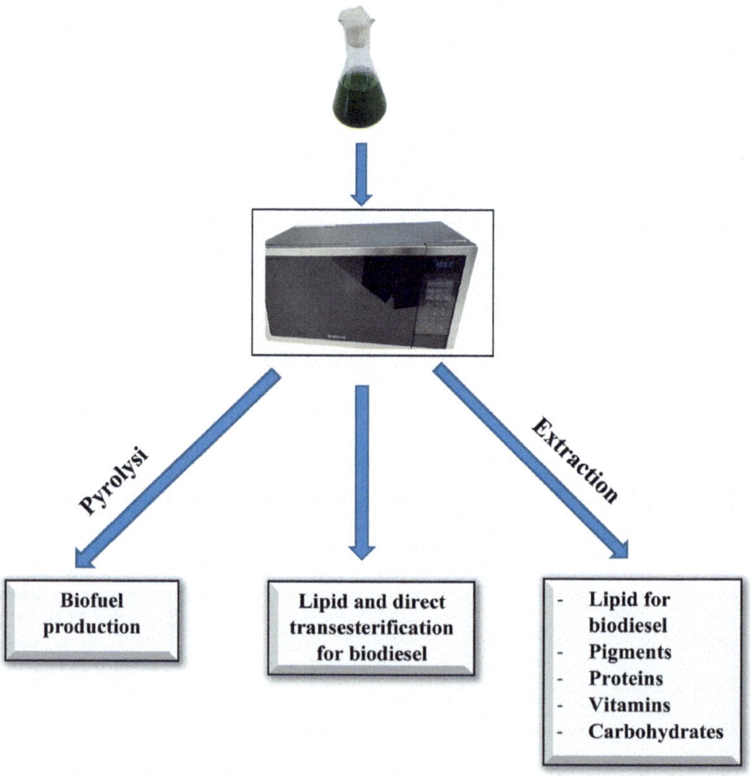

Fig. 7.2 Applications of microwave in the production of different high-value compounds and biofuels

design comprising 15 runs to assess lipid extraction using a methanol and sulfuric acid mixture. Variety of MW powers (180–600 W), heating times (2–8 min), and extraction times (3–4 h) were designated as independent variables in the experimental design. The obtained data showed a range of lipid extraction efficiencies, from 14 to 49%, with ideal conditions for maximum lipid extraction (42%) determined to be a 473 W power, 4 h extraction time, and 8 min MW time.

The yield of lipids and biodiesel is mainly affected by both, microalgae species under investigation and the operating parameters of MW extraction. *C. vulgaris* and *Nannochloropsis sp.* consistently exhibited high lipid content, reaching a maximum of 31.7% and 38.31% dry weight (DW), respectively. Enhanced lipid recoveries were generally noted when with power levels exceeding 1000 W, temperatures surpassing 65 °C, and solvent ratios, within the same solvent system. Extraction time, nevertheless, showed no significant effect [10]. Therefore, it is imperative to carefully select the optimal extraction solvent system and their respective ratios before making comparisons with conventional solvent systems. Moreover, the chosen solvent process for lipid extraction requires an assessment of its selectivity towards the intended lipid classes, aiming to minimize the extraction of non-lipid contaminants.

There is an urgent demand for more comprehensive comparisons among different solvent systems, encompassing diverse species and operating parameters of the MW (temperature, duration, and wattage), with a specific focus on wet processing. This focus is critical for promoting sustainable production on a large scale, underscoring the significance of comprehending and refining extraction conditions for practical and environmentally conscious applications.

7.2.3 Microwave-assisted Lipids Extraction and Direct Transesterification for Production of Biodiesel

Biodiesel is a safe, non-toxic, renewable, and biodegradable fuel (98% biodegradation in just a few weeks in aqueous environments). Depending on the feedstock biodiesel can also contain fewer sulfur compounds than diesel [57]. For biodiesel production, lipids need to be transesterified [58, 59]. Despite microalga being widely used as an oleous feedstock, the production cost of biodiesel from algal poses the biggest challenge in making it economically feasible [60], as drying and lipid extraction from the microalgae is energy-hungry [61]. Many works reported that the drying process alone consumes up to 70% of the total biodiesel production energy demand [62–64]. In addition, the high initial investment and operation costs required for the processing industrial plant are a major obstacle to the initiation of industrialized algal biodiesel production [65, 66]. Using direct or in situ, transesterification of wet microalgae biomass can be used to combine the two processes (extraction and transesterification) thus reducing production costs.

Technologies such as MW and ultrasound (US) can facilitate the in-situ transesterification process, especially in low-temperature reactions less than 100 °C [67]. At low temperatures, the energy essential to rupture the biomass cell wall and to enable mass transfer between the lipid and the reactant can be significantly minimized. The presence of water in the wet microalgae contributes to the MW-assisted wet in situ transesterification process because water using MW irradiation helps heat materials [68]. Through this process, the local heating happens from the inside of the microalgae cell; which results in effective heating of the overall cell compared to the normal heating process where heat is transferred by conduction and convection [69]. Because of the intermolecular fraction, temperature and pressure rapidly build up inside the microalgae cell, leading to the cell rupture, and releasing the lipid found inside it.

In contrast to convective heating, the MW-assisted process minimizes the energy and time spent on the in-situ transesterification process. In the MWE process, the amount of energy utilized in the transesterification process is only about one-third of the total energy consumed in the convection process [47]. Indeed, usually, the MW-assisted transesterification process would take only a few minutes in comparison to hours to reach completion via the convective transesterification process [70]. Microwave-assisted esterification can also result in higher biodiesel yield. Martinez-Guerra et al. [71] demonstrated 96.2% *Chlorella* lipid conversion using MW-assisted transesterification versus 78.1% conventional Bligh and Dyer process.

Microwave-assisted transesterification has some disadvantages. Johnson and Wen [67] showed that in situ transesterification using wet *Schizochytrium limacinum* microalgae revealed that the biodiesel yield from undried biomass was around 20% lower relative to the yield when dried microalgae was used. The use of MW irradiation is also known to produce local hot spots in the heated biomass [72], therefore, it still requires stirring action during the transesterification process, especially when the reaction takes a long time. An additional concern is the scalability of the apparatus when considering the mass production of biodiesel via MW-assisted transesterification which still needs more research. The explanation for this concern is the limitation of the MW irradiation's ability to penetrate effectively the biomass at a depth of more than 10–20 mm [73].

Microwave power is a non-trivial parameter: the process using dried microalgae biomass to facilitate the oil extraction from *Nannochloropsis oculata* resulted in more lipids yield using MW thermal treatment at 50% MW power (443 W) than with 100% power (943 W), although the efficacy of the cell wall destruction after 5 min using MW thermal treatment at full power was higher (70%) than that at 50% power (46%) [74]. The highest lipid yield at 943W was obtained with 1 min treatment of dry algae weight, with 17.3% cell lysis at 42.2 °C. Initially, the lipid content yield increased from 82 to 93 mgg^{-1} dry algae weight compared to the control sample. The kinetic work revealed that the rate constant for lipid extraction using MW pre-treatment was greater at 443 W compared to the controls and the extracted lipids were analyzed and found suitable for use as a biomass for biodiesel production [74].

Many factors affect the effectiveness of the MAE for cell lysis and lipids extraction, among them MW power, treatment time, final temperature, solid–liquid ratio, and the nature of the organic solvent utilized for lipids extraction, dielectric properties [38]. Balasubramanian et al. [75] showed that a continuous MW process could be used for disruption of microalgal and lipids extraction with the lipid recovery of 77% (wt/wt) from *Scenedesmus obliquus*, at higher MW frequencies and temperature of 95 °C while 30 min. Contact time and temperature were the main factors affecting the lipid yields suggesting that the MW process is likely to be effective for different microalgae strains. When MW In situ transesterification was conducted, acid catalyst concentration, reaction time, MW power, and the addition of co-solvent (n-hexane) were found to affect the yield of biodiesel, from *Chlorella* sp. with oil yield of 11.37%, and optimum biodiesel yield of 75.68%, obtained at MW power of 450 W, and 60 min incubation period [76].

7.2.4 Microwave-Assisted Extraction of Pigment

Microalgae, as photosynthetic organisms, synthesize various pigments to capture light effectively. These pigments are commonly categorized into two major classes—chlorophylls and carotenoids [77]. Chlorophyll plays a crucial role in oxygenic photosynthesis and is found in all photosynthetic microalgae [78]. Carotenoids, which can serve as both primary and secondary pigments, exhibit variations in profile based on species and growth conditions. Astaxanthin, lutein, and β-carotene are among the most well-known and commercially significant carotenoids [79].

Extracting pigments from microalgae is particularly challenging due to their rigid cell walls [80, 81], and can be significantly facilitated by MAE. MAE can produce results similar to traditional hot soaking but in a few minutes instead of 30–60 min [82]. The effectiveness of MAE extends to high mechanical resistance microalgae, such as diatoms with silica frustules. In these cases, MAE induces the permeabilization of frustule without causing cell rupturing. However, for species lacking a frustule and featuring a thick outer exopolysaccharide envelope, studies have found that conventional methods are better in terms of yield and extraction time [82].

As of now, microalgae have been a source of extracted pigments including astaxanthin, β-carotene, chlorophyll-a, chlorophyll-b, and fucoxanthin. In terms of solvents utilized for pigment extraction, while hexane has proven to be an effective lipid extractor, it is less suitable for polar pigments like fucoxanthin due to its non-polar nature [83]. An alternative to hexane is ethanol, selected for its non-toxic properties, nevertheless, acetone has demonstrated better extraction efficiency for astaxanthin, due to its lower polarity [84]. In the astaxanthin MAE, Zhao et al. [85] achieved 0.80% DW astaxanthin, but with only 37% of the above using acetic ether at a lower temperature (45°C), while using acetone, a 74% recovery of astaxanthin could be obtained [84]. These results suggest further steps are needed to enhance astaxanthin recovery, focusing on the isomers formed during MAE. Exploiting the complex life cycle of *H. pluvialis* offers another avenue, where the production of astaxanthin in the motile morphotype, lacking the thick wall of aplano-spores [86], holds the potential for higher astaxanthin recovery through MAE.

Esquivel-Hernández et al. [52], obtained 2.46±0.09 μg/g carotenoids from *A. platensis* by operating MAE at 50 °C and 400 W. Zhao et al. [53] employed MAE for *H. pluvialis* for obtaining astaxanthin, exploring optimal conditions using response surface methodology based on mathematical models. Optimal conditions were 141 W, 83 s, and a solvent of 9.8 mL, resulting in 594±3.02 μg astaxanthin from 100 mg of the dried algal sample. Esquivel-Hernández et al. [15] conducted MAE using both polar solvents (ammonium acetate and ethanol) and non-polar solvents (limonene and ethyl acetate) to extract bioactive compounds from *A. platensis.* The highest levels of pigments were obtained with elevated concentrations of non-polar solvents, increased temperature, and extended extraction times. In the limonene and ethyl acetate fractions, the amounts of C-phycocyanin and A-phycocyanin were 2.28±0.10 μg/g and 4.11±0.03 μg/g, respectively. The non-polar solvent fractions also exhibited α-tocopherol and β-carotene concentrations of 37.86±0.78 μg/g and 123.64±1.45 μg/g, respectively.

Fucoxanthin is another high-value pigment. In 2012, Xiao et al. used MAE to extract fucoxanthin from *Laminarian japonica, Undaria pinnatifida*, and *Sargassum fusiforme*. The extraction temperature was 60 °C, with a 10-min experiment time and 300W power. The obtained fucoxanthin yields from *L. japonica, U. pinnatifida*, and *S. fusiforme* were 5.13 mg, 109.3 mg, and 2.12 mg, all from 100g. HPLC analysis revealed 90% pigment purity. Additional pigments were also published for chlorophylls and β-carotene which showed a maximum yield of β-carotene (0.12% DW) using acetone [82]. Optimization of MAE conditions for pigments is essential, ensuring that they remain intact without degradation or isomer changes, ideally combined with the application of environmentally safe and cost-effective solvents.

While the method finds a wide application in the food industry [39], MAE has been acknowledged as an effective technique for recovering pigments, including different chlorophylls, phycobiliproteins, and carotenoids. However, the limitations of heat production and cost restrict its widespread use on a large scale.

7.2.5 Microwave-Assisted Extraction for High-Value Proteins, Vitamins, Carbohydrates Extract

MAE was also used to extract other compounds, e.g., proteins (phycoerythrin, phycocyanin, allophycocyanin), carbohydrates, and vitamins. Food pigmentation and fluorescent dyes such as phycoerythrin, phycocyanin, Phycobiliproteins, and allophycocyanin have high economic value. Typically, Phycobilin proteins are obtained from microalgae using a freeze–thaw process, or hot soaking [55], but solvents-free methods using MW technology are explored. Using MAE, a 0.23% DW phycocyanin was extracted from *A. platensis* applying dry biomass (60 °C, 15 min) [15] although this protein can reach 20% of the biomass DW [87]. This is considered a low yield and may be attributed to the use of low power output (400 W), along with an inefficient solvent and the use of a low solvent-to-sample ratio (7:1). High power might improve extraction but can contribute to the risk of thermal degradation/deterioration. Additional work suggested that other thermolabile products (flavonoids) were not increased with elevated power outputs [88]. A solvent-free method yielded 3.48% DW phycocyanin at 100 ∘C from *Porphyridium purpureum* DW biomass [55]. *A. platensis* is also a valuable source of allophycocyanin with up to 19.8% DW [89], but the up-to-date MAE of allophycocyanin from *A. platensis* has not been evaluated. Yields of 3.51% DW have been obtained in *P. purporeum* applying a dry process [55]. For phycoerythrin production, *P. purporeum* has been identified as a good source (3.3% DW) [90], where MAE showed a 3.48% DW yield at 40 °C in 10 s, whereas conventional soaking at the same temperature required 60 min for a comparable yield [55]. In MAE it was observed that increasing the time had no further improvements in yields of phycoerythrin, whereas an increase in temperature (50–100 °C) resulted in decreased yields, confirming thermal damage occurs above 40 °C.

MAE was used to obtain vitamins and essential micronutrients that must be acquired through the diet. Ascorbic acid (Vitamin C) and tocopherols (vitamin E) are particularly sought after for their antioxidant potential [91]. α-tocopherol, extracted from *A. platensis* at 40 °C, 400 W for 50 min, yielded very low results. B vitamins, vital for growth and reproduction, are often consumed below recommended daily levels. Microalgae present a non-chemical, vegan source of B vitamins, essential for individuals who typically obtain them from cow's milk. Notably, microalgae can serve as a natural fortification source for non-vegans. Two important B vitamins, vitamin B1 (thiamine) and (riboflavin) vitamin B2 were identified and extracted from microalgae. Thiamine is employed for treating and preventing thiamine deficiency and related problems such as beriberi, while riboflavin prevents and treats riboflavin deficiency and has been associated with migraine prevention [92]. *A. platensis* has yielded riboflavin (0.010% DW) and thiamine (0.085% DW), but in low amounts [15]. As water-soluble vitamins, their extraction efficiency was found to increase when the amount of water in the solvent was increased. The critical parameters influencing extraction using MWs were determined to be solvent ratio and extraction time. Presently, vitamin yields from MAE remain low, necessitating further optimization of operating parameters to enhance extraction efficiency.

Furthermore, methane can be derived from the residual biomass remaining after lipid extraction [93], where 307 mL of methane/g total volatile was obtained through microalgal pre-treatment. This yield demonstrated a 2.4-fold increase compared to a subsequent study by Passos et al. [94].

Additionally, soluble carbohydrates can be extracted as bulk materials from microalgae. These carbohydrates hold the potential for use as a feedstock for bioethanol, although careful consideration of energy requirements is necessary. The extraction of carbohydrates using only MW demands 34.3 MJ/kg of energy, surpassing the energy requirement for lipid extraction (1.18 MJ/kg) [94].

7.2.6 Recent Trends and Developments

MAE has predominantly been utilized for isolating specific product components from microalgae, with a primary focus on lipids and carotenoids. A shift from dry to wet process is recommended to enhance the overall cost-effectiveness of the process (remove the need for energetically expensive drying) [95]. In the context of biodiesel production, the adoption of direct transesterification, eliminating the need for a separate extraction and transesterification method, has resulted in reduced energy input and total extraction costs. While MAE has traditionally centered on lipid extraction, recent investigations have explored the feasibility of a biorefinery approach, aiming to concurrently produce lipids, carbohydrates, and proteins [96].

Although many advances have been made in lab-scale MW technology, a few cases of industrial-scale applications have been reported, such as the oil extraction from soybeans and rice bran [97] and organic compounds from Baldo leaves (*Peumus boldus*) [98]. As for microalgae, the potential for commercial applications has been widely suggested, however, reactor designs suitable for sustainable application remain a challenge [17, 99], and maintenance costs have emerged as a bottleneck for upscaling [100].

Although continuous MAE systems for rapid oil extraction from microalgae at the laboratory scale were demonstrated [75], large-scale application necessitates for comprehensive estimation of energy input and overall costs. A comprehensive report by Balasubramanian et al. [75], compared the commercial application of various extraction methods for microalgae, concluding that MAE is still moderately suitable for scaling up. Despite the potential, the high capital expenditure associated with MAE emphasizes the need for a well-established microalgal biorefinery concept to offset these costs, which, as of now, remains unexplored.

7.2.7 Microwave Pyrolysis for Biofuel Production

Microwave pyrolysis has appeared as a promising option for the utilization of microalgae biomass to produce biofuels. The pyrolysis of microalgae involves the application of MW, resulting in the production of bio-oil, biochar, and flue gases. The controlled heating provided by MW enables control over the pyrolysis process, improving the efficiency of the production of biofuel.

Various methods have been used to convert microalgae into biofuels, with thermochemical methods, especially pyrolysis, which is being widely utilized due to its capability to produce high-quality results and its cost-effectiveness compared to other methods such as convection combustion, and chemical, and biochemical methods [101, 102]. Reaction time, heating rate, and temperature within the reactor affect the quality and yield of biofuel production in the pyrolysis process [103] using MW technology can improve both the quality and yield of biofuels due to the acceleration of chemical reactions through efficient transfer of heat [104]. Optimize quality and yield of production of biofuel from microalgae using Microwave-Assisted Pyrolysis (MAP should consider several variables, including microalgae characteristics, catalyst type, type of reactor, absorbent type, size, and concentration, power of microwave, temperature, and time [105, 106].

Microalgae are a very promising approach for the production of third-generation biofuel and present advantages over previous generations [105]. Microalgae can be utilized as a potential biomass for large-scale biofuel production of biofuel without causing environmental disruption, thus addressing the limitations associated with first and second-generation biofuels [107]. The growth rate of microalgae is greater than that of other biofuel resources, facilitating greater production [103]. Importantly, their growth and production process does not need soil or to compete with the food sector for land use [107]. Although production costs are relatively high, the final cost efficiency is low for this product [108]. Therefore, microalgae have been used for biofuel production in various methods of conversion, as reported in several studies. Among these methods can be found in thermochemical conversion, gasification, combustion, pyrolysis, and liquefaction [109–116]. Among these, is pyrolysis, a process of thermal degradation of material with no oxygen. Pyrolysis can be operated at a temperature range of 400–600°C, followed by condensation of formed steam, which is a promising and economical method for biomass conversion [105]. Furthermore, pyrolysis represents a relatively recent approach to biofuel conversion. Despite the need for additional steps in the conversion to fuel, this method is simple, rapid, and effective, especially in bio-oil production. The biomass products resulting from pyrolysis conversion include gaseous fuels, solid (char), and liquid (bio-oil) [117].

As a developing and promising heating technology, MWs have gotten attention as a heating medium due to their ability to enhance and accelerate chemical reactions. In comparison to conventional pyrolysis, MW-assisted pyrolysis has some advantages, such as simple control of the pyrolysis procedure, less time, and high heating effectiveness [110]. In a work which was performed by Zainan et al. [118], they showed that the impact of catalysts on the production of bio-oil from *Chlorella vulgaris* was explored using the pyrolysis procedure, making use of a fixed bed reactor. The temperature range is 300–600 °C and variations in the ratio of catalysts to algae are at 5:1, 2:1, and 1:1. The obtained data showed that catalytic pyrolysis is a promising procedure in producing bio-oil from microalgae. The suitable pyrolysis temperature, whether catalytic or non, was found to be 500 °C. Pan et al. [111] showed that the pyrolysis process for the residues of *Nannochloropsis sp.*, with and without adding a catalyst, is possible to produce bio-oil. It has been shown that adding a catalyst provides more aromatic hydrocarbons compared to noncatalytic pyrolysis. While biomass, particularly microalgae, is considered a poor absorber of microwaves, some reports suggest that incorporating a microwave absorption enhancer enhances biomass heating, facilitating bio-oil production. Consequently, blending the raw material with a heat absorber can enhance its ability to absorb MW effectively [110, 119, 120]. Thus, additional research is warranted to explore the efficacy of combining a catalyst with a heat absorber in the production of bio-oil, serving as a crucial raw material in biofuel production.

7.3 Challenges and Future Prospects

Employing MAE as an economically viable biorefinery strategy has enabled the extraction of a diverse array of products from microalgae. While the potential for lipid extraction, a precursor for biodiesel, exists by applying the MAE, further optimization of process parameters is needed to ensure overall economic sustainability. Although traditionally employed at a laboratory scale, recent demonstrations of its successful commercial-scale application for biodiesel production from waste cooking oil suggest the scalability of MAE [121].

Direct transesterification using MW appears promising for biodiesel production but requires exploration at a commercial scale. Hexane/Isopropanol has been proposed as a suitable solvent for extracting carbohydrates and proteins from lipid-extracted algae biomass, but its feasibility for commercial applications is questionable [96]. Industrial processes have typically focused on single-product approaches, but recent efforts emphasize a biorefinery concept, aiming to derive multiple products from microalgae in a sustainable manner [81]. Ansari et al. [96] pursued a biorefinery approach, extracting lipids using MW from *S. obliquus* while employing different methods for extracting various proteins and carbohydrates. Proteins were identified as the costliest fraction, followed by lipids and carbohydrates, omitting carotenoids from the study.

For a successful biorefinery, the extraction sequence was determined as proteins first, followed by lipids and carbohydrates. However, the suitability of MAE for commercial-scale uses has not been so far explored and currently remains confined to laboratory-scale use. In developing the biorefinery platform, selecting the most commercially promising selected organism is crucial. Despite research on microalgae biofuels dating back to the seventies, particularly with the Aquatic Species Program (ASP) exploring many microalgae spp., biodiesel from microalgae has yet to achieve cost competitiveness with oil refinery platforms [122].

Exploration of three microalgal species has revealed their potential to yield commercially valuable products: *P. tricornutum* for EPA, fucoxanthin and chrysolaminarin (carbohydrate), [123], *P. purpureum* for EPA, zeaxanthin, phycobiliproteins β-carotene, and exopolysaccharide [124], and *Nannochloropsis sp.* for EPA and high-value proteins [125]. *Nannochloropsis sp.*, with its higher lipid content and higher biodiesel conversion efficiency, shows a greater potential for creating a sustainable microalgal biorefinery using MAE. This method could simultaneously produce biodiesel, EPA, pigments, and phycobiliproteins. Successful extraction of phycobiliproteins from *P. purporeum* indicates the feasibility of applying this process to *Nannochloropsis sp.*, although additional research and a comprehensive life cycle analysis are still needed.

Microwave technology holds promise for a biorefinery approach, contingent upon addressing challenges such as reducing energy requirements and using environmentally friendly solvents. A notable challenge with MWs, especially local ones, is the non-uniform distribution of radiation within the cavity, leading to non-uniform heating of samples [126]. Meanwhile, MAE has limitations in using polar solvents and is not fit for volatile target compounds, necessitating additional separation processes even after microalgal cell breakdown and release of the intracellular contents [36, 81].

7.4 Conclusion

Although there are obvious advantages to MW-assisted technologies in microalgae valorization, challenges such as optimizing the process of different parameters, scaling up production, and thus assuring economic feasibility still need further research. Upcoming research should focus on dealing with these challenges, as well as the utilization of the full potential of MW-assisted technologies in the commercial-scale production of valuable microalgae and bio-based products.

Declaration The authors have not used any AI tools during the writing of Chapter.

References

1. Udayan A, Arumugam M, Pandey A (2017) Nutraceuticals from algae and cyanobacteria. In: Algal green chemistry: recent progress in biotechnology, pp 65–89. https://doi.org/10.1016/B978-0-444-63784-0.00004-7
2. Lum KK, Kim J, Lei XG (2013) Dual potential of microalgae as a sustainable biofuel feedstock and animal feed. http://www.jasbsci.com/content/4/1/53
3. Wang HMD, Chen CC, Huynh P, Chang JS (2015) Exploring the potential of using algae in cosmetics. Biores Technol 184:355–362. https://doi.org/10.1016/J.BIORTECH.2014.12.001
4. Abdelfattah A, Ali SS, Ramadan H, El-Aswar EI, Eltawab R, Ho SH, Elsamahy T, Li S, El-Sheekh MM, Schagerl M, Kornaros M, Sun J (2023) Microalgae-based wastewater treatment: mechanisms, challenges, recent advances, and future prospects. Environ Sci Ecotechnol 13:100205. https://doi.org/10.1016/J.ESE.2022.100205
5. Goswami RK, Agrawal K, Verma P (2022a) Microalgae biomass biorefinery: a sustainable renewable energy feedstock of the future. In Micro-algae: next-generation feedstock for biorefineries: cultivation and refining processes, pp 1–29. Springer Nature Singapore, Singapore
6. Van Tuyen N, Limjuco LA, Lee K, Dang NM (2022) Integrated applications of microalgae to wastewater treatment and biorefinery: recent advances and opportunities. In: Applied chemistry for engineering, vol 33, issue 3. Korean Society of Industrial Engineering Chemistry, pp 242–257. https://doi.org/10.14478/ace.2022.1031
7. Morais Junior WG, Gorgich M, Corrêa PS, Martins AA, Mata TM, Caetano NS (2020) Microalgae for biotechnological applications: cultivation, harvesting and biomass processing. Aquaculture 528:735562. https://doi.org/10.1016/J.AQUACULTURE.2020.735562
8. Araújo R, Vázquez Calderón F, Sánchez López J, Azevedo IC, Bruhn A, Fluch S, Garcia Tasende M, Ghaderiardakani F, Ilmjärv T, Laurans M, Mac Monagail M, Mangini S, Peteiro C, Rebours C, Stefansson T, Ullmann J (2021) Current status of the algae production industry in Europe: an emerging sector of the blue bioeconomy. Front Marine Sci 7. https://doi.org/10.3389/fmars.2020.626389

9. Koyande AK, Chew KW, Rambabu K, Tao Y, Chu DT, Show PL (2019) Microalgae: a potential alternative to health supplementation for humans. Food Sci Human Wellness 8(1):16–24. https://doi.org/10.1016/J.FSHW.2019.03.001
10. Kapoore RV, Butler TO, Pandhal J, Vaidyanathan S (2018) Microwave-assisted extraction for microalgae: from biofuels to biorefinery. In: Biology 7(1):1–25. MDPI AG. https://doi.org/10.3390/biology7010018
11. Patel A, Mikes F, Matsakas L (2018) An overview of current pretreatment methods used to improve lipid extraction from oleaginous microorganisms. In: Molecules, vol 23, issue 7. MDPI AG. https://doi.org/10.3390/molecules23071562
12. Theegala CS (2015) Algal cell disruption and lipid extraction: a review on current technologies and limitations. In: Prokop A, Bajpai RK, Zappi ME (eds) Algal biorefineries. Springer, Cham, pp 419–441
13. Karim A, Amirul Islam M, Bin KZ, Faizal CKM, Khan MMR, Yousuf A (2020) Microalgal cell disruption and lipid extraction techniques for potential biofuel production. In: Microalgae cultivation for biofuels production, pp 129–147. https://doi.org/10.1016/B978-0-12-817536-1.00009-6
14. Menegazzo ML, Fonseca GG (2019) Biomass recovery and lipid extraction processes for microalgae biofuels production: a review. Renew Sustain Energy Rev 107:87–107. https://doi.org/10.1016/J.RSER.2019.01.064
15. Esquivel-Hernández DA, Rodríguez-Rodríguez J, Rostro-Alanis M, Cuéllar-Bermúdez SP, Mancera-Andrade EI, Núñez-Echevarría JE, García-Pérez JS, Chandra R, Parra-Saldívar R (2017) Advancement of green process through microwave-assisted extraction of bioactive metabolites from *Arthrospira platensis* and bioactivity evaluation. Biores Technol 224:618–629. https://doi.org/10.1016/J.BIORTECH.2016.10.061
16. Vasistha S, Khanra A, Clifford M, Rai MP (2021) Current advances in microalgae harvesting and lipid extraction processes for improved biodiesel production: a review. Renew Sustain Energy Rev 137:110498. https://doi.org/10.1016/J.RSER.2020.110498
17. Gong M, Hu Y, Yedahalli S, Bassi A (2017) Oil extraction processes in microalgae, vol 1. https://www.researchgate.net/publication/319205128
18. Seo JY, Kim MG, Lee K, Lee Y-C, Na J-G, Jeon SG, Bin PS, Oh Y-K (2017) Multifunctional nanoparticle applications to microalgal biorefinery. In Rai M, da Silva SS (eds.) Nanotechnology for bioenergy and biofuel production. Green chemistry and sustainable technology. Springer, Cham, pp 59–87
19. Bharte S, Desai K (2021) Techniques for harvesting, cell disruption and lipid extraction of microalgae for biofuel production. Biofuels 12(3):285–305. https://doi.org/10.1080/17597269.2018.1472977
20. Chen Q, Liu D, Wu C, Yao K, Li Z, Shi N, Wen F, Gates ID (2018) Co-immobilization of cellulase and lysozyme on amino-functionalized magnetic nanoparticles: an activity-tunable biocatalyst for extraction of lipids from microalgae. Biores Technol 263:317–324. https://doi.org/10.1016/J.BIORTECH.2018.04.071
21. Onumaegbu C, Alaswad A, Rodriguez C, Olabi AG (2018) Optimization of pre-treatment process parameters to generate biodiesel from microalga. Energies 11(4). https://doi.org/10.3390/en11040806
22. Pan Y, Alam MA, Wang Z, Huang D, Hu K, Chen H, Yuan Z (2017) One-step production of biodiesel from wet and unbroken microalgae biomass using deep eutectic solvent. Biores Technol 238:157–163. https://doi.org/10.1016/J.BIORTECH.2017.04.038
23. Radošević K, Cvjetko Bubalo M, Gaurina Srček V, Grgas D, Landeka Dragičević T, Redovniković RI (2015) Evaluation of toxicity and biodegradability of choline chloride based deep eutectic solvents. Ecotoxicol Environ Saf 112:46–53. https://doi.org/10.1016/J.ECOENV.2014.09.034
24. Barba FJ, Parniakov O, Pereira SA, Wiktor A, Grimi N, Boussetta N, Saraiva JA, Raso J, Martin-Belloso O, Witrowa-Rajchert D, Lebovka N, Vorobiev E (2015) Current applications and new opportunities for the use of pulsed electric fields in food science and industry. Food Res Int 77:773–798. https://doi.org/10.1016/J.FOODRES.2015.09.015

25. Han SF, Jin W, Yang Q, El-Fatah Abomohra A, Zhou X, Tu R, Chen C, Xie GJ, Wang Q (2019) Application of pulse electric field pretreatment for enhancing lipid extraction from *Chlorella pyrenoidosa* grown in wastewater. Renew Energy 133:233–239. https://doi.org/10.1016/J.RENENE.2018.10.034
26. Raso J, Frey W, Ferrari G, Pataro G, Knorr D, Teissie J, Miklavčič D (2016) Recommendations guidelines on the key information to be reported in studies of application of PEF technology in food and biotechnological processes. Innov Food Sci Emerg Technol 37:312–321. https://doi.org/10.1016/J.IFSET.2016.08.003
27. Mena-García A, Ruiz-Matute AI, Soria AC, Sanz ML (2019) Green techniques for extraction of bioactive carbohydrates. TrAC Trends Anal Chem 119:115612. https://doi.org/10.1016/J.TRAC.2019.07.023
28. Deshmukh S, Kumar R, Bala K (2019) Microalgae biodiesel: a review on oil extraction, fatty acid composition, properties and effect on engine performance and emissions. Fuel Process Technol 191:232–247. ISSN 0378-3820. https://doi.org/10.1016/j.fuproc.2019.03.013
29. Sati H, Mitra M, Mishra S, Baredar P (2019) Microalgal lipid extraction strategies for biodiesel production: a review. Algal Res 38:101413. ISSN 2211-9264. https://doi.org/10.1016/j.algal.2019.101413
30. Denery JR, Dragull K, Tang CS, Li QX (2004) Pressurized fluid extraction of carotenoids from *Haematococcus pluvialis* and *Dunaliella salina* and kavalactones from Piper methysticum. Anal Chim Acta 501(2):175–181. https://doi.org/10.1016/J.ACA.2003.09.026
31. Ryckebosch E, Muylaert K, Eeckhout M, Ruyssen T, Foubert I (2011) Influence of drying and storage on lipid and carotenoid stability of the Microalga Phaeodactylum tricornutum. J Agric Food Chem 59(20):11063–11069. https://doi.org/10.1021/jf2025456
32. Ganzler K, Salgó A, Valkó K (1986) Microwave extraction: a novel sample preparation method for chromatography. J Chromatogr A 371(C):299–306. https://doi.org/10.1016/S0021-9673(01)94714-4
33. Bagade SB, Patil M (2021) Recent advances in microwave assisted extraction of bioactive compounds from complex herbal samples: a review. In: Critical reviews in analytical chemistry, vol 51, issue 2, pp 138–149. Taylor and Francis Ltd. https://doi.org/10.1080/10408347.2019.1686966
34. Hahn T, Lang S, Ulber R, Muffler K (2012) Novel procedures for the extraction of fucoidan from brown algae. Process Biochem 47(12):1691–1698. https://doi.org/10.1016/J.PROCBIO.2012.06.016
35. Sosa-Hernández JE, Escobedo-Avellaneda Z, Iqbal HMN, Welti-Chanes J (2018) State-of-the-art extraction methodologies for bioactive compounds from algal biome to meet bio-economy challenges and opportunities. In: Molecules, vol 23, issue 11. MDPI AG. https://doi.org/10.3390/molecules23112953
36. Kadam SU, Tiwari BK, O'Donnell CP (2013) Application of novel extraction technologies for bioactives from marine algae. J Agric Food Chem 61(20):4667–4675. https://doi.org/10.1021/jf400819p
37. Nair LG, Agrawal K, Verma P (2023) Organosolv pretreatment: an in-depth purview of mechanics of the system. Bioresour Bioprocess 10(1). Springer
38. Ali M, Watson IA (2015) Microwave treatment of wet algal paste for enhanced solvent extraction of lipids for biodiesel production. Renew Energy 76:470–477. https://doi.org/10.1016/J.RENENE.2014.11.024
39. Chemat F, Rombaut N, Meullemiestre A, Turk M, Perino S, Fabiano-Tixier AS, Abert-Vian M (2017) Review of green food processing techniques. Preservation, transformation, and extraction. Innov Food Sci Emerg Technol 41:357–377. https://doi.org/10.1016/J.IFSET.2017.04.016
40. Goswami RK, Agrawal K, Upadhyaya HM, Gupta VK, Verma P (2022) Microalgae conversion to alternative energy, operating environment and economic footprint: an influential approach towards energy conversion, and management. Energy Convers Manage 269:116118
41. Morais S (2013) Ultrasonic- and microwave-assisted extraction and modification of algal components. In: Functional ingredients from algae for foods and nutraceuticals, pp 585–605. https://doi.org/10.1533/9780857098689.3.585

42. Siahaan EA, Soo Chun B (2020) Innovative alternative technology for fucoxanthin recovery. In: Kim S-K (ed) Encyclopedia of marine biotechnology. Wiley, pp 3213–3227
43. Tsubaki S, Onda A, Hiraoka M, Hiraoka M, Fujii S, Wada Y (2017) Microwave-assisted hydrothermal processing of seaweed biomass. In: Ruiz H, Hedegaard Thomsen M, Trajano H (eds.) Hydrothermal processing in biorefineries. Springer, Cham, pp 443–460
44. Zhou X, Jin W, Tu R, Guo Q, Fang HS, Chen C, Wang Q, Liu W, Jensen PD, Wang Q (2019) Optimization of microwave assisted lipid extraction from microalga Scenedesmus obliquus grown on municipal wastewater. J Clean Prod 221:502–508.https://doi.org/10.1016/J.JCLEPRO.2019.02.260
45. Quitain AT, Kai T, Sasaki M, Goto M (2013) Supercritical carbon dioxide extraction of fucoxanthin from *Undaria pinnatifida.* J Agric Food Chem 61(24):5792–5797. https://doi.org/10.1021/jf400740p
46. Qv XY, Zhou QF, Jiang JG (2014) Ultrasound-enhanced and microwave-assisted extraction of lipid from *Dunaliella tertiolecta* and fatty acid profile analysis. J Sep Sci 37(20):2991–2999. https://doi.org/10.1002/jssc.201400458
47. Wahidin S, Idris A, Shaleh SRM (2014) Rapid biodiesel production using wet microalgae via microwave irradiation. Energy Convers Manage 84:227–233. https://doi.org/10.1016/J.ENCONMAN.2014.04.034
48. Dai YM, Chen KT, Chen CC (2014) Study of the microwave lipid extraction from microalgae for biodiesel production. Chem Eng J 250:267–273. https://doi.org/10.1016/J.CEJ.2014.04.031
49. Iqbal J, Theegala C (2013) Microwave assisted lipid extraction from microalgae using biodiesel as co-solvent. Algal Res 2(1):34–42. https://doi.org/10.1016/J.ALGAL.2012.10.001
50. Pan J, Muppaneni T, Sun Y, Reddy HK, Fu J, Lu X, Deng S (2016) Microwave-assisted extraction of lipids from microalgae using an ionic liquid solvent [BMIM][HSO4]. Fuel 178:49–55. https://doi.org/10.1016/J.FUEL.2016.03.037
51. Onumaegbu C, Alaswad A, Rodriguez C, Olabi A (2019) Modelling and optimization of wet microalgae *Scenedesmus quadricauda* lipid extraction using microwave pre-treatment method and response surface methodology. Renew Energy 132:1323–1331. https://doi.org/10.1016/J.RENENE.2018.09.008
52. Esquivel-Hernández DA, López VH, Rodríguez-Rodríguez J, Alemán-Nava GS, Cuéllar-Bermúdez SP, Rostro-Alanis M, Parra-Saldívar R (2016) Supercritical carbon dioxide and microwave-assisted extraction of functional lipophilic compounds from *Arthrospira platensis.* Int J Mole Sci 17(5):658. https://doi.org/10.3390/ijms17050658
53. Zhao L, Chen G, Zhao G, Hu X (2009) Optimization of microwave-assisted extraction of astaxanthin from *Haematococcus pluvialis* by response surface methodology and antioxidant activities of the extracts. Sep Sci Technol 44(1):243–262. https://doi.org/10.1080/01496390802282321
54. Xiao X, Si X, Yuan Z, Xu X, Li G (2012) Isolation of fucoxanthin from edible brown algae by microwave-assisted extraction coupled with high-speed countercurrent chromatography. J Sep Sci 35(17):2313–2317
55. Juin C, Chérouvrier JR, Thiéry V, Gagez AL, Bérard JB, Joguet N, Kaas R, Cadoret JP, Picot L (2015) Microwave-assisted extraction of phycobiliproteins from porphyridium purpureum. Appl Biochem Biotechnol 175(1):1–15. https://doi.org/10.1007/s12010-014-1250-2
56. Krishnan S, Ghani NA, Aminuddin NF, Quraishi KS, Azman NS, Cravotto G, Leveque JM (2020) Microwave-assisted lipid extraction from Chlorella vulgaris in water with 0.5%–2.5% of imidazolium based ionic liquid as additive. Renew Energy 149:244–252. https://doi.org/10.1016/J.RENENE.2019.12.063
57. Vyas AP, Verma JL, Subrahmanyam N (2010) A review on FAME production processes. Fuel 89(1):1–9. https://doi.org/10.1016/J.FUEL.2009.08.014
58. Ma G, Hu W, Pei H, Jiang L, Song M, Mu R (2015) In situ heterogeneous transesterification of microalgae using combined ultrasound and microwave irradiation. Energy Convers Manage 90:41–46. https://doi.org/10.1016/J.ENCONMAN.2014.10.061

59. Nair LG, Agrawal K, Verma P (2022) An overview of sustainable approaches for bioenergy production from agro-industrial wastes. Energy Nexus, p100086
60. Su Y, Song K, Zhang P, Su Y, Cheng J, Chen X (2017) Progress of microalgae biofuel's commercialization. Renew Sustain Energy Rev 74:402–411. https://doi.org/10.1016/J.RSER.2016.12.078
61. Adeniyi OM, Azimov U, Burluka A (2018) Algae biofuel: current status and future applications. Renew Sustain Energy Rev 90:316–335. https://doi.org/10.1016/J.RSER.2018.03.067
62. Branco-Vieira M, Costa D, Mata TM, Martins AA, Freitas MAV, Caetano NS (2020) A life cycle inventory of microalgae-based biofuels production in an industrial plant concept. Energy Rep 6:397–402. https://doi.org/10.1016/J.EGYR.2019.08.079
63. Nazloo EK, Moheimani NR, Ennaceri H (2022) Biodiesel production from wet microalgae: progress and challenges. Algal Res 68:102902. https://doi.org/10.1016/J.ALGAL.2022.102902
64. Skorupskaite V, Makareviciene V, Gumbyte M (2016) Opportunities for simultaneous oil extraction and transesterification during biodiesel fuel production from microalgae: a review. Fuel Process Technol 150:78–87. https://doi.org/10.1016/J.FUPROC.2016.05.002
65. Venkata Mohan S, Rohit MV, Chiranjeevi P, Chandra R, Navaneeth B (2015) Heterotrophic microalgae cultivation to synergize biodiesel production with waste remediation: progress and perspectives. Biores Technol 184:169–178. https://doi.org/10.1016/J.BIORTECH.2014.10.056
66. Vlaskin MS, Chernova NI, Kiseleva SV, Popel' OS, Zhuk AZ (2017) Hydrothermal liquefaction of microalgae to produce biofuels: state of the art and future prospects. Thermal Eng 64(9):627–636.https://doi.org/10.1134/S0040601517090105
67. Johnson MB, Wen Z (2009) Production of biodiesel fuel from the microalga *Schizochytrium limacinum* by direct transesterification of algal biomass. Energy Fuels 23(10):5179–5183. https://doi.org/10.1021/ef900704h
68. Cui Y, Liang Y (2014) Direct transesterification of wet Cryptococcus curvatus cells to biodiesel through use of microwave irradiation. Appl Energy 119:438–444. https://doi.org/10.1016/J.APENERGY.2014.01.016
69. Bundhoo ZMA (2018) Microwave-assisted conversion of biomass and waste materials to biofuels. Renew Sustain Energy Rev 82:1149–1177. https://doi.org/10.1016/J.RSER.2017.09.066
70. Teo CL, Idris A (2014) Enhancing the various solvent extraction method via microwave irradiation for extraction of lipids from marine microalgae in biodiesel production. Biores Technol 171:477–481. https://doi.org/10.1016/J.BIORTECH.2014.08.024
71. Martinez-Guerra E, Gude VG, Mondala A, Holmes W, Hernandez R (2014) Extractive-transesterification of algal lipids under microwave irradiation with hexane as solvent. Biores Technol 156:240–247. https://doi.org/10.1016/J.BIORTECH.2014.01.026
72. Idris A, Bukhari A, Yusof NM, Mohd Yusof N, Tan KG (2012) A simultaneous cooling and dielectric heating: an advanced technology to improve the yield of lactides. https://www.researchgate.net/publication/262070507
73. Li J, Dai J, Liu G, Zhang H, Gao Z, Fu J, He Y, Huang Y (2016) Biochar from microwave pyrolysis of biomass: a review. Biomass Bioenerg 94:228–244. https://doi.org/10.1016/J.BIOMBIOE.2016.09.010
74. Ali M, Watson IA (2016) Microwave thermolysis and lipid recovery from dried microalgae powder for biodiesel production. Energ Technol 4(2):319–330. https://doi.org/10.1002/ente.201500242
75. Balasubramanian S, Allen JD, Kanitkar A, Boldor D (2011) Oil extraction from Scenedesmus obliquus using a continuous microwave system—design, optimization, and quality characterization. Biores Technol 102(3):3396–3403. https://doi.org/10.1016/J.BIORTECH.2010.09.119
76. Qadariyah L, Panjaitan MR, Mujaddid F, Kalsum U (2018) Biodiesel production from dry microalga biomass by microwave-assisted in-situ transesterification. In: MATEC web of conferences, 156.https://doi.org/10.1051/matecconf/201815606005

77. Richmond A, Hu Q (2013) Handbook of microalgal culture applied phycology and biotechnology. Wiley. https://doi.org/10.1002/9781118567166.fmatter
78. Masojídek J, Koblížek M, Torzillo G (2007) Photosynthesis in microalgae. In Richmond A (ed.) Handbook of microalgal culture: biotechnology and applied phycology, 2nd edn. Wiley-Blackwell
79. Guedes AC, Amaro HM, Malcata FX (2011) Microalgae as sources of carotenoids. In: Marine drugs, vol 9, issue 4. MDPI AG, pp 625–644. https://doi.org/10.3390/md9040625
80. Cooney M, Young G, Nagle N (2009) Extraction of bio-oils from microalgae. Sep Purif Rev 38(4):291–325. https://doi.org/10.1080/15422110903327919
81. Lee SY, Cho JM, Chang YK, Oh YK (2017) Cell disruption and lipid extraction for microalgal biorefineries: a review. Biores Technol 244:1317–1328. https://doi.org/10.1016/J.BIORTECH.2017.06.038
82. Pasquet V, Chérouvrier JR, Farhat F, Thiéry V, Piot JM, Bérard JB, Kaas R, Serive B, Patrice T, Cadoret JP, Picot L (2011) Study on the microalgal pigments extraction process: performance of microwave assisted extraction. Process Biochem 46(1):59–67. https://doi.org/10.1016/J.PROCBIO.2010.07.009
83. Kim SM, Jung YJ, Kwon ON, Cha KH, Um BH, Chung D, Pan CH (2012) A potential commercial source of fucoxanthin extracted from the microalga Phaeodactylum tricornutum. Appl Biochem Biotechnol 166(7):1843–1855. https://doi.org/10.1007/s12010-012-9602-2
84. Ruen-ngam D, Shotipruk A, Pavasant P (2011) Comparison of extraction methods for recovery of astaxanthin from *Haematococcus pluvialis*. Sep Sci Technol 46(1):64–70. https://doi.org/10.1080/01496395.2010.493546
85. Zhao X, Zhang X, Fu L, Zhu H, Zhang B (2016) Effect of extraction and drying methods on antioxidant activity of astaxanthin from **Haematococcus pluvialis**. Food Bioprod Process 99:197–203. https://doi.org/10.1016/J.FBP.2016.05.007
86. Butler TO, McDougall GJ, Campbell R, Stanley MS, Day JG (2018) Media screening for obtaining *Haematococcus pluvialis* red motile macrozooids rich in astaxanthin and fatty acids. Biology 7(1). https://doi.org/10.3390/biology7010002
87. Vernès L, Granvillain P, Chemat F, Vian M (2015) Phycocyanin from Arthrospira platensis. Production, extraction and analysis. Curr Biotechnol 4(4):481–491. https://doi.org/10.2174/2211550104666151006002418
88. Veggi PC, Martinez J, Meireles MAA (2012) Fundamentals of microwave extraction. In: Chemat F, Cravotto G (eds) Microwave-assisted extraction for bioactive compounds. Springer, pp 15–52
89. Spolaore P, Joannis-Cassan C, Duran E, Isambert A (2006) Commercial applications of microalgae. J Biosci Bioeng 101(2):87–96. https://doi.org/10.1263/JBB.101.87
90. Kathiresan S, Sarada R, Bhattacharya S, Ravishankar GA (2007) Culture media optimization for growth and phycoerythrin production from Porphyridium purpureum. Biotechnol Bioeng 96(3):456–463. https://doi.org/10.1002/bit.21138
91. Carocho M, Barreiro MF, Morales P, Ferreira ICFR (2014) Adding molecules to food, pros and cons: a review on synthetic and natural food additives. In: Comprehensive reviews in food science and food safety, vol 13, issue 4, pp 377–399. Blackwell Publishing Inc. https://doi.org/10.1111/1541-4337.12065
92. Graulet B, Girard CL (2017) B Vitamins in cow milk. In: Ronald Ross W, Collier RJ, Preedy VR (eds) Dairy in human health and disease across the lifespan. Academic Press, pp 211–224
93. Passos F, Solé M, García J, Ferrer I (2013) Biogas production from microalgae grown in wastewater: effect of microwave pretreatment. Appl Energy 108:168–175. https://doi.org/10.1016/J.APENERGY.2013.02.042
94. Passos F, Carretero J, Ferrer I (2015) Comparing pretreatment methods for improving microalgae anaerobic digestion: thermal, hydrothermal, microwave and ultrasound. Chem Eng J 279:667–672. https://doi.org/10.1016/J.CEJ.2015.05.065
95. Angles E, Jaouen P, Pruvost J, Marchal L (2017) Wet lipid extraction from the microalga Nannochloropsis sp.: disruption, physiological effects and solvent screening. Algal Res 21:27–34. https://doi.org/10.1016/J.ALGAL.2016.11.005

96. Ansari FA, Gupta SK, Shriwastav A, Guldhe A, Rawat I, Bux F (2017) Evaluation of various solvent systems for lipid extraction from wet microalgal biomass and its effects on primary metabolites of lipid-extracted biomass. Environ Sci Pollut Res 24(18):15299–15307. https://doi.org/10.1007/s11356-017-9040-3
97. Terigar BG, Balasubramanian S, Sabliov CM, Lima M, Boldor D (2011) Soybean and rice bran oil extraction in a continuous microwave system: from laboratory- to pilot-scale. J Food Eng 104(2):208–217. https://doi.org/10.1016/J.JFOODENG.2010.12.012
98. Petigny L, Périno S, Minuti M, Visinoni F, Wajsman J, Chemat F (2014) Simultaneous microwave extraction and separation of volatile and non-volatile organic compounds of boldo leaves from lab to industrial scale. Int J Mol Sci 15(5):7183–7198. https://doi.org/10.3390/ijms15057183
99. Lee JY, Yoo C, Jun SY, Ahn CY, Oh HM (2010) Comparison of several methods for effective lipid extraction from microalgae. Biores Technol 101(1):S75–S77. https://doi.org/10.1016/J.BIORTECH.2009.03.058
100. Ghasemi Naghdi F, González González LM, Chan W, Schenk PM (2016) Progress on lipid extraction from wet algal biomass for biodiesel production. In: Microbial biotechnology, vol 9, issue 6, pp 718–726. Wiley and Sons Ltd. https://doi.org/10.1111/1751-7915.12360
101. Voloshin RA, Rodionova MV, Zharmukhamedov SK, Nejat Veziroglu T, Allakhverdiev SI (2016) Review: biofuel production from plant and algal biomass. Int J Hydrogen Energy 41(39):17257–17273. https://doi.org/10.1016/J.IJHYDENE.2016.07.084
102. Kumar B, Verma P (2020) Life cycle assessment: blazing a trail for bioresources management. In: Energy conversion and management X. Elsevier, p 100063
103. Du Z, Li Y, Wang X, Wan Y, Chen Q, Wang C, Lin X, Liu Y, Chen P, Ruan R (2011) Microwave-assisted pyrolysis of microalgae for biofuel production. Biores Technol 102(7):4890–4896. https://doi.org/10.1016/J.BIORTECH.2011.01.055
104. Menéndez JA, Arenillas A, Fidalgo B, Fernández Y, Zubizarreta L, Calvo EG, Bermúdez JM (2010) Microwave heating processes involving carbon materials. Fuel Process Technol 91(1):1–8. https://doi.org/10.1016/J.FUPROC.2009.08.021
105. Azizi K, Keshavarz Moraveji M, Arregi A, Amutio M, Lopez G, Olazar M (2020) On the pyrolysis of different microalgae species in a conical spouted bed reactor: bio-fuel yields and characterization. Biores Technol 311:123561. https://doi.org/10.1016/J.BIORTECH.2020.123561
106. Sardi B, Ningrum RF, Ardianyah VA, Qadariyah L, Mahfud M (2022) Production of liquid biofuels from microalgae *Chlorella* sp. via catalytic slow pyrolysis. Int J Technol 13(1):147–156
107. Azizi K, Keshavarz Moraveji M, Abedini Najafabadi H (2018) A review on bio-fuel production from microalgal biomass by using pyrolysis method. Renew Sustain Energy Rev 82:3046–3059. https://doi.org/10.1016/J.RSER.2017.10.033
108. Ziolkowska JR, Simon L (2014) Recent developments and prospects for algae-based fuels in the US. Renew Sustain Energy Rev 29:847–853. https://doi.org/10.1016/J.RSER.2013.09.021
109. Fang P, Gong Z, Wang Z, Wang Z, Meng F (2019) Study on combustion and emission characteristics of microalgae and its extraction residue with TG-MS. Renew Energy 140:884–894. https://doi.org/10.1016/J.RENENE.2019.03.114
110. Menéndez JA, Domínguez A, Fernández Y, Pis JJ (2007) Evidence of self-gasification during the microwave-induced pyrolysis of coffee hulls. Energy Fuels 21(1):373–378. https://doi.org/10.1021/ef060331i
111. Pan P, Hu C, Yang W, Li Y, Dong L, Zhu L, Tong D, Qing R, Fan Y (2010) The direct pyrolysis and catalytic pyrolysis of *Nannochloropsis* sp. residue for renewable bio-oils. Bioresour Technol 101(12):4593–4599. https://doi.org/10.1016/J.BIORTECH.2010.01.070
112. Rajak U, Nashine P, Verma TN, Pugazhendhi A (2019) Alternating the environmental benefits of Aegle-diesel blends used in compression ignition. Fuel 256:115835. https://doi.org/10.1016/J.FUEL.2019.115835

113. Choudhary P, Assemany PP, Naaz F, Bhattacharya A, de Siqueira Castro J, do Couto Couto ED, Calijuri ML, Pant KK, Malik A (2020) A review of biochemical and thermochemical energy conversion routes of wastewater grown algal biomass. Sci Total Environ 726:137961. ISSN 0048-9697. https://doi.org/10.1016/j.scitotenv.2020.137961
114. Gong Z, Fang P, Wang Z, Li Q, Li X, Meng F, Zhang H, Liu L (2020) Catalytic pyrolysis of chemical extraction residue from microalgae biomass. Renew Energy 148:712–719. ISSN 0960-1481. https://doi.org/10.1016/j.renene.2019.10.158
115. Li F, Srivatsa SC, Bhattacharya S (2019) A review on catalytic pyrolysis of microalgae to high-quality bio-oil with low oxygeneous and nitrogenous compounds. Renew Sustain Energy Rev 108:481–497. ISSN 1364-0321. https://doi.org/10.1016/j.rser.2019.03.026
116. Vo TK, Ly HV, Lee OK, Lee EY, Kim CH, Seo JW, Kim J, Kim SS (2017) Pyrolysis characteristics and kinetics of microalgal *Aurantiochytrium* sp. KRS101. Energy 118:369–376. ISSN 0360-5442. https://doi.org/10.1016/j.energy.2016.12.040
117. Huang YF, Chiueh PT, Lo SL (2016) A review on microwave pyrolysis of lignocellulosic biomass. Sustain Environ Res 26(3):103–109. https://doi.org/10.1016/J.SERJ.2016.04.012
118. Zainan NH, Srivatsa SC, Li F, Bhattacharya S (2018) Quality of bio-oil from catalytic pyrolysis of microalgae *Chlorella vulgaris*. Fuel 223:12–19. https://doi.org/10.1016/J.FUEL.2018.02.166
119. Menéndez JA, Inguanzo M, Pis JJ (2002) Microwave-induced pyrolysis of sewage sludge. Water Res 36(13):3261–3264. https://doi.org/10.1016/S0043-1354(02)00017-9
120. Zuo W, Tian Y, Ren N (2011) The important role of microwave receptors in bio-fuel production by microwave-induced pyrolysis of sewage sludge. Waste Manage 31(6):1321–1326. https://doi.org/10.1016/J.WASMAN.2011.02.001
121. Tatke P, Jaiswal Y (2011) An overview of microwave assisted extraction and its applications in herbal drug research. Res J Med Plants 5(1):21–31. https://scialert.net/abstract/?doi=rjmp.2011.21.31
122. Sheehan J, Dunahay T, Benemann J, Roessler P (1978) A look back at the U.S. department of energy's aquatic species program: biodiesel from algae close-out report
123. Gao B, Chen A, Zhang W, Li A, Zhang C (2017) Co-production of lipids, eicosapentaenoic acid, fucoxanthin, and chrysolaminarin by *Phaeodactylum tricornutum* cultured in a flat-plate photobioreactor under varying nitrogen conditions. J Ocean Univ China 16(5):916–924. https://doi.org/10.1007/s11802-017-3174-2
124. Coward T, Fuentes-Grünewald C, Silkina A, Oatley-Radcliffe DL, Llewellyn G, Lovitt RW (2016) Utilising light-emitting diodes of specific narrow wavelengths for the optimization and co-production of multiple high-value compounds in *Porphyridium purpureum*. Biores Technol 221:607–615. https://doi.org/10.1016/J.BIORTECH.2016.09.093
125. Chua ET, Schenk PM (2017) A biorefinery for Nannochloropsis: Induction, harvesting, and extraction of EPA-rich oil and high-value protein. Biores Technol 244:1416–1424. https://doi.org/10.1016/J.BIORTECH.2017.05.124
126. Chen W, Sommerfeld M, Hu Q (2011) Microwave-assisted Nile red method for in vivo quantification of neutral lipids in microalgae. Biores Technol 102(1):135–141. https://doi.org/10.1016/J.BIORTECH.2010.06.076

Chapter 8
Overview of Radiation-Based Technologies and Implementation of Microwave Thermochemical Technologies in the Development of Biomass-Based Biorefinery

Ayusmita Ray, Somdutt Ravindran, and Amit Ghosh

Abstract Waste biomass valorization intends to play a crucial role in facilitating transitions toward green and sustainable development. Currently, a cascade of technologies broadly classified under thermochemical and biochemical processes are extensively studied to optimize resource efficiency and extract potentially beneficial biomolecules from multi-varied waste materials. Identification and subsequent adoption of optimal and economically viable biomass valorization methods are essential for ensuring the efficient synthesis of different classes of value-added products like biofuels, bio-oil, biochar, bioplastics, biofertilizers, biosurfactants, biomethane, biogas, etc. Currently, thermochemical processes like pyrolysis, hydrothermal liquefaction (HTL), torrefaction, and gasification are mostly used for biomass-to-biofuel technologies due to their higher efficacy and sustainability. However, instead of relying solely on conventional thermal sources to carry out biomass-to-biofuel conversion, irradiation can be employed to generate the necessary heat, resulting in a more sustainable bio-recovery. Irradiation-assisted conversion approaches offer the notable advantage of rapid and homogeneous heating while also ensuring higher product quality, thereby making them cost-effective. Additionally, single-step processes utilizing different chemical catalysts can be employed to effectively convert the products or by-products obtained from the preceding processes into chemically valuable compounds. Therefore, these valorization strategies offer direct and efficacious conversion of biomass to bio-products, adding value to the waste and

A. Ray · S. Ravindran · A. Ghosh (✉)
P.K. Sinha Centre for Bioenergy and Renewables, Indian Institute of Technology Kharagpur, Kharagpur 721302, West Bengal, India
e-mail: amitghosh@iitkgp.ac.in

A. Ghosh
School of Energy Science and Engineering, Indian Institute of Technology Kharagpur, Kharagpur 721302, West Bengal, India

© The Author(s), under exclusive license to Springer Nature Switzerland AG 2024
P. Verma and B. Kumar (eds.), *Sustainable Radiation Technologies in Waste-biomass Valorization*, Green Chemistry and Sustainable Technology,
https://doi.org/10.1007/978-3-031-63941-8_8

thereby contributing towards a circular bio-economy. In this chapter, we will comprehensively analyze the existing irradiation-based direct conversion techniques, elucidate their underlying mechanisms, and assess their effect on diverse biomasses for enhanced production of biofuel and value-added products.

Keywords Biomass · Valorization · Biofuels · Irradiation · Thermochemical · Circular bio-economy

8.1 Introduction

Worldwide economic progress and the ongoing rise in our living standards have led to a gradual but continuous rise in global energy demand [1]. The extensive exhaustion of resources and the increase in per capita waste generation have made waste management a crucial contention. The current global concern thus revolves around the effective utilization of both the biomass feedstock and any waste produced, thereby adding value to the waste. Waste biomass including agricultural wastes, municipal wastes, and industrial wastes, should be treated and recycled to produce value-added bio-compounds leading to a more economical bio-refinery [2]. Biomass-valorization involves various technologies categorized under two major classes, namely, biological and thermochemical. The thermochemical techniques operating at extremely high temperatures hold multiple advantages over biological methods for waste treatment as discussed below [3–5]

(1) Ability to handle a broad spectrum of waste biomass mainly lignocellulosic biomass (rice husk, sugarcane bagasse, June grass, etc.), which are not readily digested via biological enzymatic hydrolysis,
(2) Thermochemical processes naturally have a faster reaction kinetics as compared to biological techniques leading to rapid biomass conversion,
(3) Generation of high energy density value-added compounds via thermochemical valorization which are more suitable for energy production,
(4) The thermochemical method has reduced sensitivity towards biomass quality, hence extensive feedstock pre-treatment is not required. It is capable of successfully valorizing biomass with significant impurities as opposed to biological techniques,
(5) The efficiency/affectivity of the thermochemical technology is independent of contamination issues, or any other environmental conditions, unlike the biological method. Thermochemical processes operating at high temperatures can directly and effectively destroy hazardous contaminants.
(6) Irradiation technologies assisted with chemical treatment enhance biomass swelling leading to internal pores of higher volume, thereby increasing the biomass accessibility for enzymatic hydrolysis.
(7) It can generate a plethora of valuable chemicals from biomass conversion with biotechnological importance.

Currently, radiation-based biomass pre-treatment coupled with thermochemical valorization has emerged as a promising technique promoting an efficient circular bio-economy [6]. Optimization of any biomass valorization method will incur biomass pre-treatment as the initial step to facilitate increased accessibility of the biomass feedstock for further processing like fermentation and saccharification [7]. In this regard, non-conventional-based electron-beam-based irradiation-based pre-treatment processes have gained much attention due to their immense advantages over other conventional methods. MW radiation has emerged as one of the most promising irradiation technologies for biomass valorization leading to impressive delignification (up to 90%)as reported by studies conducted [8, 9]. Apart from MW radiation, ultrasonic waves and radioactive gamma rays have also shown immense potential biomass valorization, majorly lignocellulose [10–13] Electron beam irradiation (EBI) has demonstrated the potential to alter the chemical structure as well as the solid-state morphology of lignocellulosic wastes [12]. These irradiations facilitate the production of ions and/or radicals in the biomass triggering chemical reactions that lead to subsequent chemical bond breakage and reduced molecular mass with increased irradiation doses. Hence, non-conventional energies ensure maximum utilization of carbon without additional carbon emission, and this attribute may be exploited for a sustainable bio-refinery [12].

In the following sections, we will discuss in detail the different radiation-based valorization techniques that are adapted for efficient biomass-to-bioproduct conversion.

8.2 Analysis of Different Irradiation-Assisted Valorization Technologies

8.2.1 Microwave Irradiation

In the electromagnetic spectrum, microwave incurs a wavelength ranging between radio waves and infrared radiation. Microwave irradiation valorization operates by interfering with the recalcitrant structures of the biomass feedstock by causing selective heating on the polar parts in the aqueous environment of the biomass [14, 15]. The interaction between the feedstock and the microwave radiation leads to the movements and extensive collision of the polar molecules in the aqueous environment triggering various physicochemical reactions. In comparison with the conventional heating methods, MW irradiation involves energy transfer through the atomic or molecular, ensuring uniform heating and causing a more efficient recovery of the high-value-added chemicals [16–18]. Lignocellulosic biomasses like wheat straw, rice straw, rice hull, dried mango leaves, sugarcane bagasse, sweet sorghum bagasse, etc., have been successfully subjected to MW irradiation for improved devolatilization. Further, it also ensured reduced conversion time and higher heating rates as no time was wasted in heating the surrounding areas. MW energy is directly delivered into the biomass through the interaction of the molecules with the electromagnetic field. All in all, MW irradiation incurs various significant advantages over conventional heating methods, (1) rapid heating since the material gets heated first through dipolar rotation/ionic conduction without any initial interaction with the environment, (2) shorter reaction times with enhanced product yield, (3) direct heating of the material ensures lower energy input, (4) instant termination of electric current facilitates better control of the system, allowing for accurate calculation of the amount of power applied [19, 20].

A study reported almost 53% enhanced sugar yield from switchgrass using MW irradiation-based pre-treatment than that from conventional heating methods [12]. MW irradiation coupled with the deep eutectic solvent-assisted valorization method has also gained attention in ensuring efficient extraction of value-added chemicals [21–23]. Valorization techniques involving MW radiation-assisted chemical treatment, enzymatic hydrolysis, and eutectic solvent-based extraction for generation of high-value compounds and biofuels have been highlighted in Table 8.1 with reference to multiple lignocellulosic biomass feedstocks.

8.2.2 Ultrasonic Irradiation

This valorization method relies on the eco-friendly concept of cavitation, obtained from utilizing ultrasonic radiation on lignocellulosic materials. In the process, an ultrasonic wave is produced by an inducer, creating a pressure differential within the reaction system. As the ultrasound wave travels through the medium, it leads to the formation of cavitation bubbles filled with dissolved gas and liquid vapor [31]. These cavitation-induced shear forces act to break down the intricate network structure of lignocellulosic biomass (LCB), facilitating the extraction of targeted compounds like hemicellulose, cellulose, and/or lignin [31]. The parameters that influence the ultrasonic field include frequency, duration of reaction, choice of solvent, biomass characteristics, kinetics, and reactor configuration. Among these, optimizing the duration of ultrasonication plays a key role in the valorization process, with a maximum impact. The power of sonication is linked to its frequency and is a significant factor in the effectiveness of LCB pretreatment. A frequency range of 10–100 kHz is adequate for the cell breakage process, as beyond this range, it can have detrimental effects on the pretreatment process [32]. The utilization of low-frequency ultrasound (20–40 kHz) was initially implemented on a laboratory scale in the late 1960s [33]. In contemporary times, this technique is commonly employed at both pilot and large scales for the breakdown of sludge [34]. Ultrasonication is widely employed as a promising sustainable technology for cell disruption, and extraction of biocompounds (e.g., proteins, polysaccharides, and lipids) in food processing. Further, it was reported that the efficiency of ultrasonic pretreatment diminishes as the feedstock concentration increases from 2%–5 wt.% up to 10 wt.%, requiring significantly more power to achieve better results on LCB [35].

Table 8.1 Effects of microwave irradiation-based biomass valorisation

Feedstock	Process condition	Product formation	References
Wheat straw	Alkaline (2–5 wt.% NaOH which is about around 0.4 g/g biomass) pretreatment coupled with microwave irradiation (700 W/10–60 min; 60–140 °C)	A maximum of 73% xylan recovery was obtained. About 80% hemicellulose and 90% lignin removal in a short treatment time	Janker-Obermeier et al. [24]
Sugarcane bagasse	• MW radiation-assisted deep eutectic solvent (DES)/dimethyl sulfoxide (DMSO) system • Optimal reaction time of 40 min at 140 °C • Optimal DES type was ChCl: Acetic acid (1:2) whereas raw material: DMSO was 1:50	• Significant yields of 5-hydroxymethylfurfural, levulinic acid, and furfural: 31.29%, 28.38% and 35.65%, respectively	Amesho et al. [25]
	• Combined MW-alkali treatment at 600 W for 4 min with 1% NaOH followed by enzymatic hydrolysis • Combined microwave-alkali-acid treatment with 1% NaOH followed by 1% sulfuric acid at 450 W for 5 min	• MW-alkali treatment gave reducing sugar yield of 0.665 g/g dry biomass • Combined MW-alkali/acid treatment ensured reducing sugar yield of 0.83 g/g dry biomass which is 1.2-fold higher than that with only MW-alkali treatment • Resulted almost 90% of lignin removal from the bagasse	Binod et al. [9]
	• Optimal MW-assisted acid treatment includes a reaction temperature and acid-soaking time are 180 °C for 30 min, respectively, • Optimal acid (sulfuric acid) concentration is in the range of 0.005 M	• Bagasse degradation of 40–44% (wt) with 80–98% hemicellulose removal	Wei Hsin Chen et al. [26]

(continued)

Table 8.1 (continued)

Feedstock	Process condition	Product formation	References
Rice straw	• Optimal condition for MW-alkali coupled treatment with alkali concentration of 2.75%, followed by an MW irradiation time (200–800 W) of 22.50 min • 22.50 min with 30 g/L substrate concentration	• 65–72% rice straw saccharification. Optimal MW-irradiation condition—700 W for 6 min • 57–62% rice hull saccharification • Highest enzyme activities of Cellulases and xylanase reached the 6th day of fermentation • Increased hydrolysis rate with combined MW-alkali pretreatment	Singh et al. [27]
Chestnut shell waste (CSW)	• MW irradiation coupled deep eutectic solvent (DES) extraction system (MWDA) • Optimal delignification with 25% water in ChCl–oxalic acid dihydrate DES extraction for 30 min at 65 °C • Optimal MW irradiation (10, 20, 30 W) for 60 min at around 85 °C MW-irradiation	• 28.6±1.1% (w/w) Yield of polyphenols were extracted substantially improved with lignin reduction from 21 to 17%	González-Rivera et al. [22]
Coconut coir	• Microwave irradiation-assisted deep eutectic solvent (DES) based treatment: • The experiment was performed in an oil bath at 150 °C for 20 min using 0.5 g coconut coir and 15 g DES (ChCl:lactic acid = 1:4)	• Bio-oil yield from DES-extracted lignin was almost 38.8% • Highest lignin yield was noted to be about 82% with > 95% purity which is 6 times higher than that from conventional heating. The yield of polyphenols was substantially improved	Mankar et al. [28]
Grape pomace	• Microwave-assisted extraction (MAE): irradiation of 1000 W for 10 min	• 45% anthocyanin recovery	Da Rocha and Noreña [29]

(continued)

Table 8.1 (continued)

Feedstock	Process condition	Product formation	References
Grasses and fallen leaves, garden biomass	• MW-assisted natural deep eutectic solvent (NADES) system • NADES dilution of 49% maximizes phenolic extraction • Optimal MW irradiation of 800 W for 0.84 min	• MW-NADES displayed the highest polyphenolic recovery 762.11±12.61 μg/ 100 mg FWB with almost intact biological activities	Abdelrahman et al. [21]
Grasses and fallen leaves, garden biomass	• Mild acid pretreatment (5% H_2SO_4) with MW-heating (700 W) at 200 °C with a reaction time of 15 min	• Mild acid pretreatment coupled with MW-irradiation gave a cellulose recovery of 53.95% with a reducing sugar extraction of 46.97%	Gabhane et al. [30]

However, it was reported that ultrasound pre-treatment coupled with chemical reagents to the biomass change improves the solution system shear stress and mass transfer by altering the solution viscosity [36]. Ultrasound-assisted NaOH pre-treatment of corn straw displayed the highest biogas production of 510 mL/g with 6 wt.% CaO and 2 wt.% NaOH alkali and ultrasound irradiation at 50 °C for 30 min [37]. Furthermore, lignin degradation was increased by almost 18% with NaOH-coupled ultrasound- pre-treatment as compared to other conventional chemical pre-treatments without radiation [38]. Ultrasonic radiation has also been applied coupled with microwave irradiation for the isolation of hydrosols, solid residues, and essential oils, from plants such as lavender and rosemary [39]. Nevertheless, various research studies have highlighted certain constraints associated with ultrasound technology. The substantial energy released by ultrasound waves can lead to the destruction or depolymerization of some bioactive components, reducing their molar mass [40]. Additionally, undesirable aqueous-soluble components may be leached into the solvent, necessitating additional steps to eliminate impurities in subsequent processing [40]. Table 8.2 summarizes effects of ultrasonic radiation-assisted valorization of waste biomass.

8.2.3 Gamma-Ray Irradiation

Gamma rays, also regarded as high-energy photon rays generated by the radioactive decay of ^{60}Co and ^{137}Cs, are preferably utilized for biomass pre-treatment as compared to other radioactive isotopes [46]. It stands out as the most energetic electromagnetic radiation, characterized by an exceptionally short wavelength and high frequency (Fig. 8.1). The gamma rays work by transferring their energy to the atoms in the biomass feedstock, causing scission in the intra- and intermolecular bonds, lowering the polymerization degree of cell wall components, cellulose

Table 8.2 Effects of ultrasonic radiation-assisted valorization of waste biomass

Feedstock	Process conditions	Product formation	References
Glycerol	• Ultrasound-assisted activated clay catalyst-mediated • Rapid two-minute ultrasonic cavitation • 5% w/w catalyst loading (cyclohexanone)	90% conversion of glycerol into economically valuable hyacinth fragrance along with furan-based fuel additives	Pawar et al. [41]
Grape pomace	• Ultrasound-assisted extraction (UAE) for 10 min in aqueous ethanol	• A UAE resulted in a high yield of polyphenols (9.57 mg GAE (gallic acid equivalents)/g of dry pomace) • It is to be noted that conventional extraction procedures required almost 100 times longer extraction times	Goula et al. [42]
Pumpkin seeds	• Optimal ultrasound-assisted three-phase partitioning (UTPP)-PSO, PSP, PSPS • Salting by $(NH_4)_2SO_4$ addition of 30 g/100 mL, a t-butanol to slurry ratio of 1.0:1.0 (v/v) with a pH 5 • An ultrasonic power of 118 W, for 20 min, and a duty cycle of 60%	• Optimal extraction yields of Pumpkin seed oil, pumpkin seed protein (PSP), and pumpkin seed polysaccharides (PSPS) were 39.79, 14.30, and 1.97% with ultrasonication at 118 W for 20 min	Wang et al. [13]
Tomato processing wastes	• Optimal ultrasound-assisted extraction (UAE)—50 W, 65 W, 90 W for 1–30 min	• Maximum β-carotene of 4.42–4.90 mg/kg-Dw under UAE of 90W for 30 min • Lycopene 52.21–57.19 mg/kg-DwUAE of 90 W for 15 min	Yilmaz et al. [43]
Wheat bran	• Optimal ultrasound-assisted enzymatic extraction: at 50 °C for 70 min; and ultrasonic power 180 W	• Arabinoxylan yield was 42.6 ± 0.17 mg/g of wheat bran	Wang et al. [44]

(continued)

Table 8.2 (continued)

Feedstock	Process conditions	Product formation	References
Vegetable waste	• Optimal ultrasound and microwave-assisted Ternary deep eutectic solvents (US-MW-TDES) • US—30 min and MW-irradiation for 20 min-assisted TDES pretreatment	• Significant lignin removal from garlic skin (GS) and green onion root (GOR), reaching 90.14% and 92.34%, respectively • Cellulose content of GOR and GS was increased by 50.28% and 48.52%, respectively	Ji et al. [10]
Cassava starch	Optimal ultrasound radiation for 60 min at 40% of amplitude	• The highest sugar (reducing) concentration obtained was 0.114 mg/mL from cassava starch	Puspasari et al. [45]

and lignin, along with partial disruption of lignocellulosic complex materials [47]. Scanning electron microscopy (SEM) facilitated morphological studies comparing LCBs obtained from rice straw, bagasse, and corn stalk after subjecting them to gamma irradiation (dosage of 1000 kGy) along with non-irradiated sources [47]. The irradiated biomasses exhibited a broader distribution of lignin, increased porosity, and were consisting of small droplets, indicating structural changes in lignin at the 1000 kGy irradiation dose. Despite the interference of lignin droplets, the SEM analysis revealed that all three LCBs displayed uneven surfaces and significant reactive areas after exposure to gamma irradiation, crucial for further valorization steps. The radiation being transferred to the biomass through molecular collision leads to the formation of various cracks and irregularities on the matrix surface [48]. This gamma irradiation pre-treatment process addresses a significant challenge in industrial-scale bioethanol production from lignocellulose by degrading fibers and increasing the surface area necessary to access cellulose [49]. It was also reported by various studies that coupling gamma irradiation with an acid hydrolysis step for biomass pre-treatment resulted in an enhanced fermentable sugar yield (Glucose, xylose, arabinose, etc.) from the lignocellulosic feedstocks [50, 51]. The generation of gamma rays relies on radioactive isotopes, posing a constant risk of nuclear leakage and catastrophic events. Additionally, the use of nuclear-based technology may incur substantial expenses, this economic burden could render the adoption of such technology impractical and financially unsustainable. However, protective shields or suits are mandatory while handling such radioactive rays to ensure human safety. Biological shields which are radiation-absorbing materials are placed around the nuclear reactors to ensure maximized reduction in human exposure to the radiation. Examples of various waste biomass valorization facilitated by gamma-ray irradiation have been summarised in Table 8.3.

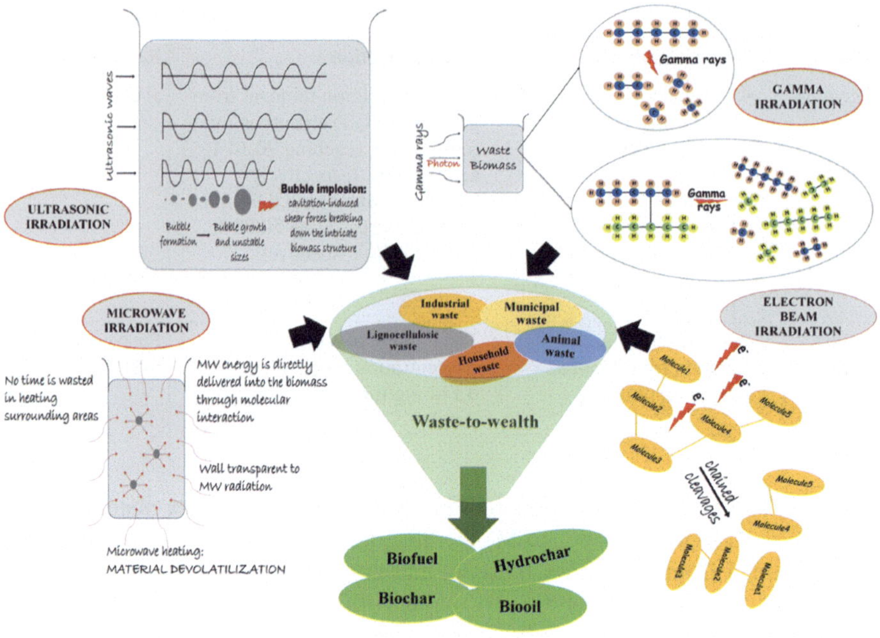

Fig. 8.1 Different radiation-assisted waste feedstock valorisation techniques

8.2.4 *Electron Beam Irradiation*

In electron beam irradiation, biomass feedstocks are exposed to rays of accelerated electrons, where on collision energy transfers between the electrons to the atoms in the biomass feedstocks (Fig. 8.1). The interaction of biomass with the accelerated electrons causes majorly chained cleavages in polymeric substances, where polymeric chains are cleaved at random positions in the backbone into at least two still highly molecular fragments [46]. However, the electron beam is majorly characterized by lower penetrative ability and simultaneous high doses, with a commercial ionizing energy range of 3–12 meV (million electron volts) [12].

Wood properties under electron beam irradiation have been analyzed with EBI ranging from 25–100 kGy doses which showed an increased wood hardness, on the contrary, a higher dose of 200 kGy resulted in an initial degradation process [53]. Spruce wood samples on exposure to 2 MGy EBI dose led to an almost 90% cellulose conversion to simple sugars [52]. Again, 80 kGy EBI pretreated rice straw at 0.12 mA and 1 meV upon enzymatic hydrolysis led to a glucose yield of 52.1% of the theoretical maximum after 132 h. The yield from pretreated rice was considerably higher than that obtained from untreated rice straw [54]. EBI has been evaluated for pretreatment of multiple grass varieties which has been further enzymatically hydrolyzed using *Trichoderma reesei* for 144 h. Significant improvements in the reducing sugar yield of almost 79% from 250 kGy EBI exposed samples were released within a few hours of hydrolysis [55]. It is also to be noted that the crystallinity level of the LCB also decreased with EBI treatment and simultaneously showed an indirect correlation between crystallinity degree and yield of glucose post enzymatic hydrolysis [47]. Recently, EBI was employed to generate composite biofilms from biopolymers such as polyvinylchloride, banana peels, and carboxymethyl cellulose-acting fillers, thereby promoting the bioremediation of dyes from wastewater. EBI (2.5 kGy/pass) was also studied for sugarcane bagasse valorization in comparison with gamma-ray

Table 8.3 Effects of gamma-ray irradiation-assisted valorization of waste biomass

Feedstock	Process condition	Product formation	References
Oil waste	• Gamma-ray assisted recovery via *Serratia marscens* N2 • Irradiation intensity 1000 Gy	• High yield of 19.2 g/L for biosurfactant (detergent). It was about onefold higher than that from non-radiated samples • 83.3% valorization of pre-used cooking oil was achieved after 6 days	Elkenawy and Gomaa [50]
Fish waste	• Direct gamma irradiation • 10 kGy gamma radiation	Biodegradable film production for food packaging by protein dispersion in fish waste	Mishra et al. [11]
Cellulose materials	• Direct gamma irradiation • Gamma-irradiation between 1.7–2 × 10^6 Gy dose levels • High efficacy when conducted before enzymatic cellulose conversion into simple sugars	60–80% glucose yield from the cellulose content after 42 h of enzymatic hydrolysis	Ardica et al. [52]
Rice straw	• Gamma ray irradiation (GRI)assisted alkali treatment coupled with microalgal growth • Radiation dose of 100 kGy with dose rate of 10 kGy/h	• Glucose and xylose saccharification yield from GRI-alkali treated biomass increased up to 91.65% and 98.84%, respectively • Maximum biomass and FAME concentrations of microalgae grown on rice straw hydrolysates were 6.51 g/L and 2.95 g/L, respectively	Joe et al. [51]
Sugarcane bagasse	• Direct gamma irradiation • Co^{60} radiation at 2.5 kGy/h for optimal output	Cellulose content reduced from 48 to 36% Hemicellulose content decreased from 31 to 16%. However, no considerable change in lignin content	Kapoor et al. [47]

irradiation, where it was concluded that gamma-ray assisted valorization facilitated higher biorecovery [47].

8.3 Different Microwave-Assisted Thermochemical Processes for Biomass to Biofuel Technology

8.3.1 Microwave-Assisted Pyrolysis

Microwave-assisted pyrolysis (MAP) involves the application of an electromagnetic wave's electric field component to heat biomass. During MAP it is not necessary for the feedstock to be dried externally and the heat transmission during the process takes place directly within the feedstock [56]. The absorption of microwave radiation by the feedstock is greatly influenced by its moisture content [57]. The efficiency of microwave pyrolysis heating is significantly higher compared to traditional pyrolysis methods that rely on electric resistance heating. This is due to the rapid and efficient volumetric heating achieved through microwave technology. Therefore, to enhance the overall effectiveness and productivity of bio-oil by increasing heating rates, it is necessary to utilize microwave ovens that operate at lower frequencies and have higher microwave output power [58]. The primary advantage of MAP technology is its ability to facilitate rapid thermochemical conversion reactions in large biomass materials. This technology ensures fast and uniform internal heating of the feedstock, enabling immediate responses during start-up and shut-down processes. Notably, MAP technology does not necessitate agitation for effective heat transfer [59–63]. MAP shows great potential as a technology in comparison to conventional pyrolysis processes. It offers a more gentle and medium-speed pyrolytic process [57]. Nevertheless, to commercialize this technology there are still some scale-up challenges that need further research. The bio-oil production during biomass MAP is affected by several parameters, such as the microwave output power and the ratio of feedstock to microwave absorber. In some reaction systems, a microwave absorber is necessary due to the limited ability of that biomass to absorb microwaves effectively. It is crucial to find the minimal quantity of absorber required to attain an ideal pyrolysis condition, since higher absorber-to-biomass ratio results in increased energy consumption per unit of biomass [60]. Also, the biochar attained from the pyrolysis of the biomass is known for its exceptional ability to absorb MW, making it a highly promising candidate for partial recycling. By reusing the produced char, heat can be recovered, leading to improved MW absorption and overall energy efficiency [60]. However, alternative MW absorbers to char include sulfuric acid, ionic liquids, activated carbons, SiC, metallic oxides, and ionic liquids have been studied. These absorbers are often used in conjunction with biomass to increase the yield or quality of bio-oil [64].

For example, the MAP of several feedstocks was conducted (corn stover, bamboo leaves, rice husk, rice straw, sugarcane peel, sugarcane bagasse, and waste coffee grounds), interestingly, the solid yields of all the above-mentioned feedstocks were found to be quite similar [65]. The type of feedstock has a notable influence on yields of liquids and gases, highlighting its importance to explore further in that aspect. The maize stover exhibited the least amount of liquid yield, while boasting the greatest gas yield of 40% [66]. On the other hand, rice husk demonstrated a significantly higher liquid yield (48 wt.%) compared to other materials, although its gas yield was relatively lower (30 wt.%) [67]. Additionally, it was also noticed that larger and bulkier biomass, such as wood, can undergo direct pyrolysis using MAP instead of requiring prior shredding into smaller fragments and subsequent pyrolysis under the conventional method. This is mainly due to the difference in the method used. In contrast to the traditional approach, where heat is transferred from the outer surface of the material to the interior via convection and/or conduction, MAP creates hotspots within the material [68]. Previous studies using microalgal feedstock for MAP have investigated the following parameters: temperature range: 450–825 °C, power supply range: 390–2250 W, and absorber: biomass (microalgae) ratios: 5–130% [59]. Under those circumstances, the bio-oils that were produced had their yield reaching up to 59 wt.%. A study was conducted to investigate the impact of temperature and the ratio of the absorber to biomass on the bio-oil fraction of *Chlorella* sp. and *Nannochloropsis* sp. using SiC (microwave absorber) and zeolite HZSM-5 MAP [67, 69]. The highest bio-oil yields achieved for *Chlorella* sp. and *Nannochloropsis* sp. were 57 wt.% and 59 wt.%, at temperatures of 550 °C and 500 °C, respectively, *Chlorella* sp. achieved this result without the need for an adsorbent, whereas *Nannochloropsis* sp. required a catalyst/biomass ratio was 0.5. The use of Zeolite HZSM-5 increased the overall water content of the bio-oil, leading to a decreased higher heating value (HHV). Similarly, in a study wherein *Chlorella* sp. was subjected to MAP, it resulted in a 28.6% bio-oil yield when the microwave output was set at 750 W and the final temperature reached approximately 480 °C [60]. The experiment revealed that the maximum char yield of 28% was attained using a power output of 500 W (362 °C). Whereas, the highest gas yield of 35% was obtained when using a power output of 1250 W (around 540 °C). The biochar yield ranged from 25 to 28%, while the yields of other products were 22–28.6% for bio-oil, 18–21% for the water phase, and 24–35% for gases. The GC–MS analysis revealed that the bio-oil consisted of 22.18%, of aliphatic and aromatic hydrocarbons. These hydrocarbons are highly sought after and are similar to the ones found in gasoline and diesel. The bio-oil also exhibited the following characteristics as pH level of 9.7, a density of 0.98 kg/L, and an HHV of 30.7 MJ/kg. In addition, it exhibited a lower oxygen content of 16.5 wt.% compared to bio-oil derived from LCB (33–39 wt.%) [60]. However, before utilizing it as engine fuel, it would be necessary to extract both N and O from it. This suggests that microwave-assisted pyrolysis has the potential to be a highly efficient technique for producing biochar.

Further, in a subsequent study, the sewage was subjected to a high heating temperature, in the presence of graphite as a microwave absorbent. The same conditions were also maintained in conventional pyrolysis for comparison. The bio-oil products derived through microwave pyrolysis exhibit an elevated calorific value ranging from

33.9 to 36.8 MJ/kg, along with a low concentration of polycyclic aromatic hydrocarbon (PAHs) [70]. Nevertheless, the bio-oil obtained by employing traditional high-temperature heating methods led to an elevated level of PAHs. The bio-oil obtained by treating sewage sludge with MAP exhibited significantly higher purity in comparison to the bio-oil derived through traditional pyrolysis [71]. The improved quality of bio-oil is due to the efficient heating and cooling capabilities of microwave pyrolysis [72]. Despite its numerous advantages over traditional fast pyrolysis, the MAP still faces difficulties with respect to the heating consistency, energy efficiency, and capital expenses. To overcome some of the heating-related challenges effective catalysts can be utilized. These catalysts enhance the absorption of microwaves, leading to more efficient heat distribution. Ultimately, this could contribute to the development of a more energy-efficient system. For example in a recent study, a spent lithium-ion battery was employed as a microwave-absorbing catalyst to further improve the production of biochar from rice straw. During this Microwave assisted catalytic pyrolysis (MACP), it was discovered that the solid product had a higher specific surface area, whilst the liquid product's had a higher calorific value. Overall, it was postulated that this process contributes to the development of a circular economy in which two waste items are recycled together synergistically to synthesize valuable products [73]. In another study, Techno-Economic Analysis (TEA) comparison was drawn between conventional biochar production (CPC) and microwave-assisted biochar production (MWPC). The findings revealed that the minimum selling price for CPC ranged from €436/tonne to €863/tonne, whereas for MWPC, it fell between €564/tonne and €979/tonne. Despite the slightly higher cost for MWPC, a 20% price increase, attributed to its higher quality, suggests that prioritizing aspects such as quality and other parameters over quantity could render microwave-assisted biochar production more economically viable [74]. Overall, additional research and development are required to tackle these challenges to make this technology highly commercial.

8.3.2 Hydrothermal Processes

8.3.2.1 Microwave Assisted Hydrothermal Liquefaction (MW-HTL)

Considering the advantages of microwave heating and the high absorption of water in microwave energy, the use of microwave-based heating to conduct the hydrothermal reactions for feedstock valorization have the potential to emerge as an industrially scalable promising technology (Fig. 8.2). Water possesses excellent microwave energy absorption properties and also serves as both the reaction medium and catalyst from a hydrothermal perspective. In regards to the first point, water at subcritical or supercritical (temperature and pressure > 374 °C and > 22.1 MPa respectively) conditions has a remarkable ability to dissolve organic compounds, making it ideal for biomass conversion [75, 76]. This technology uses water as an environmentally friendly processing agent combined with microwave irritation as an effective heating tool along with biomass as a renewable source of raw material. This combination of green chemistry and sustainable bio-refinery have the potential to make microwave-assisted hydrothermal liquefaction (MW-HTL) a promising method for producing biocrude. The properties of MW-HTL that have been identified include elevated conversion efficiency, lower processing temperatures, and the capability to handle wet and aqueous raw materials. The integration of hydrothermal conditions with microwave-assisted heating is considered a potent and efficient technology for achieving targeted and sustainable biomass conversion [77, 78].

In a study by Yang et al. [78], MW-HTL, focused on investigating the biomass model components to analyse the generation of biocrude (acids, alcohol, aldehyde, amide, amine, ester, furan, hydrocarbon, ketone, and phenol) under MW irradiation. The study used four biomass model components; protein, saccharide, lignin, and lipid which was statistically tailored established to analyse the intricate relationships among these elements. The highest percentage of biocrude was produced by lipids, followed by proteins, as single components. When saccharide and lipid were combined, they synergistically produced a higher amount of biocrude under the influence of MW radiation. This outcome was likely influenced by the presence of lipid decomposition intermediates. The biocrude yield was lower for saccharide but for other components the yield was comparable to the conventional method. In saccharide binary mixtures, the overall MW irradiation resulted in a lower production of biocrude, but a higher amount of solid residue (SR). MW irradiation may be the reason behind higher intensity of repolymerization of intermediates decomposed from saccharide, resulting in the formation of more SR. The biocrude yield of the protein-lipid mixture remained unaffected by the heating method. However, when it came to lignin-lipid, MW-HTL resulted in a higher biocrude yield but a lower

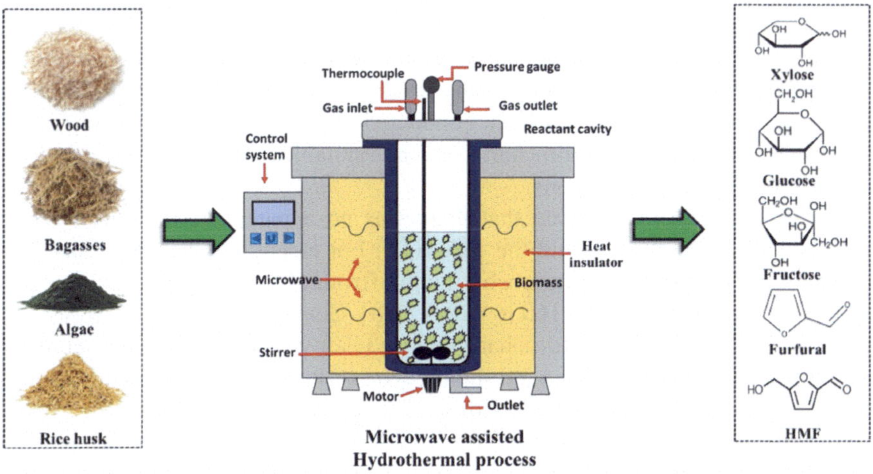

Fig. 8.2 Microwave assisted hydrothermal process for the transformation of the feedstock into value added chemicals

SR yield compared to traditional HTL. The heating method had minimal impact on both the model components and actual feedstock as far as of chemical composition of biocrude. In general, the impact of the heating method was closely linked to the substrates [78]. Currently analysis of these studies does not state a clear preference for one type of heating method over the other and more research is required optimize and identify unique set of conditions for using different feedstock.

8.3.2.2 Microwave Assisted Hydrothermal Treatment Carbonisation (MA-HTC)

Apart from MA-HTL, there are another microwave-assisted hydrothermal treatment technologies known as MA-HT carbonisation (MA-HTC), that facilitate the transformation of biomass into liquid and solid biofuels, as well as chemically enhanced products [79]. Microwave-assisted hydrothermal carbonization (MA-HTC) has garnered significant interest in recent times owing to its numerous benefits in biomass conversion. These attributes include efficient conversion, lower processing temperatures, and the ability to handle moist and water-based biomass [80]. The MA-HTC entails subjecting biomass to moderate temperatures and pressures for a short reaction duration. Also, it was seen that within the substrate throughout this process, there was a reduction in the proportions of hydrogen and oxygen. Simultaneously, the ratio of carbon (C) and higher heating value (HHV) rises, resulting in the formation of a solid substance called hydrochar, which is both high in energy content and carbon content [81–83]. Hydrochar derived from biomass possesses advantageous characteristics including a high energy density, elevated carbon content, strong mechanical properties, and a non-fibrous structure [84]. Hence, it possesses versatility for various potential applications such as absorbents, capacitors, fuels, catalysts, and filter aids. MA-HTC, in comparison to conventional hydrothermal carbonisation, offers advantages such as increased speed, enhanced controllability, environmental friendliness, time saving, energy efficiency, and technological efficiency [85, 86]. Owing to these features, extensive research has been conducted to employ this technology in the conversion of biomass into hydrochar, a resource suitable for both energy generation and various material applications.

Further, researchers observed that there was no apparent correlation between biomass and temperature within the MA-HTC. For example, when rice husk was utilized, a consistent drop in hydrochar output (from 58.6 to 49.5%) was found as the temperature climbed from 160 to 220 °C [83]. Conversely, when fish waste was employed, an initial rise in hydrochar yield (~34%) was noted at higher temperatures. Subsequently, it reached a threshold beyond which it gradually declined with further temperature increases [87]. However, in a specific case, researchers found that raising the temperature between 160 and 200 °C did not have a significant effect on hydrochar production when converting rice straw through the process of MA-HTC [88]. However, the discrepancies noted may stem from the diverse chemical compositions of the raw materials, underscoring the significance of biomass composition in influencing the process. A similar correlation between biomass chemical composition and reaction time was also observed with the MA-HTC [77]. Therefore, it is recommended to determine the reaction time experimentally for each specific biomass. In general, the optimal reaction time for most processes was found to be between 20 and 60 min.

In addition to temperature, catalyst, time, and the composition of the biomass, the size of the biomass particles also plays a crucial role in hydrochar production. It was observed that the hydrochar enhanced with an increase in the size of biomass particles. This is attributed to larger particles which encounter a reduction in the water surrounding their surface, leading to a more effective exposure to microwave radiation. Furthermore, during FTIR studies, it was observed that the hydrochar generated through MA-HTC of biomass exhibited better chemical and functional characteristics in comparison to the hydrochar produced through traditional hydrothermal methods [80]. In summary, this technology looks promising and in the long run with proper research, it can be commercialized.

8.3.2.3 Microwave-Assisted Hydrothermal Treatment Hydrolysis (MA-HTT)

MA-HTT presents an alternative approach for synthesizing value-added chemicals and biomaterials. The advantages of microwave-based heating, such as reduced time and energy consumption and increased product output, have been established. During the process, biomass undergoes decomposition, resulting in the formation of water-soluble chemicals with significant commercial value [77]. Liginocellulse-derived sugar-based bio-alcohol production has gathered a significant attention due to its inherent advantages, including renewability, cost-effectiveness, and environmental friendliness [89]. While biological fermentation of sugars is a potential method for bioenergy production, the recalcitrant structure of the feedstock poses significant challenge for the direct fermentation of biomass. The biomass has undergone a pre-treatment step to ensure that the polymeric sugar is readily available for the process. Various pre-treatment technologies have been utilized in recent years, including physical, chemical, physicochemical, and biological treatments [90]. Every approach possesses its own perks and drawbacks. In 1984, the feedstock underwent the MA-HTT process for the first time, serving as first stage in the process of breaking down of hardwoods and softwoods into smaller components [91]. Subsequently, further research has been carried out on this technology due to its efficacy and selection of products when compared with alternative technologies. In addition, it is more controlled, and energy and time-efficient process [76]. The remarkable control capability of this process enables the optimization of sugar yield while minimizing the presence of inhibitory molecules that hinder fermentation, such as furfural, HMF, acetic and formic acids, and phenolic compounds. This approach has been employed to generate saccharide-rich aqueous solution which can be used for fermentative operations.

Temperature plays a dual role in the process therefore it requires careful control and monitoring. On one hand, elevated temperatures facilitate depolymerization of polysaccharides into fermentable sugars. Nevertheless, these high temperatures also encourage further breakdown of the polysaccharides into secondary products, including furfural, HMF, and carboxylic acids, while promoting lignin depolymerization, leading to the production of phenolic compounds. Unfortunately, these components are known inhibitors for most yeasts which employ saccharide fermentation. Also, temperature management and maintenance is crucial for achieving an optimal yield of fermentable sugars. When brewer's spent grains (BSG) were subjected to MA-HTT to produce sugar it was revealed that the temperature of 192.7 °C and within 5.4 min were ideal. Under these conditions, it was feasible to recover 64w% of total hemicellulose sugars and 70% of the retrievable glucose in the enzymatic hydrolysate [92].

Similarly, the application of MA-HTT to aspen wood at 195 °C for 20 min resulted in the highest xylan yield, approximately 66.1. In another study the effect of MA-HTT on eucalyptus biomass catalysts was explored. An increment in the temperature had a positive impact on the xylose monomer yielding up to 49.58 mg g^{-1} from 2.52 mg g^{-1} when the temperature was risen to 190 °C from 150 °C. However, temperatures above 200 °C had a detrimental impact due to the gradual degradation of xylose. Despite this, they achieved an 89.2% of glucose conversion with a 64.9% solid recovery [93]. Further, in another study, sugarcane bagasse was subjected to MA-HTT. The process was carried out for 7 min at 170 ± 5 °C, with a biomass loading of 1:40 m/m under acidic catalysis, and they were able to achieve a sugar yield of 86% [9]. The carnauba leaves, macauba pulp, macauba shell, and pine nutshell were exposed to elevated temperatures (up to 200 °C) in the presence of a TiO_2 catalyst, resulting in an HMF/furfural yield of 53.24% [94]. During MW-HTT, corn stover was subjected to specific conditions, including temperature of 200 °C, time of 28 min, and biomass loading ratio of 1:100 m/m. in the presence of maleic acid, yielded 61% furfural [95]. These observations indicate that in MW-HTT, prolonged elevated temperatures are influencing the synthesis of furfurals from the released saccharides.

8.3.2.4 Techno-Economic Aspect of Microwave Radiation-Based Thermochemical Technologies

Techno-economic analysis (TEA) enables the evaluation of the economic viability of a technology and may also assist in identifying which part or/and material of the process is more sensitive to price [96]. Keeping this in mind many strategies can be implemented to make the process economical and profitable. Several considerations must be taken into account before constructing a biomass valorization unit based on MW. For instance, feedstock are significant recurring expense in these MW-thermochemical processes; therefore, it is critical to ensure while establishing the valorization unit the necessary feedstock varieties are readily available in large quantities throughout the year at a cheaper price to mitigate the overall process cost. In addition to the cost of acquiring feed material, there is another recurring expense associated with obtaining fresh catalysts. However, this cost can be substantially reduced by recovering and regenerating the used catalysts. Therefore, it is more advisable to go for an efficient, inexpensive catalyst with a higher regeneration or reusability capability. In a study, techno-economic analysis on MW-co-pyrolysis of food waste and low-density polyethylene was performed, wherein they estimated that the fixed cost would be 10,00,000 INR/kg and the monthly operational cost would be 8248 INR/kg at a lab scale. In addition, they estimated that the initial setup cost at a commercial scale (100 kg/day) would be 53,12,585 INR/kg, while the operational expenditures would amount to 24,420 INR/kg. Overall, they proposed that establishing a commercial MW-co-pyrolysis facility would result in improved conversion efficiency, and by selling biochar and bio-oil for 40/kg and achieving an internal recovery rate of 7%, the plant would reach its breakeven point within 4 years [97]. In another study, a techno-economic study was conducted on a small-scale microwave aided ex situ catalytic pyrolysis (MAECP) system, which indicated a potential annual profit of up to 135,494 USD [98]. The proposed hypothesis suggests that the thermochemical process, along with catalytic upgradation, transforms the feedstock into aromatic hydrocarbons. This presents an appealing opportunity for large-scale implementation, since pure fractions of aromatic hydrocarbons have higher value in comparison to mixed compounds. In addition, a sensitivity analysis was conducted, during which it was discovered that the return on investment is heavily dependent on bio-oil yield, bio-oil quality, selling price and feedstock price. One more re-occurring cost apart from feedstock is of procurement of the new catalysts which can be partially reduced by recovery and regeneration of the used catalysts. It was observed that the utilization of the MAP results in a substantial reduction in biomass preparation cost. However, the fluctuation in the initial cost of feedstock does have an impact on the total price. Moreover, it was indicated by the analysis that the overall production and quality of the bio-oil would have a substantial positive impact on the Return on investment (ROI) [98]. Further, an extensive analysis was carried out to examine the techno-economic feasibility aspects of a commercial-scale MAP for the production of activated carbon in the UK utilizing agricultural pea waste as a feedstock [99]. During this, they considered the locations that could potentially reduce transport, storage, and fixed and recurring costs. With an initial investment of £100 million, it was estimated that the total cost could be recovered within three years. The net present value was calculated to be £4,476,137,297.79, with the internal rate of return found to be 55% and the ROI being 52%. Within this context, the emphasis was placed on activated carbon rather than bio-oil, stating that bio-oil production is not viable based on baseline productions and that there are also issues regarding bio-oil pricing [99]. This indicates that irradiation-based technologies can not only be used for biomass valorisation for biofuel production but also be used for producing other valuable compounds, and it is advisable to prioritize more than one product in order to increase profitability. Nevertheless, the available studies on this topic are both limited and scattered. To gain a deeper understanding, it is essential that more extensive research be conducted, incorporating techno-economic aspects.

8.4 Thermochemical Products and Their Applications

8.4.1 Biofuels

Biofuels and various value-added fuel additives are majorly obtained from food, agricultural and industrial waste valorisation. Studies have revealed that microwave pretreatment (100–175 °C) assisted anaerobic digestion of food and kitchen wastes leads to biomethane generation, which is further used as a sustainable and green bioenergy fuel. Further, on coupling chemical/biological biomass (food waste) pretreatment with dark fermentation carried out by a diverse group of fermentative bacteria leads to the production of biohydrogen, the cleanest form of biofuel [100]. Moreover, agricultural wastes such as bagasse, corn cob when subjected to thermochemical pretreatment combined with hydrolysis also yields impressive amounts of biomethane, biohydrogen and bioethanol [101]. Microwave-assisted natural deep eutectic solvent extraction from spent coffee grounds yielded significant amounts of biobutanol along with other valuable chemicals such as gallic acid, catechin, etc.

8.4.2 Bio-char

Bio-char is a carbon-rich solid product (65–90% carbon) generated from thermal degradation of organic wastes under oxygen-limited or anaerobic conditions. Biochar acts as a low-cost, eco-friendly adsorbent for heavy metals, pesticides, dyes, microbial contaminants, where the properties may be improved with customization of pyrolysis process parameters and type of biomass used as its feedstock [102]. It is stated that biochar yields from aquatic biomass is considerably higher than that from terrestrial biomass as aquatic biomass incur more fixed carbon. Biochar has extensive applications in wastewater treatment, industrial wastewater bioremediation via adsorption of priority contaminants, catalytic degradation processes, and surface redox reactions. Biochar includes mechanisms such as electrostatic interactions, ion exchange, functional group exchange between the heavy metals or contaminants and the biosorbents in the biochar. Further, this leads to surface complexation or surface precipitation promoting the removal of environmental contaminants. The biochar characteristics may be customized depending on factors such as feedstock type, reaction time, heat transfer rate and temperature, however, irrespective of the biomass used, an increased temperature facilitates significantly increased surface area of the biochar. However, at a temperature above 900 °C, pore expansion and/or collision with the adjacent pores may result in reduced number of overall pores [103] They may be further manipulated via diverse physical or chemical activation treatments to artificially alter the morphology and composition of the biological cells, and significantly enhance its adsorption ability [103]. Physical activation includes treatment of raw biochar with steam or gas, improving its pore structure, size, volume and diverse chemical characteristics like hydrophobicity, polarity, etc. On the contrary, chemical activation of raw biochar involves the participation of chemicals (acids, alkalis, oxidants) resulting in a completely different form of biochar with new chemical groups on the surface [103]. Moreover, biochar apart from its role in wastewater treatment, also has been reported to be used as an additive in the biomethanation process to facilitate enhanced biological activity in the microorganisms [104]. Physically or chemically modified biochar are also studied extensively for its role as a catalyst in wide range of energy production reactions; cracking, transesterification, electrocatalysis etc., as well as in producing value-added chemicals which incurs oxidation, hydrolysis, isomerization, etc.

8.4.3 Bio-oil

Bio-oil is generated from majorly flash and fast pyrolysis of biomass, which is a dark brown liquid mixture of organic matters including simple sugar monomers and oligomers obtained from hydrolysed cellulose-hemicellulose, various sugar derivatives such as alcohols, carboxylic acids, ketones, esters, aldehydes, pyrolytic lignin and lignin derived phenolic compounds, along with high content of water [105]. However, it is noteworthy that moisture content of bio-oil is dependent on pyrolytic conditions and type of biomass feedstocks used where usually on an intact wet basis, water content is present between 50 and 60% (prior to drying) [105].

The composition and purity of bio-oil is dependent on; (i) biomass feedstock, (ii) heat transfer rate, (iii) final pyrolysis char temperature, (iv) residence time, (v) char removal efficiency, (vi) moisture content in biomass, (vii) duration and temperature of storage, and (viii) exposure to contamination during storage [106]. Bio-oil is derived from diverse biomass including, agro-wastes, bagasse, rice hull, wheat-rice straw, switchgrass, woods, etc., and has multivarious applications; it may act as an effective fuel oil or diesel substitute in engines, furnaces, and turbines for power generation. Furthermore, bio-oil can serve as a feedstock to produce many value-added biochemicals such as resins, adhesives, etc. Additionally, bio-oil may be utilized as a petrochemical alternative automobile fuel (after upgradation) and in the synthesis of anhydro-sugars (levoglucosan) which play a critical role in manufacturing biopolymers and pharmaceuticals [107–109].

8.4.4 Pyrolysis Gases

Pyrolysis promoting thermal decompositions of biomass at extensively higher temperatures results in the formation of various non-condensable toxic gases, majorly carbon monoxide (CO) apart from CO_2, H_2, CH_4 [106]. The degradation of lignocellulosic materials facilitates the breakdown of carbonyl and carboxyl functional groups in the three lignocellulosic components; lignin, cellulose, and hemicellulose. Lignin treatment at around 200 and 450 °C leads to rearrangements of its aromatic hydrocarbons leading to H_2 and CO_2 formation. On the other hand, cellulose pyrolysis at 325 and 400 °C undergoes demethylation causing CH_4 generation [106, 110]. Simultaneously, the degradation of the benzene ring in cellulose gives rise to CO and H_2. Other than these, C_2H_6 and C_1–C_4 hydrocarbons also constitute major components of the pyrolysis gas.

The type of product formation and its distribution majorly is dependent on the feedstock type, reaction conditions, and the conversion process. For instance, flash pyrolysis favors bio-oil production (liquid fraction), whereas slow pyrolysis is mainly employed for biochar (solid fraction) formation. However, formation of the gaseous fraction is significantly monitored by the reaction temperature and elemental composition of the feedstock used. Pyrolysis gases can be recycled into the pyrolysis reactor or may be used as an additional energy source to aid the fluidization process for bio-oil. The typical lower heating value (LHV) of pyrolysis gases ranges between 10 and 20 MJ/N m^3. Hence, it may be concluded that coupling the pyrolysis gas with the fluidizing stream within the pyrolysis reactor may facilitate increased hydrocarbon content in the produced bio-oil [106].

8.4.5 Hydrochar

Hydrochar is a valuable product of hydrothermal carbonization. It shares many similarities with biochar, but the main distinction lies in its production process and conditions, which involve a different pre-treatment. The biomass undergoes a complex series of chemical reactions, often resulting in the breakdown and decomposition of the intricate polymeric structure of the biomass-yielding hydrochar [111]. As the name implies, the composition of hydrochar consists of a combination of liquid (water) and solid char, resulting in a slurry-like consistency. In addition, hydrochar is composed of lower levels of alkali and alkaline earth elements, as well as heavy metals, while also having a HHV [112]. The carbonization process typically yields hydrochar, bio-oil, and gases, which can be utilized as biofuels or serve as precursors for other value-added chemicals. Hydrochar, the highly sought-after product of HTC, can yield anywhere from 40–70% [113]. Several studies have focused on optimizing various catalytic-mediated HTCs to increase the hydrochar yield and make the process more cost-effective by reducing the necessary temperature and pressure. The hydrochar surface is highly aromatized and contains numerous oxygen-containing groups. It is well-known that these oxygen-conducting groups have a strong affinity for water. Hydrochar has the potential to be used as a material for retaining water in the soil, also it has been studied to release nutrients gradually into the soil [114, 115]. Further exploration is needed towards this property in order to develop a hydrochar capable of progressively releasing water to sustain soil moisture over time. Furthermore, hydrochar typically does not auto-ignite because of the presence of these oxygen groups. In addition, hydrochar produced at higher temperatures, specifically 260 °C, exhibits combustion properties comparable to coal [116]. However, the production of this particular grade of hydrochar necessitates a significant amount of water, making the process less economically viable. It is observed that in hydrochar, the solid yield and H/C–O/C ratio decrease as the energy density increases, which can be attributed to longer reaction times and higher temperatures [117]. The solid-state $_{13}$C Nuclear-Magnetic-Resonance (NMR) technique provides valuable insight into the composition of hydrochar. In this case, it is revealed that approximately 60% of the carbon atoms in the sample are cross-linked directly through sp^2 or sp^3 carbon groups, resembling the structure of a furan ring [118]. The hydrochars particles have a porous structure, allowing them to be effectively combined with other furfurals to enhance the overall energy density. In general, hydrochar is a promising energy density product. However, further research is required to make it economical.

8.5 Future Prospects

Biorefineries derived from sustainable 2G LCBs have currently emerged as a research hotspot. The employment of physical pretreatment methods for biomass which includes exposure to different radiations (microwave, ultrasound, electron beam, etc.) coupled with thermochemical processes like pyrolysis, hydrothermal liquefaction, gasification, etc. for biomass conversion have proved to be the most promising strategy for an efficient waste to wealth conversion. Radiation assisted waste valorization helps in reducing the crystalline index of the cellulose in the LCB, thereby increasing accessibility to the active sites or the fermentable sugars. Furthermore, reports also showed that combining radiation with a chemical pretreatment, involving the use of chemical factors such as acids, alkalis, alkaline and metal salts, ionic liquids (ILs) and DES (deep eutectic solvent) facilitates improved degradation of lignocellulosic components in an aquatic medium and higher fractionation efficiency with reduced energy cost [119]. Unfortunately, implementation of these advance solvent systems may pose a significant financial challenge for involving high energy required for product extraction and purification, as well as solvent recycling. Therefore, the development of a cost-effective product separation technology is a pre-requisite for a sustainable biorefinery. It is also to be noted that, among the diverse thermochemical biomass conversion processes, pyrolysis has really emerged as one of the advantageous methods for waste treatment. Several innovative alterations in the process time (temporal), mass and heat transfer (thermodynamic), and functional domains may be explored for further process intensification. However, the working efficacy of these complex systems depends on various surrounding factors which makes the whole process time-consuming and labour-intensive. In this scenario, the machine learning approach has emerged as a powerful platform due to the gradual advancement of AI in solving non-linear problems (classification and regression). Presently, ML-algorithm-based predictive models enable the identification of the correlation patterns to optimize biofuel and value-added chemical production, by determining the optimal process parameters, efficient feedstock selection, and improving the overall production yields. Moreover, ML-algorithms also help in evaluating the purity of the products formed expressed in terms of its elemental composition. Further, to ensure a green refinery, studies are performed to explore microorganisms whose innate metabolic versatility can be exploited to valorize waste biomasses consisting of lignin, hemicellulose, etc. Fungi, bacteria, yeasts are studied extensively for their capability to produce hydrolyzing enzymes such as laccase, cellulase which can hydrolyze the complex sugars into simpler and fermentable forms which makes biofuel production easier. Moreover, metabolic engineering approaches to genetically alter the wild type strains with overexpressing target genes is one of the leading research fields focusing on eco-friendliness as well as cost-effectiveness [120–122]. Along with synthetic biology comes the integration of systems biology, mainly genome-scale metabolic modelling aiding in designing super strains for consolidated bioprocessing. Most importantly, to initiate a circular economy, incorporation of the utilization of the by-products from these radiation-assisted thermochemical processes into the life cycle assessment (LCA) studies is a prerequisite to evaluate these processes for a carbon-efficient and carbon–neutral resource circulation. Therefore, achievements of the prospects may enable LCB to drive a circular bio-economy.

8.6 Conclusion

Comprehensive knowledge regarding the constituents of various feedstocks and biomasses has been accumulated in the past few decades. As a consequence of this understanding and the transition towards green technology for biofuel or value added molecule production the researchers are now turning their attention towards the technologies that facilitate the complete bioconversion of biomass into biofuels or other valuable commodity chemicals. Among the emerging and promising technologies in this field, irradiation-based thermochemical technologies stand out. This process has the ability to achieve targeted and controllable yield. Given these promising features, this book chapter review analyzed different microwave irritation based thermochemical technologies and applications. This includes microwave-assisted pyrolysis, hydrothermal liquefaction, carbonization, and the treatment of various biomass for its conversion into value-added products such as hydrochar, bio-oil, saccharide solution, HMF, furfurals, and more. However, more research is required to establish industrial-scale irradiation-based biorefinary technology that can effectively compete with the existing traditional technologies. In the future, the biggest breakthrough would be the development of ML modulated self-regulating irradiation based single bioreactor technology capable of valorising all types of biomass on a continuous mode at an industrial scale.

Acknowledgements The authors are thankful to the Department of Science and Technology-GoI (Grant No. CRG/2020/002080), Department of Biotechnology-GoI (Grant No. BT/PR37958/GET/119/297/2020), and Scheme for Promotion of Academic and Research Collaboration (SPARC), MHRD-GoI (Grant No. SPARC/2019–2020/P1991/SL).

Declaration The authors have not used any AI tools during the writing of the Chapter.

References

1. Agrawal K, Bhardwaj N, Kumar B, Chaturvedi V, Verma P (2019) Process optimization, purification and characterization of alkaline stable white laccasse from *Myrothecium verrucaria* ITCC-8447 and its application in delignification of agroresidues. Int J Biol Macromol 125:1042–1055. https://doi.org/10.1016/j.ijbiomac.2018.12.108
2. Kumar B, Verma P (2020) Biomass-based biorefineries: an important architype towards a circular economy. Elsevier, Fuel, p 119622
3. Bhatt M, Chakinala AG, Joshi JB, Sharma A, Pant KK, Shah K, Sharma A (2021) Valorization of solid waste using advanced thermo-chemical process: a review. J Environ Chem Eng 9(4):105434. https://doi.org/10.1016/j.jece.2021.105434
4. Durak H (2023) Comprehensive assessment of thermochemical processes for sustainable waste management and resource recovery. Processes 11(7). https://doi.org/10.3390/pr11072092
5. Okolie JA, Epelle EI, Tabat ME, Orivri U, Amenaghawon AN, Okoye PU, Gunes B (2022) Waste biomass valorization for the production of biofuels and value-added products: a comprehensive review of thermochemical, biological and integrated processes. Process Saf Environ Prot 159:323–344. https://doi.org/10.1016/j.psep.2021.12.049
6. Venkatesh G (2022) Circular bio-economy—paradigm for the future: systematic review of scientific journal publications from 2015 to 2021. In: Circular economy and sustainability, vol 2. https://doi.org/10.1007/s43615-021-00084-3
7. Kumar B, Bhardwaj N, Agrawal K, Chaturvedi V, Verma P (2020) Current perspective on pretreatment technologies using lignocellulosic biomass: an emerging biorefinery concept. Fuel Process Technol 199:106244. https://doi.org/10.1016/j.fuproc.2019.106244
8. Bhardwaj N, Kumar B, Verma P (2020) Microwave-assisted pretreatment using alkali metal salt in combination with orthophosphoric acid for generation of enhanced sugar and bioethanol. Biomass Convers Biorefinery. https://doi.org/10.1007/s13399-020-00640-1
9. Binod P, Satyanagalakshmi K, Sindhu R, Janu KU, Sukumaran RK, Pandey A (2012) Short duration microwave assisted pretreatment enhances the enzymatic saccharification and fermentable sugar yield from sugarcane bagasse. Renew Energy 37(1):109–116. https://doi.org/10.1016/j.renene.2011.06.007
10. Ji Q, Yu X, Yagoub AEGA, Chen L, Zhou C (2020) Efficient removal of lignin from vegetable wastes by ultrasonic and microwave-assisted treatment with ternary deep eutectic solvent. Ind Crops Prod 149(February):112357. https://doi.org/10.1016/j.indcrop.2020.112357
11. Mishra PK, Gautam RK, Kumar V, Kakatkar AS, Chatterjee S (2021) Synthesis of biodegradable films using gamma irradiation from fish waste. Waste Biomass Valorization 12(5):2247–2257. https://doi.org/10.1007/s12649-020-01143-w
12. Singh R, Krishna BB, Kumar J, Bhaskar T (2016) Opportunities for utilization of non-conventional energy sources for biomass pretreatment. Biores Technol 199:398–407. https://doi.org/10.1016/j.biortech.2015.08.117
13. Wang H, Chen K, Cheng J, Jiang L, Yu D, Dai Y, Wang L (2021) Ultrasound-assisted three phase partitioning for simultaneous extraction of oil, protein and polysaccharide from pumpkin seeds. Lwt 151(July):112200. https://doi.org/10.1016/j.lwt.2021.112200

14. Foong SY, Liew RK, Yang Y, Cheng YW, Yek PNY, Wan Mahari WA, Lam SS, et al (2020) Valorization of biomass waste to engineered activated biochar by microwave pyrolysis: progress, challenges, and future directions. Chem Eng J 389(February):124401. https://doi.org/10.1016/j.cej.2020.124401
15. Kumar B, Bhardwaj N, Verma P (2019) Pretreatment of rice straw using microwave assisted $FeCl_3$–H_3PO_4 system for ethanol and oligosaccharides generation. Bioresour Technol Rep 7(2019):100295. https://doi.org/10.1016/j.biteb.2019.100295
16. Arpia AA, Chen WH, Lam SS, Rousset P, de Luna MDG (2021) Sustainable biofuel and bioenergy production from biomass waste residues using microwave-assisted heating: a comprehensive review. Chem Eng J 403(April 2020):126233. https://doi.org/10.1016/j.cej.2020.126233
17. Bhardwaj N, Kumar B, Verma P (2022) Microwave-assisted pretreatment using alkali metal salt in combination with orthophosphoric acid for generation of enhanced sugar and bioethanol. Biomass Convers Biorefinery 12:1069–1076. https://doi.org/10.1007/s13399-020-00640-1
18. Kumar B, Verma P (2019b) Optimization of microwave-assisted pretreatment of rice straw with $FeCl_3$ in combination with H_3PO_4 for improving enzymatic hydrolysis. Adv Plant Microb Biotechnol 41–48.https://doi.org/10.1007/978-981-13-6321-4_7
19. Kumar B, Bhardwaj N, Verma P (2020) Microwave assisted transition metal salt and orthophosphoric acid pretreatment systems: generation of bioethanol and xylo-oligosaccharides. Renew Energy 158:574–584. https://doi.org/10.1016/j.renene.2020.05.006
20. Usmani Z, Sharma M, Tripathi M, Nizami AS, Gong L, Nguyen QD, Reddy MS, Thakur VK, Gupta VK (2023) Converting biowaste streams into energy–leveraging microwave assisted valorization technologies for enhanced conversion. J Energy Inst 107(November 2022):101161. https://doi.org/10.1016/j.joei.2022.101161
21. Abdelrahman R, Hamdi M, Baba, WN, Hassan HM, Maqsood S (2023) Synergistic combination of natural deep eutectic solvents and green extraction techniques for the valorization of date palm leaves: optimization of the solvent-biomass interaction. Microchem J 195(September):109503. https://doi.org/10.1016/j.microc.2023.109503
22. González-Rivera J, Mero A, Husanu E, Mezzetta A, Ferrari C, D'Andrea F, Bramanti E, Pomelli CS, Guazzelli L (2021) Combining acid-based deep eutectic solvents and microwave irradiation for improved chestnut shell waste valorization. Green Chem 23(24):10101–10115. https://doi.org/10.1039/d1gc03450b
23. Rachiero GP, Berton P, Shamshina J (2022) Deep eutectic solvents: alternative solvents for biomass-based waste valorization. Molecules 27(19):1–14. https://doi.org/10.3390/molecules27196606
24. Janker-Obermeier I, Sieber V, Faulstich M, Schieder D (2012) Solubilization of hemicellulose and lignin from wheat straw through microwave-assisted alkali treatment. Ind Crops Prod 39(1):198–203. https://doi.org/10.1016/j.indcrop.2012.02.022
25. Amesho KTT, Cheng PC, Chang KL, Peng YP, Jhang SR, Lin YC (2022) Microwave-assisted deep eutectic solvents/dimethyl sulfoxide system for efficient valorization of sugar bagasse waste into platform chemicals: a biorefinery approach for circular bioeconomy. Bioresour Technol 363(September):127969. https://doi.org/10.1016/j.biortech.2022.127969
26. Chen WH, Ye SC, Sheen HK (2012) Hydrolysis characteristics of sugarcane bagasse pretreated by dilute acid solution in a microwave irradiation environment. Appl Energy 93:237–244
27. Singh A, Tuteja S, Singh N, Bishnoi NR (2011) Enhanced saccharification of rice straw and hull by microwave-alkali pretreatment and lignocellulolytic enzyme production. Bioresour Technol 102(2):1773–1782. https://doi.org/10.1016/j.biortech.2010.08.113
28. Mankar AR, Pandey A, Pant KK (2022) Microwave-assisted extraction of lignin from coconut coir using deep eutectic solvents and its valorization to aromatics. Bioresour Technol 345(October 2021):126528. https://doi.org/10.1016/j.biortech.2021.126528
29. Da Rocha CB, Noreña CPZ (2020) Microwave-assisted extraction and ultrasound-assisted extraction of bioactive compounds from grape pomace. Int J Food Eng 16(1–2):1–10. https://doi.org/10.1515/ijfe-2019-0191

30. Gabhane J, Prince William SPM, Vaidya AN, Mahapatra K, Chakrabarti T (2011) Influence of heating source on the efficacy of lignocellulosic pretreatment—A cellulosic ethanol perspective. Biomass Bioenergy 35(1):96–102. https://doi.org/10.1016/j.biombioe.2010.08.026
31. Flores EMM, Cravotto G, Bizzi CA, Santos D, Iop GD (2021) Ultrasound-assisted biomass valorization to industrial interesting products: state-of-the-art, perspectives and challenges. Ultrason Sonochem 72:105455. https://doi.org/10.1016/j.ultsonch.2020.105455
32. Purkait MK, Haldar D (2021) Conventional pretreatment methods of lignocellulosic biomass. In: Lignocellulosic biomass to value-added products, pp 31–46.https://doi.org/10.1016/b978-0-12-823534-8.00009-0
33. Swinwood JF, Waite TD, Kruger P, Rao SM (1994) Radiation technologies for waste treatment: a global perspective. IAEA Bull 36(1):11–15
34. Liu C, Xiao B, Dauta A, Peng G, Liu S, Hu Z (2009) Effect of low power ultrasonic radiation on anaerobic biodegradability of sewage sludge. Biores Technol 100(24):6217–6222. https://doi.org/10.1016/j.biortech.2009.07.001
35. Ma S, Li Y, Li J, Yu X, Cui Z, Yuan X, Zhu W, Wang H (2022) Features of single and combined technologies for lignocellulose pretreatment to enhance biomethane production. Renew Sustain Energy Rev 165(May):112606. https://doi.org/10.1016/j.rser.2022.112606
36. Bussemaker MJ, Zhang D (2013) Effect of ultrasound on lignocellulosic biomass as a pretreatment for biorefinery and biofuel applications. Ind Eng Chem Res 52(10):3563–3580. https://doi.org/10.1021/ie3022785
37. You Z, Pan SY, Sun N, Kim H, Chiang PC (2019) Enhanced corn-stover fermentation for biogas production by NaOH pretreatment with CaO additive and ultrasound. J Cleaner Prod 238:117813. https://doi.org/10.1016/j.jclepro.2019.117813
38. Elalami D, Carrere H, Abdelouahdi K, Garcia-Bernet D, Peydecastaing J, Vaca-Medina G, Oukarroum A, Zeroual Y, Barakat A (2020) Mild microwaves, ultrasonic and alkaline pretreatments for improving methane production: Impact on biochemical and structural properties of olive pomace. Bioresour Technol 299(December 2019):122591. https://doi.org/10.1016/j.biortech.2019.122591
39. Gonzalez-Rivera J, Campanella B, Pulidori E, Bramanti E, Tiné MR, Bernazzani L, Onor M, Bàrberi P, Duce C, Ferrari C (2023) From volatiles to solid wastes: towards the full valorization of lavender and rosemary by simultaneous in situ microwaves and ultrasounds irradiation extraction. Ind Crops Prod 194(February). https://doi.org/10.1016/j.indcrop.2023.116362
40. Liu Y, Liu X, Cui Y, Yuan W (2022) Ultrasound for microalgal cell disruption and product extraction: a review. Ultrason Sonochem 87(May):106054. https://doi.org/10.1016/j.ultsonch.2022.106054
41. Pawar RR, Gosai KA, Bhatt AS, Kumaresan S, Lee SM, Bajaj HC (2015) Clay catalysed rapid valorization of glycerol towards cyclic acetals and ketals. RSC Adv 5(102):83985–83996. https://doi.org/10.1039/c5ra15817f
42. Goula AM, Thymiatis K, Kaderides K (2016) Valorization of grape pomace: drying behavior and ultrasound extraction of phenolics. Food Bioprod Process 100:132–144. https://doi.org/10.1016/j.fbp.2016.06.016
43. Yilmaz T, Kumcuoglu S, Tavman S (2017) Ultrasound-assisted extraction of lycopene and β-carotene from tomato-processing wastes. Ital J Food Sci 29(1):186–194
44. Wang J, Sun B, Liu Y, Zhang H (2014) Optimisation of ultrasound-assisted enzymatic extraction of arabinoxylan from wheat bran. Food Chem 150:482–488. https://doi.org/10.1016/j.foodchem.2013.10.121
45. Puspasari F, Airlangga B, Trisanti PN (2019) Effect of shear rate in high shear mixing process on the structure of cassava starch granule and reducing sugar product. In: IOP conference series: materials science and engineering, vol 673, no 1. IOP Publishing, p 012135
46. Basak B, Kumar R, Bharadwaj AVSLS, Kim TH, Kim JR, Jang M, Jeon BH, et al (2023) Advances in physicochemical pretreatment strategies for lignocellulose biomass and their effectiveness in bioconversion for biofuel production. Bioresour Technol 369(September 2022):128413. https://doi.org/10.1016/j.biortech.2022.128413

47. Kapoor K, Tyagi AK, Das M, Kumar V (2023) Comparative analysis of morphological and structural changes in gamma and electron beam irradiated sugarcane bagasse. Cellul Chem Technol 57(1–2):61–70. https://doi.org/10.35812/CelluloseChemTechnol.2023.57.06
48. Butnaru E, Darie-Niţă RN, Zaharescu T, Balaeş T, Tănase C, Hitruc G, Vasile C, et al (2016) Gamma irradiation assisted fungal degradation of the polypropylene/biomass composites. Radiat Phys Chem 125:134–144. https://doi.org/10.1016/j.radphyschem.2016.04.003
49. Chen WH, Pen BL, Yu CT, Hwang WS (2011) Pretreatment efficiency and structural characterization of rice straw by an integrated process of dilute-acid and steam explosion for bioethanol production. Bioresour Technol 102(3):2916–2924. https://doi.org/10.1016/j.biortech.2010.11.052
50. Elkenawy NM, Gomaa OM (2022) Valorization of frying oil waste for biodetergent production using *Serratia marcescens* N2 and gamma irradiation assisted biorecovery. Microb Cell Fact 21(1):1–20. https://doi.org/10.1186/s12934-022-01877-3
51. Joe MH, Kim JY, Lim S, Kim DH, Bai S, Park H, Choi JIl, et al (2015) Microalgal lipid production using the hydrolysates of rice straw pretreated with gamma irradiation and alkali solution. Biotechnol Biofuels 8(1):1–9.https://doi.org/10.1186/s13068-015-0308-x
52. Ardica S, Calderaro E, Cappadona C (1985) Radiation pretreatments of cellulose.pdf, 26
53. Schnabel T, Huber H, Grünewald TA, Petutschnigg A (2015) Changes in mechanical and chemical wood properties by electron beam irradiation. Appl Surf Sci 332:704–709. https://doi.org/10.1016/j.apsusc.2015.01.142
54. Bak JS, Ko JK, Han YH, Lee BC, Choi IG, Kim KH (2009) Improved enzymatic hydrolysis yield of rice straw using electron beam irradiation pretreatment. Bioresour Technol 100(3):1285–1290. https://doi.org/10.1016/j.biortech.2008.09.010
55. Karthika K, Arun AB, Rekha PD (2012) Enzymatic hydrolysis and characterization of lignocellulosic biomass exposed to electron beam irradiation. Carbohydr Polym 90(2):1038–1045. https://doi.org/10.1016/j.carbpol.2012.06.040
56. Kurian V, Gill M, Dhakal B, Kumar A (2022) Recent trends in the pyrolysis and gasification of lignocellulosic biomass. In: Biofuels and bioenergy. https://doi.org/10.1016/B978-0-323-90040-9.00028-X
57. Bi Z, He BB (2022) 26—Biochar from microalgae. 3rd Generation Biofuels. Elsevier Inc. https://doi.org/10.1016/B978-0-323-90971-6.00025-5
58. Fermoso J, Coronado JM, Serrano DP, Pizarro P (2017) Pyrolysis of microalgae for fuel production. In: Microalgae-based biofuels and bioproducts. Elsevier Ltd. https://doi.org/10.1016/B978-0-08-101023-5.00011-X
59. Cabral F, Du Z, Xie Q, Otávio J, Cheng Y, Wan Y, Ruan R, et al (2014) Fast microwave assisted pyrolysis of biomass using microwave absorbent. Bioresour Technol 156:267–274.https://doi.org/10.1016/j.biortech.2014.01.038
60. Du Z, Li Y, Wang X, Wan Y, Chen Q, Wang C, Ruan R, et al (2011) Microwave-assisted pyrolysis of microalgae for biofuel production. Bioresour Technol 102(7):4890–4896.https://doi.org/10.1016/j.biortech.2011.01.055
61. Fu W, Dai J, Zhang Y, Guang M, Liu Y, Li B (2021) Heating performances of high density polyethylene (HDPE) plastic particles in a microwave chamber. Sustain Energy Technol Assess 48(October):101581. https://doi.org/10.1016/j.seta.2021.101581
62. Li S, Li C, Shao Z (2022) Microwave pyrolysis of sludge: a review
63. Zhou N, Dai L, Lv Y, Li H, Deng W, Guo F, Ruan R, et al (2021) Catalytic pyrolysis of plastic wastes in a continuous microwave assisted pyrolysis system for fuel production. Chem Eng J 418(December 2020):129412. https://doi.org/10.1016/j.cej.2021.129412
64. Salema AA, Ani FN (2012) Microwave-assisted pyrolysis of oil palm shell biomass using an overhead stirrer. J Anal Appl Pyrol 96:162–172. https://doi.org/10.1016/j.jaap.2012.03.018
65. Huang Y, Pina A, Ferrão P, Fournier J, Lacarrière B, Corre OL (2018) Product distribution and heating performance of lignocellulosic biomass pyrolysis using microwave heating. Energy Proc 152:910–915. https://doi.org/10.1016/j.egypro.2018.09.092
66. Babich IV, Van Der Hulst M, Lefferts L, Moulijn JA, Connor PO, Seshan K (2011) Catalytic pyrolysis of microalgae to high-quality liquid. Biomass Bioenerg 35(7):3199–3207.https://doi.org/10.1016/j.biombioe.2011.04.043

67. Cabral F, Xie Q, Min M, Antônio L, Muniz R, Farenzena M, Ruan R, et al (2014) Fast microwave-assisted pyrolysis of microalgae using microwave absorbent and HZSM-5 catalyst. Bioresour Technol 166:518–526.https://doi.org/10.1016/j.biortech.2014.05.100
68. Prashanth PF (2022) Microwave-assisted torrefaction and pyrolysis of rice straw pellets for bioenergy (February):2964–2977. https://doi.org/10.1049/rpg2.12445
69. Abubakar Z, Ani FN (2013) Microwave-assisted pyrolysis of oil palm shell biomass (36):19–30
70. Menendez JA (2003) Gas chromatographic—mass spectrometric study of the oil fractions produced by microwave-assisted pyrolysis of different sewage sludges, 1012:193–206
71. Domínguez A, Menéndez JA, Inguanzo M, Pis JJ (2006) Production of bio-fuels by high temperature pyrolysis of sewage sludge using conventional and microwave heating. Bioresour Technol 97:1185–1193.https://doi.org/10.1016/j.biortech.2005.05.011
72. Asomaning J, Haupt S, Chae M, Bressler DC (2018) Recent developments in microwave-assisted thermal conversion of biomass for fuels and chemicals. Renew Sustain Energy Rev 92(March):642–657. https://doi.org/10.1016/j.rser.2018.04.084
73. He M, Zhao J, Wang D, Liang Q, Wang T, Zhao S, Ma W (2024) Microwave-assisted catalytic pyrolysis of biomass with biochar materials derived from spent lithium-ion batteries: microwave absorption and pyrolysis characteristics. J Environ Chem Eng 12(2):112099. https://doi.org/10.1016/j.jece.2024.112099
74. Haeldermans T, Campion L, Kuppens T, Vanreppelen K, Cuypers A, Schreurs S (2020) Bioresource technology a comparative techno-economic assessment of biochar production from different residue streams using conventional and microwave pyrolysis. Biores Technol 318(June):124083. https://doi.org/10.1016/j.biortech.2020.124083
75. Antonetti C, Bonari E, Licursi D, Nassi N, Maria A, Galletti R (2015) Hydrothermal conversion of giant reed to furfural and levulinic acid: optimization of the process under microwave irradiation and investigation of distinctive agronomic parameters, 21232–21253. https://doi.org/10.3390/molecules201219760
76. Tsubaki S, Oono K, Onda A, Yanagisawa K, Azuma J (2012) Bioresource technology microwave-assisted hydrothermal hydrolysis of cellobiose and effects of additions of halide salts. Biores Technol 123:703–706. https://doi.org/10.1016/j.biortech.2012.07.086
77. Gao Y, Remón J (2021) Microwave-assisted hydrothermal treatments for biomass valorisation: a critical review, pp 3502–3525. https://doi.org/10.1039/d1gc00623a
78. Yang J, Sophia Q, Niu H, Dalai A, Corscadden K, Zhou N (2020) Microwave-assisted hydrothermal liquefaction of biomass model components and comparison with conventional heating. Fuel 277(April):118202. https://doi.org/10.1016/j.fuel.2020.118202
79. Lorente A, Remo J, Clark JH (2020) Sustainable production of solid biofuels and biomaterials by microwave-assisted, hydrothermal carbonization (MA-HTC) of Brewers' spent grain (BSG). https://doi.org/10.1021/acssuschemeng.0c06853
80. Gao Y, Liu Y, Zhu G, Xu J, Yuan Q, Zhu Y, Ji L, et al (2018) Microwave-assisted hydrothermal carbonization of dairy manure: chemical and structural properties of the products. Energy 165:662–672.https://doi.org/10.1016/j.energy.2018.09.185
81. Funke A, Ziegler F, Berlin TU (2010) Hydrothermal carbonization of biomass : a summary and discussion of chemical mechanisms for process engineering, pp 160–177. https://doi.org/10.1002/bbb
82. Hoekman SK, Broch A, Robbins C (2011) Hydrothermal carbonization (HTC) of lignocellulosic biomass, pp 1802–1810
83. Nizamuddin S, Baloch HA, Siddiqui M, Mubarak N, Tunio M, Bhutto A, Jatoi AS, et al (2019) Microwave hydrothermal carbonization of rice straw: optimization of process parameters and upgrading of chemical, fuel, structural and thermal properties. Mat 12(3):403. https://doi.org/10.3390/ma12030403
84. Elaigwu SE, Greenway GM (2016) Microwave-assisted and conventional hydrothermal carbonization of lignocellulosic waste material: comparison of the chemical and structural properties of the hydrochars. J Anal Appl Pyrol 118:1–8. https://doi.org/10.1016/j.jaap.2015.12.013

85. Hessien M (2022) Microwave-assisted hydrothermal carbonization of pomegranate peels into hydrochar for environmental applications. Energies
86. Wu D, Hakkarainen M (2014) A closed-loop process from microwave-assisted hydrothermal degradation of starch to utilization of the obtained degradation products as starch plasticizers
87. Kannan S, Gariepy Y, Raghavan GSV (2017) Optimization and characterization of hydrochar produced from microwave hydrothermal carbonization of fish waste. Waste Manage 65:159–168. https://doi.org/10.1016/j.wasman.2017.04.016
88. Li Y, Tsend N, Li T, Liu H, Yang R, Gai X, Wang H (2019) Bioresource technology microwave assisted hydrothermal preparation of rice straw hydrochars for adsorption of organics and heavy metals. Bioresour Technol 273(October 2018):136–143. https://doi.org/10.1016/j.biortech.2018.10.056
89. Klein M, Neel I, Perkas N, Gedanken A (2016) Bioethanol production from *Ficus religiosa* leaves using microwave irradiation 177:20–25.https://doi.org/10.1016/j.jenvman.2016.03.050
90. Tseng K, Shiao Y, Chang R, Yeh Y (2013) Optimization of microwave-based heating of cellulosic biomass using taguchi method, 3404–3419. https://doi.org/10.3390/ma6083404
91. Ooshima H, Aso K, Harano Y, Yamamoto T (1984) Microwave treatment of cellulosic materials for their enzymatic hydrolysis. Biotechnol Lett 6(5):289–294
92. López-linares JC, Campillo V, Coca M, Lucas S, García-cubero MT (2020) Microwave-assisted deep eutectic solvent extraction of phenolic compounds from brewer's spent grain raw material. https://doi.org/10.1002/jctb.6565
93. Xu J, Liu B, Hou H, Hu J (2017) Pretreatment of eucalyptus with recycled ionic liquids for low-cost biorefinery. Biores Technol 234:406–414. https://doi.org/10.1016/j.biortech.2017.03.081
94. Da V, Lacerda S, Benito J, Sotelo L, Correa-guimarães A, Martín-ramos P, Navas-gracia LM, et al (2016) Efficient microwave-assisted acid hydrolysis of lignocellulosic materials into total reducing sugars in ionic liquids 50:761–770
95. Bernal HG, Maria A, Galletti R (2014) Furfural from corn stover hemicelluloses. A mineral acid-free approach, pp 3734–3740.https://doi.org/10.1039/c4gc00450g
96. Yew S, Chai W, Jit F, Phang F, Yeo LS, Ngu LH, How BS (2022) Future era of techno-economic analysis: insights from review. Front Sustainab
97. Neha S, Prasanna K, Ramesh K, Remya N (2022) Techno-economic analysis and life cycle assessment of microwave co-pyrolysis of food waste and low-density polyethylene. Sustain Energy Technol Assess 52(PD):102356. https://doi.org/10.1016/j.seta.2022.102356
98. Wang L, Lei H, Ruan R (2015) Techno-economic analysis of microwave-assisted pyrolysis for production of biofuels BT—production of biofuels and chemicals with microwave. In: Fang Z, Smith Richard J, Qi X (eds.). Springer Netherlands, Dordrecht, pp 251–263. https://doi.org/10.1007/978-94-017-9612-5_12
99. Zein SH, Antony A (2022) Techno-economic analysis and feasibility of industrial-scale activated carbon production from agricultural pea waste using microwave-assisted pyrolysis: a circular economy approach. Processes.https://doi.org/10.3390/pr10091702
100. Qureshi F, Yusuf M, Kamyab H, Vo DVN, Chelliapan S, Joo SW, Vasseghian Y (2022) Latest eco-friendly avenues on hydrogen production towards a circular bioeconomy: currents challenges, innovative insights, and future perspectives. Renew Sustain Energy Rev 168(September):112916. https://doi.org/10.1016/j.rser.2022.112916
101. Raud M, Kikas T, Sippula O, Shurpali NJ (2019) Potentials and challenges in lignocellulosic biofuel production technology. Renew Sustain Energy Rev 111(July 2018):44–56. https://doi.org/10.1016/j.rser.2019.05.020
102. Xie Y, Wang L, Li H, Westholm LJ, Carvalho L, Thorin E, Skreiberg Ø, et al (2022) A critical review on production, modification and utilization of biochar. J Anal Appl Pyrol 161(November 2021). https://doi.org/10.1016/j.jaap.2021.105405
103. Li Y, Xing B, Ding Y, Han X, Wang S (2020) A critical review of the production and advanced utilization of biochar via selective pyrolysis of lignocellulosic biomass. Biores Technol 312(March):123614. https://doi.org/10.1016/j.biortech.2020.123614

104. Song J, Wang Y, Zhang S, Song Y, Xue S, Liu L, Yang G, et al (2021) Coupling biochar with anaerobic digestion in a circular economy perspective: a promising way to promote sustainable energy, environment and agriculture development in China. Renew Sustain Energy Rev 144(February):110973. https://doi.org/10.1016/j.rser.2021.110973
105. Mohan D, Pittman CU, Steele PH (2017) Pyrolysis of wood/biomass for Bio-oil. Prog Energy Combust Sci 62(4):848–889
106. Vuppaladadiyam, AK, Vuppaladadiyam SSV, Sahoo A, Murugavelh S, Anthony E, Bhashkar T, Leu SY, et al (2023) Bio-oil and biochar from the pyrolytic conversion of biomass: a current and future perspective on the trade-off between economic, environmental, and technical indicators. Sci Total Environ 857(September 2022):159155. https://doi.org/10.1016/j.scitotenv.2022.159155
107. Czernik S, French R, Feik C, Chornet E (2002) Biomass thermoconversion processes. Ind Eng Chem Res 41:4209–4215
108. Solantausta Y, Nylund NO, Westerholm M, Koljonen T, Oasmaa A (1993) Wood-pyrolysis oil as fuel in a diesel-power plant. Biores Technol 46(1–2):177–188. https://doi.org/10.1016/0960-8524(93)90071-I
109. Xiu S, Shahbazi A (2012) Bio-oil production and upgrading research: a review. Renew Sustain Energy Rev 16(7):4406–4414. https://doi.org/10.1016/j.rser.2012.04.028
110. Kan T, Strezov V, Evans TJ (2016) Lignocellulosic biomass pyrolysis: a review of product properties and effects of pyrolysis parameters. Renew Sustain Energy Rev 57:1126–1140. https://doi.org/10.1016/j.rser.2015.12.185
111. Kambo HS, Dutta A (2015) A comparative review of biochar and hydrochar in terms of production, physico-chemical properties and applications. Renew Sustain Energy Rev 45:359–378. https://doi.org/10.1016/j.rser.2015.01.050
112. Cheng S, Chen T, Xu W, Huang J, Jiang S, Yan B (2020) Application research of biochar for the remediation of soil heavy metals contamination: a review. Molecules 1–21
113. Yan W, Hastings JT, Acharjee TC, Coronella CJ, Victor RV (2010) Mass and energy balances of wet torrefaction of lignocellulosic biomass (13):4738–4742. https://doi.org/10.1021/ef901273n
114. Abel S, Peters A, Trinks S, Schonsky H, Facklam M, Wessolek G (2013) Geoderma impact of biochar and hydrochar addition on water retention and water repellency of sandy soil. Geoderma 202–203:183–191. https://doi.org/10.1016/j.geoderma.2013.03.003
115. Khosravi A, Yuan Y, Liu Q, Zheng H, Hashemi M, Tang Y (2024) Hydrochars as slow—release phosphorus fertilizers for enhancing corn and soybean growth in an agricultural soil. Carbon Res. https://doi.org/10.1007/s44246-023-00086-w
116. Aliyu M, Iwabuchi K, Itoh T (2023) Upgrading the fuel properties of hydrochar by co-hydrothermal carbonisation of dairy manure and Japanese larch (*Larix kaempferi*): product characterisation, thermal behaviour, kinetics and thermodynamic properties. Biomass Convers Biorefinery 13(13):11917–11932. https://doi.org/10.1007/s13399-021-02045-0
117. Libra JA, Ro KS, Kammann C, Funke A, Berge ND, Neubauer Y, Berge ND et al (2014) Hydrothermal carbonization of biomass residuals: a comparative review of the chemistry, processes and applications of wet and dry, 7269. https://doi.org/10.4155/bfs.10.81
118. Baccile N, Laurent G, Babonneau F, Fayon F, Titirici M, Antonietti M (2017) Structural characterization of hydrothermal carbon spheres by advanced solid-state MAS 13 C NMR investigations HAL Id : hal-00393439 structural characterization of hydrothermal carbon spheres by advanced solid-state MAS 13 C NMR investigations
119. Nair LG, Agrawal K, Verma P (2022) An insight into the principles of lignocellulosic biomass-based zero-waste biorefineries: a green leap towards imperishable energy-based future. In: Biotechnology and genetic engineering reviews, pp 1–51
120. Pandey C, Sharma P, Gupta N (2023) Engineering to enhance thermostability of xylanase: for the new era of biotechnology. J Appl Biol Biotechnol 11(2):41–54. https://doi.org/10.7324/JABB.2023.110204
121. Procópio DP, Lee JW, Shin J, Tramontina R, Ávila PF, Brenelli LB, Basso TO, et al (2023) Metabolic engineering of *Saccharomyces cerevisiae* for second-generation ethanol production

from xylo-oligosaccharides and acetate. Sci Rep 13(1):1–14. https://doi.org/10.1038/s41598-023-46293-8

122. Yang J, Yue HR, Pan LY, Feng JX, Zhao S, Suwannarangsee S, Zhao XQ, et al (2023) Fungal strain improvement for efficient cellulase production and lignocellulosic biorefinery: current status and future prospects. Bioresour Technol 385(May):129449. https://doi.org/10.1016/j.biortech.2023.129449

Chapter 9
Ultrasound-Based Technologies for Agricultural Waste Biomass Valorization: Mechanism, Recent Technical Advances, Limitations, and Future Aspect

Cunshan Zhou, Lei Zhang, Jiakang Liang, and Manni Ren

Abstract Bioenergy attracts great attention as a renewable energy source; on the one hand, environmental pollution caused by agricultural waste biomass can be solved; on the other hand, the additional values of agricultural products will be boosted. Hence, efficient processing technologies are essential to agricultural waste biomass valorization. Ultrasound has been termed a novel technology and proved to be a very effective physical processing method. It is closely related to ultrasonic propagation properties and processing mechanisms. Deep eutectic solvent (DES) can promote agricultural waste biomass valorization because of high extracting efficiency from different components of plant-based materials with other exceptional advantages, including low value, simple synthesis, tiny volatility, preferable solubility, better biodegradability, and little toxicity. The combined utilization of ultrasound and DES can make the high-value applications of agricultural waste biomass and is conducive to the practicability of commercial-scale extraction. This chapter will present the fundamentals of ultrasound-based processing, including the processing mechanism, cavitation bubble behavior, and affecting factors. Subsequently, recent ultrasound-based technical advances in the fields, such as pretreatment, extraction, hydrolyzation, homogenization, fermentation, nanocrystallization, and catalysis, will be introduced, analyzed, and concluded. Finally, limitations and future perspectives on ultrasound-based technologies and DES will be predicted for further agricultural waste biomass valorization.

Keywords Ultrasound · DES · Agricultural waste biomass · Valorization

C. Zhou (✉) · L. Zhang · J. Liang · M. Ren
School of Food and Biological Engineering, Jiangsu University, Zhenjiang 212013, China
e-mail: cunshanzhou@163.com

© The Author(s), under exclusive license to Springer Nature Switzerland AG 2024

P. Verma and B. Kumar (eds.), *Sustainable Radiation Technologies in Waste-biomass Valorization*, Green Chemistry and Sustainable Technology,
https://doi.org/10.1007/978-3-031-63941-8_9

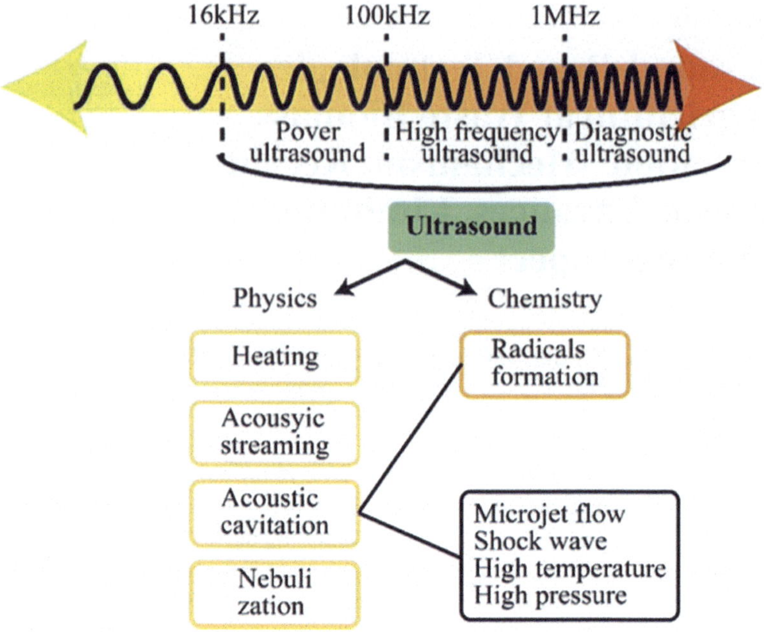

Fig. 9.1 Frequency range and action mode of ultrasound waves

9.1 Ultrasound-Based Processing Fundamentals

9.1.1 Ultrasonic Processing Mechanism

Sound waves with mechanical properties are known as ultrasound when their frequency exceeds 16 kHz. There are three kinds of ultrasound based on frequency. The scope of low-frequency ultrasound is from 16 to 100 kHz, that of high-frequency ultrasound is from 100 to 1000 kHz, and that of diagnostic ultrasound is from 1000 to 10,000 kHz [5, 21]. In biomass conversion processes, almost all ultrasound frequencies used are in the low-frequency range. Ultrasound acting on a liquid medium causes physicochemical variations of the medium (heating, acoustic flow, cavitation, atomization, generation of free radicals, etc.) (Fig. 9.1); due to these capabilities, ultrasound has great potential for application in biomass conversion processes [21].

9.1.2 Ultrasonic Cavitation Bubble Behavior

The cavitation effect is the most important mode of action of ultrasound [48, 104]. When ultrasonic waves propagate in a liquid medium at a specific pressure, periodic compression and sparse circulation are produced.

The solvent molecules are stacked together during the compression cycle; conversely, the solvent molecules are pulled apart during the sparse cycle. If there is enough energy in this alternating cycle, cavities are formed as the bonding of solvent molecules is overcome, and cavities are filled with solvent vapor and gases in the solvent [15]. During sparse circulation, negative pressure increases the volume of the bubble. When acoustic waves shift to a compression cycle, the bubble volume is decreased because of the positive pressure. It is considered that the ratio of expansion and compression is almost equal; however, the bubble compression is slightly less than its expansion cycle, thus bubbles can continue to grow. When the size of the bubble is small, it is comparatively stable, the bubble size still gets bigger, and, eventually, the bubble collapses, causing the release of energy, triggered by the ultrasound bubble formation, growth, and collapse process known as the cavitation effect [42, 85]. Because of the cavitation effect as micro-bubbles collapse, violent physical reactions, including micro-jets, shock waves, local high temperature, and pressure, will occur in the medium, which promotes the decomposition of water molecules and results in a strong oxidative free radical (OH, $O_2^{-}\cdot$, etc.) (Fig. 9.1) [7, 21].

9.1.3 Ultrasonic Affecting Factor

9.1.3.1 Frequency

A significant factor influencing the cavitation effect of ultrasound is frequency. Generally speaking, low-frequency ultrasound tends to produce large cavitation bubbles with high energy, and the bubbles quickly reach their maximum size and rupture; whereas high-frequency ultrasound, diagnostic ultrasound usually transmits only a small amount of energy, resulting in a limited cavitation effect [42]. Due to the high-intensity cavitation effect brought about by low-frequency ultrasound, low-frequency ultrasound is used in several industries such as cleaning, auxiliary extraction, auxiliary drying, and so on [81].

The application of multi-frequency ultrasound is also an excellent way to control the ultrasonic cavitation effect. Multi-frequency ultrasound introduces the concept of simultaneous or alternating use of ultrasound waves at different frequencies, which can overcome the shortcomings of a single ultrasound field distribution. In addition, the synergy of multiple-frequency ultrasound waves can provide more accurate control of the cavitation effect [104]. Ma et al. [60] used triple-frequency ultrasound to assist DES for the maize kernel fractionation and the cavitation ability of triple-frequency ultrasound was successfully utilized to fully penetrate the solvent into the plant cell wall structure, and the linkage bonds were increased significantly between lignin and carbohydrates and ether bonds of the lignin, thereby enhancing the fractionation efficiency. By combining ethanol and DES with triple-frequency ultrasound, 90.14% and 85.78% of the lignin in garlic skin and shallot root, respectively, could be removed [40].

9.1.3.2 Power (Power Density)

The power of ultrasound affects the formation, growth, collapse, and energy release of bubbles produced by cavitation effects. Increment of ultrasonic power usually results in the formation of more and larger bubbles. The energy released as bubbles collapse also increases with increasing ultrasonic power. In addition, high power favors a higher threshold for cavitation effects, i.e. high power makes cavitation effects more likely to occur [42]. It should be noted that the effect of ultrasonic treatment cannot become more obvious as ultrasonic power gets bigger,as the power beyond a certain range, the cavitation effect will be weakened due to the disruption of bubble dynamics, and the expansion cycle of the bubble's abnormal growth is diminished [40, 41].

9.1.3.3 Intermittent Ratio

Intermittent ratio between ultrasonic switching on and off is very important in the trial ultrasonic work, e.g. switching on for 3 s and stopping for 2 s. Bubbles induced by ultrasound are usually not instantaneous from generation to collapse; therefore, a long switch-on time means that the bubbles have a longer time to grow, and collapse and therefore produce more denser bubbles and stronger cavitation [11].

The increment of medium temperature can be affected by high intermittency ratios. Considering processing temperature requirements, the thermal effect caused by ultrasound is often reduced by controlling the intermittency ratio [98].

9.1.4 Application of DES for Biomass

9.1.4.1 Overview of DES

DES, an eutectic mixture, includes hydrogen bond donors (HBD) and hydrogen bond acceptors (HBA), usually with halide anions and quaternary ammonium cations. Hydrogen bonding formed between HBD and HBA promotes charge delocalization, resulting in the DES melting point being smaller than that of individual elements, and the stronger the hydrogen bonding, the faster the melting point decreases [44]. The ability to form hydrogen bonding connections between Cl^- in choline chloride (ChCl) and hydroxyl hydrogens in HBD can be seen in Fig. 9.2. The general formula Cat^+X^-zY can be used to illustrate DES; Cat^+ generally refers to phosphorus, ammonium, or sulfonium salt cation; X^- is the Lewis base, commonly halogen anion; z refers to the ratio of the number of molecules between Y and the anion X^-; Y represents Lewis or Brønsted acid [82].

DES is categorized into four classes. The first class of DES is composed of metal halides, including CuCl3, and quaternary ammonium salts, including ChCl, like the metal halide/imidazolium salt system. Because non-hydrated metal halides have a

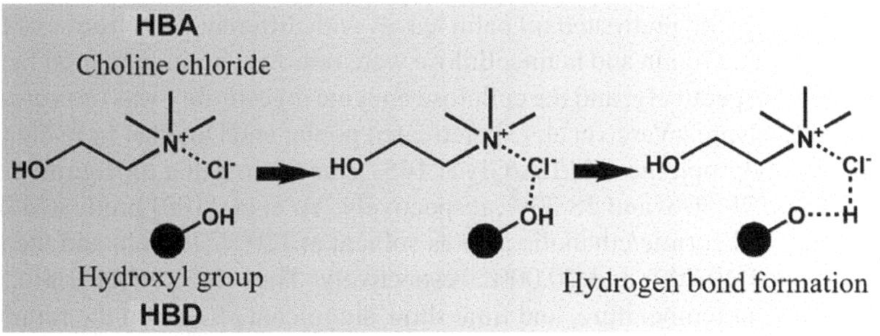

Fig. 9.2 Mutual effect between HBD and HBA (ChCl)

high melting point, they have been less studied in biomass [82]. The second class is formed from ChCl with hydrated metal halides (e.g. $AlCl_3 \cdot 6H_2O$), which are less costly and insensitive to air and moisture, and, therefore, more suitable for large-scale industrial production. The third class of DES is composed of ChCl with amides, carboxylic acids, alcohols, ethanol, etc. These DES have the advantages of low cost, biodegradable, not easily reacting with water, wide electrochemical window, and wide range of HBD inclusion, and, therefore, are more widely used. The fourth class of DES is the co-melts formed by $ZnCl_2$-like metal halides with ethylene glycol (EG), acetamide, urea, and 1,6-hexanediol. Most of the DES reported so far are based on choline cations, which are inexpensive and non-toxic compared to pyridine and imidazole and are categorized as vitamins in Europe for use as feed for megafauna [38].

The advantages of DES are low value, simple synthesis, tiny volatility, preferable solubility, better biodegradability, little toxicity, etc. It has the advantages of ILs and makes up for the disadvantages of ILs [27]. The selective deconstruction, separation, and transformation of biomass components can be realized by DES with recycling ability [80]. The lignin of biomass is dissolved by DES also with upgrading properties. Separating high-purity lignin from biomass can realize lignin appreciation [54]. At present, during the pretreatment, separation, and transformation of biomass, DES is broadly utilized.

9.1.4.2 Advances in DES Pretreatment of Biomass

DES has excellent lignin extraction properties. The β-O-4 bonds in lignin are sensitive to acidic protons; therefore, acidic DES can significantly improve lignin removal. Cronin et al. [20] utilized ChCl-lactic acid (LA) DES to obtain lignin from fermentation residues of corn stalks and found that the purity and yield of lignin were 94.7% and 75.0%, respectively. Zhang et al [102] used ChCl-oxalic acid (OA)/LA DES to pretreat corncobs and found that the lignin removal reached 98.50% and 93.10%, respectively.

Zulkefli et al. [106] pretreated oil palm leaves with different molar ratios of ChCl-LA; it was seen that lignin and hemicellulose were removed by an increased factor of 1.36 and 3.78, respectively; and the cellulose enzyme digestibility was improved by a factor of 1.42. Alvarez-Vasco et al. [3] pretreated poplar and Douglas-fir using ChCl-acetic acid/acetylpropionic acid/LA/Gly at 145 °C and found that the lignin removal of both reached 78.50% and 58.20%, respectively. Yu et al. [100] pretreated herbal residue with ChCl-formic/ethanolic acid as solvent at 120 °C for 8 h, and the lignin removal reached 40.70% and 60.00%, respectively. Thus, DES molar ratio, HBD type, pretreatment temperature, and time show significant effects on the removal of biomass lignin. In addition, Gly as HBD synthesized DES has tiny volatility and better biodegradability, as well as biocompatibility for enzymes and microorganisms [17], and for the biodiesel industry, it is a low-value accessory substance. It can be considered that Gly-based DES is an extensively used green solvent applied in the pretreatment of biomass. Xu et al. [96] used ChCl-Gly as a solvent for biomass lignin removal in a pretreatment process at 180 °C for 2 h pretreated corn stalk and found that lignin removal reached 60.0%. Despite the great advantages of ChCl-Gly, this harsh pretreatment condition limited its viability. For purpose of the excellent pretreatment efficiency with reduced pretreatment temperature, Chen et al. [17] developed an acidified DES pretreatment by adding 0.90 wt% sulfuric acid to ChCl-Gly, which caused lignin removal of 76.60% in milder conditions of 120 °C for 1 h. In addition, studies have been conducted to synthesize DES using phenolics as HBD to pretreat biomass. Kim et al. [47] attempted to synthesize DES using ChCl and lignin-derived phenol p-coumaric acid (PCA) to pretreat biomass and found that the removal rate of lignin was 60.80% after a reaction at 160 °C for 3 h. The results of this study are summarized in Table 9.1. In addition, Liang et al. [55] used betaine and amino acids as cations and anions, respectively, to synthesize novel DES for the pretreatment of corncobs, and found that the lignin removal rate reached 49.06% after pretreatment, and the enzymatic efficiency was also significantly improved.

Ho and Wu [37] found a lignin removal of 55.14% after pretreating oil palm leaves with a novel type II DES, ChCl: copper (II) chloride dihydrate ($CuCl_2 \cdot 2H_2O$). In addition to binary DES, ternary or poly-DES pretreatment systems are gradually being developed. Xia et al. [95] added $AlCl_3 \cdot 6H_2O$ to ChCl-Gly to form a ternary DES (TDES) to pretreat poplar wood and found that this TDES was able to remove lignin almost completely from poplar wood. Chen et al. [16] used a polyol-based TDES (guanidine hydrochloride-EG-p-toluenesulfonic acid (PTSA)) to pretreat switchgrass, and the removal of xylan and lignin was 79.00% and 82.00%, respectively, after reaction at 120 °C for 6 min. The above studies illustrate that DES is widely used for biomass pretreatment; however, due to the wide variety of biomass species and the large differences in their compositional structures, the fractionation of biomass by DES is also very different, so it is difficult to find a specific DES for the pretreatment of all biomass.

Table 9.1 Comparison of some literature on ultrasound-assisted DES extraction of bioactive compounds

Technique	Matrix	Targeted compound	Combination impact	Refs.
US-ChCl/glycerol	Sacha inchi meal biomass	Protein	• Efficiently extract a high total of crude protein with increased solubility at alkaline pH • Improved in vitro digestibility	[78]
US-microwave-PEG 200/ChCl	Pumpkin Seed	Protein	• Rapid and effective extraction process	[57]
US-ChCl/OA with 25% of water	Grape skin	Phenolics	• Enhanced extraction efficiency	[12]
US-ChCl/Malic acid	*Carya cathayensis* Sarg peels	Phenolics	• Boosted antioxidant activity	[28]
US-LA/ammonium acetate	Dittany, fennel, marjoram, mint, sage	Phenolics; Flavonoid	• Increased concentration of polyphenols • Leading to stronger antiradical activity and reducing power	[10]
US-betaine/acetic acid with 40% of water	*Scutellaria baicalensis*	Flavonoid	• A shorter extraction time yields higher results and anti-inflammatory activities	[33]
US-ChCl/ 1,6-hexanediol	Tea	Polysaccharides	• Improved extraction yield with increased total carbohydrate content and a lower molecular weight • Superior alpha-glucosidase inhibitory and hypoglycemic activity	[94]

9.2 Recent Technical Advances

9.2.1 Ultrasound-Assisted Biomass Pretreatment

Ultrasound energies are recognized as a beneficial method for enhancing industrialized processes, with effective applications in the valorization of agricultural waste biomass [66]. These applications encompass pretreating biomass, degrading lignin, converting cellulose, and increasing sugar production via biological and chemical processes. The major impacts of employing ultrasound energy in biomass treatment include increased convective forces, sonic cavitation, and reactive radicals. Biomass pretreatment is a methodology that utilizes physical, chemical, or biological processes to separate cellulose, lignin, and inorganic compounds for valorization and processing [2, 8, 18, 19]. It is found that methods, such as alkaline/acid hydrolysis, biomass extrusion, and biological treatments, are primarily used as pretreatments but with challenges of substantial used reagents, prolonged pretreatment durations, and rugged conditions. Thus, ultrasound is an extensive alternative for improving efficiency during pretreated stages. Physical effects, including turbulence, shock waves, microjets, and mixing, lead to the pretreatment effectiveness by the ultrasonic-assisted method. Hence, the biomass becomes more susceptible to further processing. Additionally, the effect of the sonochemical method also corresponds to free radicals due to the implosion of cavitation bubbles. In ultrasound-assisted pretreatments, lignin removal has been a primary focus and is the first step during any separation processing of lignocellulosic materials, which can be attributed to ultrasonic physical effects [24, 35]. Therefore, the removal is very efficient as time and temperature decrease than traditional methods. Although the simultaneous separation of lignin, cellulose, and hemicellulose fractions is found [24, 35], the processing times are long with high operational costs and unsuitable for industrialized adoption [49]. Thus, ultrasound-assisted hybrid methods have been used to improve the pretreated efficiency. The lignocellulosic biomass sonication and cellulose fraction crystallinity are decreased. When ultrasound interacts with the cellulose fraction, intermolecular hydrogen bonds are weakened or even delinked. The organization of cellulose sheets is reduced, which is beneficial to the hydrolytic process with the improvement of accessibility, hydrophilicity, and reactivity [24, 35, 65]. Although ultrasound-assisted pretreatment steps have been closely associated with lignocellulosic fraction separation, the concept range of biomass pretreatment encompasses is broader [24, 35, 65].

In recent years, extensive research has been conducted on possible substitutive materials derived from raw food. These "second-generation materials" are typically wastes or accessory substances in conventional processes from the food industry and agricultural waste, all of which have very low value [30, 39. 86]. Ultrasound-assisted biomass pretreatments are characterized by the reduction of processing time, temperatures (potentially even atmospheric temperature), and the ability to operate the reactor at atmospheric pressure. The integration of hybrid systems, which combine ultrasound with conventional treatments, could offer a viable alternative for enhancing pretreatment efficiency [30, 39, 86]. Notably, these hybrid systems for process scale-up have been studied a little. This highlights the fact that significant advancements are still needed to establish ultrasound-assisted pretreatment as a standard procedure in industrialized applications.

9.2.2 Ultrasound-Assisted DES Extraction

Currently, green and efficient DES has been widely used to extract natural products, including phenols, flavonoids, proteins, polysaccharides, anthraquinone proteins, polysaccharides, anthraquinones, terpenoids, alkaloids, etc., instead of organic solvents [9, 13, 21, 36, 43, 99]. Ultrasound-assisted extraction (UAE) as a new and effective extraction technique, and a new and effective extraction technique, can not only reduce solvent and energy consumption of solvent and energy but also shorten the extraction time; therefore, the UAE of DES extraction has become an important tool for the efficient extraction of functional components from plant tissues [92]. Therefore, the extraction of DESs by UAE has become an important tool for the efficient extraction of functional components from plant tissues [72, 84, 89, 93].

9.2.3 Ultrasound-Assisted Protein Enzymatic Hydrolysis

9.2.3.1 Effect of Ultrasound on Protein Structure

The chemical bonds that maintain the conformation of protein molecules include the van der Waals force, hydrogen bond, hydrophobic interaction, ionic bond, and disulfide bond. The ultrasonic modification mechanism of protein is mainly that cavitation effects cause physicochemical changes in the protein. There into, protein polar groups with absorbed energy induce their aggregation or unfolding. In addition, the protein conformational is changed because of ultrasonic treatments, and free radicals caused by the cavitation effect can destroy the interaction between protein molecules and change the primary, secondary, tertiary, or quaternary structure of the protein, thus affecting its physical and chemical properties. The effect of the ultrasonic field changed the chemical bonds in protein, changed the physical and chemical properties of protein and the site of enzymatic hydrolysis, and then affected the enzymatic hydrolysis effect.

The changes in protein structure caused by ultrasound can also be analyzed from the microstructure and nano-mechanical properties. Scanning electron microscope can directly show the particle size and shape of protein before and after ultrasonic treatment. For example, it has been found that ultrasonic treatment destroys the macromolecular aggregates of soybean protein isolate, making them dispersed and fine, but these fine aggregates will recombine, which has the same effect when [56, 58, 64] ultrasonic treatment is applied to pea or rice protein isolates [70].

We can also study the changes in the apparent morphology and mechanical properties caused by the action of ultrasound on protein from the nano-scale. With the help of an atomic force microscope, Wen et al. [90] found that 20/40 kHz dual-frequency ultrasound decreased the particle size of protein molecules in watermelon seeds and increased the dispersion degree of molecules. Ma et al. [61] obtained the changes in surface roughness and Young's modulus of elasticity caused by ultrasonic treatment of wheat gluten by calculating and analyzing the scanned images of an atomic force microscope. Therefore, it is beneficial to further study the mechanism of ultrasonic-induced enzymatic hydrolysis by analyzing the microstructure changes of the protein.

9.2.3.2 Effect of Ultrasound on Biological Activity of Protein Enzymatic Hydrolysates

Protein enzymatic hydrolysis assisted by ultrasonic pretreatment can improve the ACE inhibition rate, antioxidant activity, antibacterial activity, and immune activity of the enzymatic hydrolysate [56, 58, 64, 92]. At present, high-intensity ultrasound and other emerging technologies are applied to change the structure of the protein, improve the enzymatic hydrolysis rate, and obtain high-yield peptide products with strong biological activity which has become a hot spot in the field of food science [73]. Studies have shown that the influence of ultrasound on the structure of a protein may expose more cleavage sites and increase the hydrolysis rate. Because of ultrasonic treatment, electrostatic interaction is destroyed with the increment of surface hydrophobicity; the release of C-terminal ACE peptide containing hydrophobic residues is promoted, and the site originally in the spherical hydrophobic core of whey protein is exposed [6]. To improve ACE inhibition, it is believed that the active site of ACE is more matched with peptides with hydrophobic residues [1, 34, 52]. In addition, ultrasonic wave improves the antioxidant activity of wheat gluten hydrolysate [105], and when it is applied to the hydrolysis of milk (concentrated milk protein, concentrated whey protein, and casein), β-lactoglobulin, α-whey protein, walnut protein, and lactoferrin, the hydrolysate has higher antibacterial, immunomodulatory, and antithrombotic abilities [59, 87, 88].

9.2.4 Ultrasound-Based Nanocrystallization

For example, cellulose nanocrystals (CNCs) or lignin-containing nano-fibrillated cellulose (LCNFs) have considerable availability in the fields of foods, pharmaceuticals, and even cosmetics because of their excellent mechanical properties, environmental protection, and low value. Ultrasound is an important tool for the preparation of these materials by changing the structure and characteristics of cellulose through acoustic energy. The power of ultrasound has an important effect on the nanocrystalization of cellulose. Karimi et al.'s [45] study showed that ultrasonication below 200W reduced the crystallinity of microcrystalline cellulose (MCC) by only 1%, whereas high-power ultrasonication (>200W) significantly reduced the crystallinity of MCC. The increment of ultrasound power does not consistently lead to the reduction of cellulose crystallinity, since ultrasound acts mainly through cavitation, and the effective cavitation effect is only maximized at a specific power level [91]. Prolonged sonication at low power did not result in a significant effect on cellulose [77]. Recently, it was found the duration of ultrasound treatment had a greater effect on cellulose particle size than the change in ultrasound power in the power range of 150–600 W. Ni et al. [69] analyzed that it was due to the increase in power density, as the increment of ultrasound powers from 150 to 600 W, the power density increased from 330 to 605 J/mL for 5 min of ultrasound treatment, whereas by extending the ultrasound time at 150W, the power density could be increased to 1980 J/mL. Levanic et al. [51] added continuous sonication to the process of Tempo-mediated oxidation of cellulose and obtained similar conclusions, with an increase in the proportion of fibrous fines in the system as the number of sonication cycles was increased and the Tempo oxidation time was increased, while finer particles were generating at 60 min.

The combination of ultrasound and solvent was able to disrupt the hydrogen bonding in cellulose more easily, and the structure of the hydrogen bonds in cellulose significantly affected its crystalline properties [67]. Malyar et al. [62] used sulfuric acid and ultrasound together to treat cellulose obtained from spruce wood after delignification and succeeded in obtaining MCC at a power of only 80W at room temperature. Ji et al. [40] pretreated bagasse using ternary low eutectic solvent (TDES) in combination with ultrasound and obtained low crystallinity LCC while successfully removing lignin, with a sugar conversion of more than 90% under the combined effect of Lewis acid in the TDES composition and ultrasound. Ren et al. [75] applied DES synthesized with ChCl and oxalic acid in combination with ultrasound to extract LCNFs from sugarcane bagasse, and the size distribution (5.05–6.88 nm in diameter) of LCNFs with good freestanding structure is uniform, which enabled multidisciplinary applications. Amornnopparattanakul et al. [4] obtained cellulose with highly crystalline and thermally stable characteristics from sugarcane bagasse through a combination of enzymatic digestion and sonication. In the process of obtaining cellulose from biomass feedstocks by ultrasound-assisted sonication, prolonged sonication leads to a decrement of the cellulose crystallinity.

Kwiczak-Yigitbasi et al. [50] concluded that the energy of ultrasound mainly acts on the amorphous regions of cellulose because XRD spectra of the cotton fabric cellulose crystalline regions change less during the whole ultrasonic treatment. The amorphous regions of cellulose are first acted upon under the action of ultrasound. Thus, the short ultrasonic treatment can remove the amorphous part of the cellulose [74]. Prolonged sonication may lead to entanglement between cellulose structures resulting in agglomeration and disorganization of crystalline regions [32, 101]. In a recent experiment, the crystallinity of α-cellulose improved to 70.76% at an ultrasound energy of 12 kJ/g treatment, whereas the crystallinity reduced to 64.61% and 61.95% at 30 and 60 kJ/g treatments, respectively [76]. The study attributed the increase in cellulose crystallinity by low-energy ultrasound treatment to the hydrolysis of cellulose indeterminate regions, while cellulose I was transformed to low-crystallinity cellulose II altering the overall stability and thus leading to a decrease in crystallinity by high-energy ultrasound. Subsequently, it was demonstrated by grain size and average cross-sectional area where ultrasonication breaks interaction bonds between pristine cells, loosens the crystalline structure of cellulose, and leads to the complexation of two neighboring cells, which leads to a decrease in cellulose crystallinity [76].

Overall, ultrasound provides high shear force, microjet effect, and shock wave as effective means to achieve cellulose nanocrystallization. Ultrasound mainly acts on the indeterminate region of cellulose, and specific nanocrystallization results can be achieved by controlling the ultrasound parameters.

9.2.5 Ultrasound-Based Catalysis

Many scholars have worked on the ultrasound-assisted production of biomass oil; both homogeneous and non-homogeneous catalysts can be operated by ultrasound to change the yield [71]. The ultrasound-assisted reaction significantly reduces the time required for the ester exchange reaction and reduces the temperature and pressure required for the reaction. Chai et al. [14] successfully synthesized 2,5-furandicarboxylic acid (FDCA) from rice straw feedstock using ultrasound-assisted dilute acid pretreatment catalyzed by MnO2, which disrupts the lignocellulose structure and thus promotes the catalytic reaction. Struhs et al. [83] designed an acoustic catalytic transfer hydrogenation procedure (SCTH) to produce fuel blends and compounds from lignocellulosic biomass in a greener means by breaking and restructuring the molecular bonds of bio-oils through the phenomena of cavitation, thinning, and hydrogenation. The depolymerization process of lignin can also be assisted by ultrasound to increase the yield of lignin depolymerization products. Xu et al. [97] used ultrasound-assisted calcite-based catalysts to depolymerize corn stover lignin, and the assistance of ultrasound resulted in a phenolic monomer yield of 11.46 wt% and a biomass oil yield of 68.00 wt%. During the transesterification reaction for biodiesel production, ultrasound can significantly promote the transesterification reaction through the microbubbles generated by cavitation [68]. Khanian-Najaf-Abadi et al. [46] also demonstrated that the use of an ultrasonic reactor reduces reaction time, catalyst usage, and reagent usage. The kinetic study of this transesterification process revealed that the activation energies of the transesterification reaction under ultrasound were only 36.71 kJ/mol, which was about half of the activation energy of the reaction under mechanical stirring (62.19 kJ/mol). In addition, ultrasound enhanced mass transfer and emulsification during the reaction and improved the interfacial zone between alcohol-oil-catalyst [78].

Fallah et al. [25] also conclude that compared with the traditional mechanically stirred ester exchange method, many generated tiny bubbles by the ultrasonication method increased the interfacial area with the liquid phase of reactants and thus improved the reaction efficiency. It is worth mentioning that the by-products (alkanes, olefins, CO) produced in the pyrolysis of biomass collect and coking on the surface of the catalyst can result in catalyst deactivation, whereas ultrasonication effectively inhibits coking and converts biomass to the target aromatic hydrocarbons [53, 103].

Ultrasound has also been widely developed for the photocatalytic conversion of biomass. The use of ultrasonic assistance during the photocatalytic process increases the biomass conversion rate by several times. Palmitoleic acid hydrolysis to produce xylose has been reported to be more than two times higher with the help of ultrasonic cavitation effects. In addition, there are several explanations for the enhanced photocatalytic effect (multifactorial). During photocatalysis, photogenerated holes and photogenerated electrons are usually generated on the catalyst surface to form biomass-converting reactive species such as ROS and free radicals, and ultrasound facilitates mass transfer, which greatly facilitates the binding of reactants to substrates [23]. In addition, since ultrasound can break the hydration layer on the catalyst surface, water can be used as a solvent in the photocatalytic reaction, which meets more requirements of green conversion [31]. Dhar et al. [22] used TiO2 as a photocatalyst to increase lignin depolymerization efficiency by a factor of 2.7 compared to photocatalysis alone under ultrasound-assisted action. Some photocatalysts are photoresponsive with an accompanying piezoelectric capability such as MoS2 and ZnO, which can convert mechanical energy into electrical energy, generating piezoelectric fields to change the direction of carrier flow in heterojunctions during photocatalysis as well as to enhance ROS generation, thereby improving catalytic ability [79]. Few studies have been conducted combining biomass conversion and piezocatalysis, and the ability of piezocatalysis combined with photocatalytic processes to exhibit efficient ROS generation predicts the great potential that combined acousto-optic approaches can have in biomass conversion [29].

Overall, ultrasound is an effective alternative to improve the rate of reaction and the reaction yield, but reduces the energies required for reactions in the thermal cracking conversion of biomass. The mechanisms of the synergistic effect of ultrasound in the reaction process are all widely studied, including the cavitation effect, enhanced mass transfer, microbubbles, and so on. Very little research has been done on ultrasonic reactors [63] and ultrasonic reactors for large-scale production need to be further developed.

9.3 Limitations and Future Directions

In the previous section, we reviewed ultrasound applied in biomass conversion; ultrasonic treatments can be utilized in several stages of the biomass valorization process such as the pretreatment stage, extraction stage, hydrolysis stage, nanocrystallization, and catalytic depolymerization stage. In this section, we discuss the drawbacks, limitations, and future directions of ultrasound in these applications.

(1) Pretreatment stage: ultrasonic waves, based on their cavitation ability, destroy the connection of the components in the biomass feedstock to achieve the separation of the components. Regarding the parameters of ultrasound in the pretreatment process, researchers have carried out a large number of studies; most of the current studies remove lignin to obtain cellulose as the main goal, hemicellulose is often neglected in the treatment process, and the conversion of this component is equally important for biomass conversion. The choice of ultrasonic media for the pretreatment process is equally important, as solvents such as DES, alkalis, and acids are used for pretreatment, and these solvents are a challenge for the requirements of industrial ultrasonic equipment. In addition, the energy consumption of ultrasound needs to be considered and the development of economically viable and sustainable means of ultrasonic pretreatment is still a need for future research.
(2) Extraction: ultrasound improves the extraction of high-value substances from biomass by increasing mass transfer efficiency or destroying cellular tissues. It has been proved that the extraction efficiency can be significantly improved by ultrasonication, but a large number of solvents inevitably need to be used as a medium in the extraction process, but the effective separation of the extracted products from the medium is still a problem to be solved, such as the use of DES for the extraction of phenolics of biomass raw materials, and a large amount of water is often needed to be used for the precipitation of polyphenols after the completion of the extraction, which does not quite comply with the principle of green production. In addition, industrial applications are faced with the problem of ultrasound equipment design, where the data obtained from small ultrasound equipment used in laboratories may not apply to large ultrasound equipment used in production.
(3) Hydrolysis: ultrasound can significantly increase the efficiency of enzymatic hydrolysis. If the biomass conversion efficiency is also improved with the reduction of energy consumption, the potential practical application of ultrasound is great. In addition, the molecular changes in enzymes brought about by ultrasound need to be studied in more detail, and to realize the utilization of the whole biomass components, the ultrasound-assisted enzymatic hydrolysis of cellulose, lignin, and hemicellulose should be carried out at the same time. The homogenization effect brought by ultrasound is significant and can improve the uniformity and efficiency of the reaction in many biomass conversion processes. However, there is little literature on the homogenization of ultrasound as a separate research object. It usually only shows that the homogenization effect of ultrasound can improve the biomass conversion ability because of one reason. In fact, for different materials and reactors, the homogenization effect is different, which needs to be specialized in the study of ultrasonic researches.

In the process of biomass hydrolysis by microorganisms, ultrasound is often used to improve the efficiency of the hydrolysis process. Microorganisms exposed to ultrasound are susceptible to cell breakage or reduced microbial viability due to the shear and cavitation effects of ultrasound. Intermittent ultrasound ensures that the conversion efficiency is increased while the damage to microorganisms is reduced, and the ultrasound parameters used in fermentation systems are often not universal, so the optimization of the parameters as well as the protection of the microbial viability is necessary in practical applications.

(1) Nanocrystallization: in the production of cellulosic materials, ultrasound can reduce the crystallinity of cellulose and improve the homogeneity to achieve high-performance cellulosic materials. The crystallinity of cellulose material is positively correlated with its thermal stability. To be more specific, thermal stability of cellulose nanomaterials prepared by ultrasound is reduced, because ultrasound destroys the glycosidic bonds in cellulose, the crystallinity is reduced and thus the hydrophilicity of cellulose increases, which is detrimental to the application of cellulose in some packaging materials, such as food, medical and other applications. Although ultrasound has significant potential in the preparation of cellulose materials, the reason why its side effects can be reduced is still needed. Although ultrasound has significant potential in the preparation of cellulose materials, research is still needed on how to minimize the impact of its side effects.

(2) Catalysis: ultrasonic technology has more potential than mechanical or magnetic stirring during the preparation of biomass oil. Ultrasound can remove mass transfer limitations during catalytic reactions, as well as reduce the need for stirring due to localized high temperatures and microjet formation, and remove the large amount of coke that collects on the catalyst surface due to high temperature and anaerobic conditions. Various parameters regarding the use of ultrasound in the biomass oil production process such as catalyst loading and reactor volume. still need to be investigated for large-scale (industrial) application of ultrasound-assisted production of biomass oil. Most of the current studies use simple ultrasound equipment that is readily available in the laboratory, which may not be efficient in large-scale scenarios of commercial production. In addition, some energy calculations based on laboratory ultrasound equipment may not be meaningful for practical purposes and should be explored for cost calculations in scaled-up production.

In photocatalytic depolymerization, organic solvents, with their good solubility of biomass, are often used as the medium in the photocatalytic treatment of biomass, which is not suitable from the economic as well as sustainable point of view. Ultrasound can realize good dispersion of biomass in water, so ultrasound-assisted photocatalysis with water as the medium is a hot research topic in the future. The simultaneous acousto-optic process enhances the degree of biomass depolymerization and increases the diversity and complexity of the products, and the selective acquisition of the products in this case needs to be investigated. Ultrasound, especially high-power one, has an unavoidable function on the morphology of catalysts, and how to avoid the deactivation of catalysts due to the destruction of active sites or the changes of catalysts in the acousto-optic photocatalytic process caused by ultrasound still needs to be studied. In addition, some scholars have tried to introduce ultrasound into the synthesis process of photocatalysts in order to obtain uniform, nano-sized catalysts with good heterojunction. From the current literature, the energy consumption of ultrasound used in the acousto-photocatalytic process still needs to be explored, i.e. whether the use of ultrasound can achieve the purpose of green conversion of biomass still needs to be explored, and due to the action of ultrasound the elements such as O, H, N, and so on, dissolved in the solvents (e.g. water and acetonitrile), which play the key role of breaking the biomass in the conversion of biomass in some photocatalytic biomass conversion processes, are unavoidably changed. The key role of chemical bonds in some photocatalytic biomass conversion processes is paid little attention, so the changes brought by the introduction of ultrasound still needs to be explored.

9.4 Conclusion

In the chapter, it can be concluded that ultrasound can be applied in biomass conversion. First of all, ultrasound itself plays an important role; hence, the fundamentals of ultrasound-based processing, including the processing mechanism, cavitation bubble behavior, and affecting factors, have been explained. Ultrasound acting on a liquid medium caused physical and chemical changes with cavitation effect and generation of free radicals, etc. The ultrasonic effect was decided by the affecting factors, such as ultrasonic frequency, power density, intermittency ratios, and so on.

Second, DES is widely used for biomass pretreatment; however, due to the wide variety of biomass species and the large differences in their compositional structures, the fractionation of biomass by DES is also very different, which can hardly find a specific DES for the pretreatment of all biomass. Ultrasound energies are recognized as a beneficial method for enhancing industrialized processes, with effective applications in the valorization of agricultural waste biomass. Thus, recent ultrasound-based technical advances reviewed several stages of the biomass valorization process during pretreatment, extraction, hydrolyzation, nanocrystallization, and catalysis. The combined utilization of ultrasound and DES can make the high-value applications of agricultural waste biomass and is instrumental for exploring the feasibility of commercial-scale extraction.

Finally, the drawbacks, limitations, and future directions of ultrasound in the combined utilization of ultrasound and DES have been concluded and can be predicted for further research and application on agricultural waste biomass valorization.

Declaration: The authors provide the declaration that there is no use of AI in the writing process.

References

1. Abadia-Garcia L, Castano-Tostado E, Ozimek L, Romero-Gomez S, Ozuna C, Amaya-Llano SL (2016) Impact of ultrasound pretreatment on whey protein hydrolysis by vegetable proteases. Innov Food Sci Emerg Technol 37:84–90. https://doi.org/10.1016/j.ifset.2016.08.010
2. Agrawal K, Bhardwaj N, Kumar B, Chaturvedi V, Verma P (2019) Process optimization, purification and characterization of alkaline stable white laccasse from *Myrothecium verrucaria* ITCC-8447 and its application in delignification of agroresidues. Int J Biol Macromol 125:1042–1055. https://doi.org/10.1016/j.ijbiomac.2018.12.108
3. Alvarez-Vasco C, Ma RS, Quintero M, Guo M, Geleynse S, Ramasamy KK, Wolcott M, Zhang X (2016) Unique low-molecular-weight lignin with high purity extracted from wood by deep eutectic solvents (DES): a source of lignin for valorization [Article]. Green Chem 18(19):5133–5141. https://doi.org/10.1039/c6gc01007e
4. Amornnopparattanakul N, Yingkamhaeng N, Meesupthong R, Pinmanee P, Suwanprateep J, Nimchua T, Sukyai P (2023) Structure features of sugarcane bagasse under ultrasonic with xylanase and laccase treatment. Sugar Tech 25(4):893–905. https://doi.org/10.1007/s12355-023-01259-y
5. Arefi-Oskoui S, Khataee A, Safarpour M, Orooji Y, Vatanpour V (2019) A review on the applications of ultrasonic technology in membrane bioreactors. Ultrason Sonochem 58, Article 104633. https://doi.org/10.1016/j.ultsonch.2019.104633
6. Arzeni C, Martinez K, Zema P, Arias A, Perez OE, Pilosof AMR (2012) Comparative study of high intensity ultrasound effects on food proteins functionality. J Food Eng 108(3):463–472. https://doi.org/10.1016/j.jfoodeng.2011.08.018
7. Ashokkumar M (2011) The characterization of acoustic cavitation bubbles—an overview [Article; Proceedings Paper]. Ultrason Sonochem 18(4):864–872. https://doi.org/10.1016/j.ultsonch.2010.11.016
8. Badiei M, Asim N, Jahim JM, Sopian K (2014) Comparison of chemical pretreatment methods for cellulosic biomass. APCBEE Proc 9:170–174. https://doi.org/10.1016/j.apcbee.2014.01.030
9. Bajkacz S, Adamek J (2017) Evaluation of new natural deep eutectic solvents for the extraction of isoflavones from soy products. Talanta 168:329–335. https://doi.org/10.1016/j.talanta.2017.02.065
10. Bakirtzi C, Triantafyllidou K, Makris DP (2016) Novel lactic acid-based natural deep eutectic solvents: efficiency in the ultrasound-assisted extraction of antioxidant polyphenols from common native Greek medicinal plants. J Appl Res Med Arom Plants 3(3):120–127. https://doi.org/10.1016/j.jarmap.2016.03.003
11. Bang JH, Suslick KS (2010) Applications of ultrasound to the synthesis of nanostructured materials. Adv Mater 22(10):1039–1059. https://doi.org/10.1002/adma.200904093
12. Bubalo MC, Curko N, Tomasevic M, Ganic KK, Redovnikovic IR (2016) Green extraction of grape skin phenolics by using deep eutectic solvents. Food Chem 200:159–166. https://doi.org/10.1016/j.foodchem.2016.01.040
13. Cao J, Yang M, Cao FL, Wang JH, Su EZ (2017) Well-designed hydrophobic deep eutectic solvents as green and efficient media for the extraction of artemisinin from artemisia annua leaves. Acs Sustain Chem Eng 5(4):3270–3278. https://doi.org/10.1021/acssuschemeng.6b03092
14. Chai YZ, Yang HC, Bai M, Chen AW, Peng L, Yan BH, Zhao DY, Qin PF, Peng C, Wang XQ (2021) Direct production of 2,5-Furandicarboxylicacid from raw biomass by manganese dioxide catalysis cooperated with ultrasonic-assisted diluted acid pretreatment. Biores Technol 337, Article 125421. https://doi.org/10.1016/j.biortech.2021.125421
15. Chen F, Zhang M, & Yang C-h (2020) Application of ultrasound technology in processing of ready-to-eat fresh food: a review. Ultrason Sonochem 63, Article 104953. https://doi.org/10.1016/j.ultsonch.2019.104953

16. Chen Z, Jacoby WA, Wan CX (2019) Ternary deep eutectic solvents for effective biomass deconstruction at high solids and low enzyme loadings. Biores Technol 279:281–286. https://doi.org/10.1016/j.biortech.2019.01.126
17. Chen Z, Reznicek WD, Wan CX (2018) Deep eutectic solvent pretreatment enabling full utilization of switchgrass. Biores Technol 263:40–48. https://doi.org/10.1016/j.biortech.2018.04.058
18. Cherubini F (2010) The biorefinery concept: Using biomass instead of oil for producing energy and chemicals. Energy Convers Manage 51(7):1412–1421. https://doi.org/10.1016/j.enconman.2010.01.015
19. Cherubini F, Strømman AH (2011) Chemicals from lignocellulosic biomass: opportunities, perspectives, and potential of biorefinery systems. Biofuels, Bioproducts Biorefining 5(5):548–561. https://doi.org/10.1002/bbb.297
20. Cronin DJ, Chen XW, Moghaddam L, Zhang X (2020) Deep eutectic solvent extraction of high-purity lignin from a corn stover hydrolysate. Chemsuschem 13(17):4678–4690. https://doi.org/10.1002/cssc.202001243
21. Das AK, Sharma M, Mondal D, Prasad K (2016) Deep eutectic solvents as efficient solvent system for the extraction of kappa-carrageenan from *Kappaphycus alvarezii*. Carbohyd Polym 136:930–935. https://doi.org/10.1016/j.carbpol.2015.09.114
22. Dhar P, Teja V, Vinu R (2020) Sonophotocatalytic degradation of lignin: Production of valuable chemicals and kinetic analysis. J Environ Chem Eng 8(5), Article 104286. https://doi.org/10.1016/j.jece.2020.104286
23. Djellabi R, Aboagye D, Galloni MG, Andhalkar VV, Nouacer S, Nabgan W, Rtimi S, Constanti M, Cabello FM, Contreras S (2023) Combined conversion of lignocellulosic biomass into high-value products with ultrasonic cavitation and photocatalytic produced reactive oxygen species-A review. Bioresour. Technol 368, Article 128333. https://doi.org/10.1016/j.biortech.2022.128333
24. Fakayode OA, Akpabli-Tsigbe NDK, Wahia H, Tu SS, Ren MN, Zhou CS, Ma HL (2021) Integrated bioprocess for bio-ethanol production from watermelon rind biomass: Ultrasound-assisted deep eutectic solvent pretreatment, enzymatic hydrolysis and fermentation. Renew Energy 180:258–270. https://doi.org/10.1016/j.renene.2021.08.057
25. Fallah A, Kamran-Pirzaman A, Zanjani NG (2023) The sono-physical effect of cavitation bubbles on homogeneous catalyzed biodiesel production. Biomass Convers Biorefinery 13(6):5339–5351. https://doi.org/10.1007/s13399-021-01760-y
26. Flores EMM, Cravotto G, Bizzi CA, Santos D, Iop GD (2021) Ultrasound-assisted biomass valorization to industrial interesting products: state-of-the-art, perspectives and challenges. Ultrason Sonochem 72:105455. https://doi.org/10.1016/j.ultsonch.2020.105455
27. Francisco M, van den Bruinhorst A, Kroon MC (2012) New natural and renewable low transition temperature mixtures (LTTMs): screening as solvents for lignocellulosic biomass processing. Green Chem 14(8):2153–2157. https://doi.org/10.1039/c2gc35660k
28. Fu XZ, Wang D, Belwal T, Xu YQ, Li L, Luo ZS (2021) Sonication-synergistic natural deep eutectic solvent as a green and efficient approach for extraction of phenolic compounds from peels of Carya cathayensis Sarg. Food Chem 355:12, Article 129577. https://doi.org/10.1016/j.foodchem.2021.129577
29. Fu Y, Wang Y, Zhao H, Zhang Z, An B, Bai C, Ren Z, Wu J, Li Y, Liu W, Li P, Ma J (2021) Synthesis of ternary ZnO/ZnS/MoS_2 piezoelectric nanoarrays for enhanced photocatalytic performance by conversion of dual heterojunctions. Appl. Surface Sci. 556, Article 149695. https://doi.org/10.1016/j.apsusc.2021.149695
30. Ghisellini P, Cialani C, Ulgiati S (2016) A review on circular economy: the expected transition to a balanced interplay of environmental and economic systems. J Clean Prod 114:11–32. https://doi.org/10.1016/j.jclepro.2015.09.007
31. Giannakoudakis DA, Nair V, Khan A, Deliyanni EA, Colmenares JC, Triantafyllidis KS (2019) Additive-free photo-assisted selective partial oxidation at ambient conditions of 5-hydroxymethylfurfural by manganese (IV) oxide nanorods. Appl. Catal. B Environ. 256, Article 117803. https://doi.org/10.1016/j.apcatb.2019.117803

32. Han Z, Zhu H, Cheng J-H (2022) Structure modification and property improvement of plant cellulose: Based on emerging and sustainable nonthermal processing technologies. Food Res Int 156, Article 111300. https://doi.org/10.1016/j.foodres.2022.111300
33. Hao C, Chen LZ, Dong HJ, Xing WG, Xue FM, Cheng Y (2020) Extraction of flavonoids from scutellariae radix using ultrasound-assisted deep eutectic solvents and evaluation of their anti-inflammatory activities. ACS Omega 5(36):23140–23147. https://doi.org/10.1021/acsomega.0c02898
34. Haque E, Chand R (2008) Antihypertensive and antimicrobial bioactive peptides from milk proteins. Eur Food Res Technol 227(1):7–15. https://doi.org/10.1007/s00217-007-0689-6
35. Hassan SS, Williams GA, Jaiswal AK (2018) Emerging technologies for the pretreatment of lignocellulosic biomass. Biores Technol 262:310–318. https://doi.org/10.1016/j.biortech.2018.04.099
36. Hernandez-Corroto E, Plaza M, Marina ML, Garcia MC (2020) Sustainable extraction of proteins and bioactive substances from pomegranate peel (*Punica granatum* L.) using pressurized liquids and deep eutectic solvents. Innov. Food Sci Emerg Technol 60:11, Article 102314. https://doi.org/10.1016/j.ifset.2020.102314
37. Ho MC, Wu TY (2020) Sequential pretreatment with alkaline hydrogen peroxide and choline chloride: copper (II) chloride dihydrate—synergistic fractionation of oil palm fronds. Biores Technol 301:10, Article 122684. https://doi.org/10.1016/j.biortech.2019.122684
38. Hong S, Shen XJ, Xue ZM, Sun ZH, Yuan TQ (2020) Structure-function relationships of deep eutectic solvents for lignin extraction and chemical transformation. Green Chem 22(21):7219–7232. https://doi.org/10.1039/d0gc02439b
39. Jenkins BM, Baxter LL, Miles TR, Miles TR (1998) Combustion properties of biomass. Fuel Process Technol 54(1):17–46. https://doi.org/10.1016/S0378-3820(97)00059-3
40. Ji Q, Yu X, Yagoub AEA, Chen L, Mustapha AT, Zhou C (2021) Enhancement of lignin removal and enzymolysis of sugarcane bagasse by ultrasound-assisted ethanol synergized deep eutectic solvent pretreatment. Renew Energy 172:304–316. https://doi.org/10.1016/j.renene.2021.03.050
41. Ji QH, Yu XJ, Wu PW, Yagoub AA, Chen L, Mustapha AT, Zhou CS (2021) Pretreatment of sugarcane bagasse with deep eutectic solvents affect the structure and morphology of lignin. Ind Crops Prod 173, Article 114108. https://doi.org/10.1016/j.indcrop.2021.114108
42. Jiang Q, Zhang M, Xu B (2020) Application of ultrasonic technology in postharvested fruits and vegetables storage: a review. Ultrason Sonochem 69, Article 105261. https://doi.org/10.1016/j.ultsonch.2020.105261
43. Jiang ZM, Wang LJ, Gao Z, Zhuang B, Yin Q, Liu EH (2019) Green and efficient extraction of different types of bioactive alkaloids using deep eutectic solvents. Microchem J 145:345–353. https://doi.org/10.1016/j.microc.2018.10.057
44. Kalhor P, Ghandi K (2019) Deep eutectic solvents for pretreatment, extraction, and catalysis of biomass and food waste. Molecules 24(22):37, Article 4012. https://doi.org/10.3390/molecules24224012
45. Karimi M, Jenkins B, Stroeve P (2014) Ultrasound irradiation in the production of ethanol from biomass. Renew Sustain Energy Rev 40:400–421. https://doi.org/10.1016/j.rser.2014.07.151
46. Khanian-Najaf-Abadi M, Ghobadian B, Dehghani-Soufi M, Heydari A (2023) Experimental evaluation of simultaneous variations in biodiesel yield and color using choline hydroxide catalyst in an ultrasonic reactor. J Clean Prod 382, Article 134767. https://doi.org/10.1016/j.jclepro.2022.134767
47. Kim KH, Dutta T, Sun J, Simmons B, Singh S (2018) Biomass pretreatment using deep eutectic solvents from lignin derived phenols. Green Chem 20(4):809–815. https://doi.org/10.1039/c7gc03029k
48. Kumar B, Bhardwaj N, Agrawal K, Chaturvedi V, Verma P (2020) Current perspective on pretreatment technologies using lignocellulosic biomass: an emerging biorefinery concept. Fuel Process Technol 199:106244. https://doi.org/10.1016/j.fuproc.2019.106244

49. Kumar B, Verma P (2020) Biomass-based biorefineries: An important architype towards a circular economy. Elsevier, Fuel, p 119622
50. Kwiczak-Yigitbasi J, Demir M, Ahan RE, Canli S, Seker UOS, Baytekin B (2020) Ultrasonication for environmentally friendly preparation of antimicrobial and catalytically active nanocomposites of cellulosic textiles. Acs Sustain Chem Eng 8(51):18879–18888. https://doi.org/10.1021/acssuschemeng.0c05493
51. Levanic J, Svedstrom K, Liljestrom V, Sernek M, Crnivec IGO, Ulrih NP, Haapala A (2022) Cellulose fiber and nanofibril characteristics in a continuous sono-assisted process for production of TEMPO-oxidized nanofibrillated cellulose. Cellulose 29(17):9121–9142. https://doi.org/10.1007/s10570-022-04845-7
52. Li GH, Le GW, Shi YH, Shrestha S (2004) Angiotensin I-converting enzyme inhibitory peptides derived from food proteins and their physiological and pharmacological effects. Nutr Res 24(7):469–486. https://doi.org/10.1016/j.nutres.2003.10.014
53. Li K, Zhang G, Wang Z-x, Hu B, Lu Q (2020). Calcium formate assisted catalytic pyrolysis of pine for enhanced production of monocyclic aromatic hydrocarbons over bimetal-modified HZSM-5. Biores Technol 315, Article 123805. https://doi.org/10.1016/j.biortech.2020.123805
54. Li WX, Xiao WZ, Yang YQ, Wang Q, Chen XH, Xiao LP, Sun RC (2021) Insights into bamboo delignification with acidic deep eutectic solvents pretreatment for enhanced lignin fractionation and valorization. Ind Crops Products 170, 11, Article 113692. https://doi.org/10.1016/j.indcrop.2021.113692
55. Liang Y, Duan WJ, An XX, Qiao YY, Tian YY, Zhou HF (2020) Novel betaine-amino acid based natural deep eutectic solvents for enhancing the enzymatic hydrolysis of corncob. Biores Technol 310:9, Article 123389. https://doi.org/10.1016/j.biortech.2020.123389
56. Liu BH, Wang H, Hu T, Zhang PP, Zhang Z, Pan SY, Hu H (2017) Ball-milling changed the physicochemical properties of SPI and its cold-set gels. J Food Eng 195:158–165. https://doi.org/10.1016/j.jfoodeng.2016.10.006
57. Liu RL, Yu P, Ge XL, Bai XF, Li XQ, Fu Q (2017) Establishment of an aqueous PEG 200-based deep eutectic solvent extraction and enrichment method for pumpkin (*Cucurbita moschata*) seed protein. Food Anal Methods 10(6):1669–1680. https://doi.org/10.1007/s12161-016-0732-y
58. Liu YF, Oey I, Bremer P, Carne A, Silcock P (2018) Bioactive peptides derived from egg proteins: a review. Crit Rev Food Sci Nutr 58(15):2508–2530. https://doi.org/10.1080/10408398.2017.1329704
59. Luz C, Izzo L, Graziani G, Gaspari A, Ritieni A, Manes J, Meca G (2018) Evaluation of biological and antimicrobial properties of freeze-dried whey fermented by different strains of Lactobacillus plantarum. Food Funct 9(7):3688–3697. https://doi.org/10.1039/c8fo00535d
60. Ma QN, Ji QH, Chen L, Zhu ZL, Tu SS, Okonkwo CE, Out P, Zhou CS (2022) Multimode ultrasound and ternary deep eutectic solvent sequential pretreatments enhanced the enzymatic saccharification of corncob biomass. Ind Crops Prod 188, Article 115574. https://doi.org/10.1016/j.indcrop.2022.115574
61. Ma X, Wang D, Chen W, Ismail BB, Wang W, Lv R, Ding T, Ye X, Liu D (2018) Effects of ultrasound pretreatment on the enzymolysis of pectin: kinetic study, structural characteristics and anti-cancer activity of the hydrolysates. Food Hydrocolloids 79:90–99. https://doi.org/10.1016/j.foodhyd.2017.12.008
62. Malyar YN, Sudakova IG, Borovkova VS, Chudina AI, Mazurova EV, Vorobyev SA, Fetisova OY, Elsufiev EV, Ivanov IP (2023) Microfibrillated cellulose with a lower degree of polymerization; synthesis via sulfuric acid hydrolysis under ultrasonic treatment. Polymers 15(4), Article 904. https://doi.org/10.3390/polym15040904
63. Meza-Ramirez DB, Hernandez-Benitez C, Contreras-Arias A, Godinez LA, Rodriguez-Valadez FJ (2021) Optimization of an ultrasonic reactor on a semi-pilot scale for biodiesel production. Fuel 288, Article 119645. https://doi.org/10.1016/j.fuel.2020.119645
64. Mine Y, Ma FP, Lauriau S (2004) Antimicrobial peptides released by enzymatic hydrolysis of hen egg white lysozyme. J Agric Food Chem 52(5):1088–1094. https://doi.org/10.1021/jf0345752

65. Mosier N, Wyman C, Dale B, Elander R, Lee YY, Holtzapple M, Ladisch M (2005) Features of promising technologies for pretreatment of lignocellulosic biomass. Biores Technol 96(6):673–686. https://doi.org/10.1016/j.biortech.2004.06.025
66. Nair LG, Agrawal K, Verma P (2022) An insight into the principles of lignocellulosic biomass-based zero-waste biorefineries: a green leap towards imperishable energy-based future. Biotechnol Genetic Eng Rev 1–51
67. Nair LG, Agrawal K, Verma P (2023). Organosolv pretreatment: an in-depth purview of mechanics of the system. Biores Bioprocess 1–29
68. Ngaosuwan K, Ritprasert P, Sookman C, Kiatkittipong W, Wongsawaeng D, Mens-Appamana W, Rokhum SL, Assabumrungrat S (2023) In-situ synergistic effect of ultrasonic irradiation and solvent addition for enhancing CaO catalyzed transesterification. Fuel 338, Article 127305. https://doi.org/10.1016/j.fuel.2022.127305
69. Ni Y, Li J, Fan L (2021) Effects of ultrasonic conditions on the interfacial property and emulsifying property of cellulose nanoparticles from ginkgo seed shells. Ultrason Sonochem 70, Article 105335. https://doi.org/10.1016/j.ultsonch.2020.105335
70. O'Sullivan J, Arellano M, Pichot R, Norton I (2014) The effect of ultrasound treatment on the structural, physical and emulsifying properties of dairy proteins. Food Hydrocoll 42:386–396. https://doi.org/10.1016/j.foodhyd.2014.05.011
71. Okolie JA, Escobar JI, Umenweke G, Khanday W, Okoye PU (2022) Continuous biodiesel production: A review of advances in catalysis, microfluidic and cavitation reactors. Fuel 307, Article 121821. https://doi.org/10.1016/j.fuel.2021.121821
72. Ozturk B, Parkinson C, Gonzalez-Miquel M (2018) Extraction of polyphenolic antioxidants from orange peel waste using deep eutectic solvents. Sep Purif Technol 206:1–13. https://doi.org/10.1016/j.seppur.2018.05.052
73. Ozuna C, Paniagua-Martinez I, Castano-Tostado E, Ozimek L, Amaya-Llano SL (2015) Innovative applications of high-intensity ultrasound in the development of functional food ingredients: production of protein hydrolysates and bioactive peptides. Food Res Int 77:685–696. https://doi.org/10.1016/j.foodres.2015.10.015
74. Qu R, Tang M, Wang Y, Li D, Wang L (2021) TEMPO-oxidized cellulose fibers from wheat straw: effect of ultrasonic pretreatment and concentration on structure and rheological properties of suspensions. Carbohyd Polym 255, Article 117386. https://doi.org/10.1016/j.carbpol.2020.117386
75. Ren MN, Kong FG, Zhou CS, Fakayode OA, Liang JK, Li HX, Zhou M, Fan XY (2023) Green, one-pot biomass hierarchical utilization strategy for lignin-containing cellulose nanofibrils and fractionated lignin preparation. Ind Crops Prod, 203, Article 117193. https://doi.org/10.1016/j.indcrop.2023.117193
76. Shao BQ, Han ZB, Pang RR, Wu D, Xie B, Su YL (2023) The crystalline structure transition and hydrogen bonds shift determining enhanced enzymatic digestibility of cellulose treated by ultrasonication. Sci Total Environ 876, Article 162631. https://doi.org/10.1016/j.scitotenv.2023.162631
77. Sharma S, Pradhan R, Manickavasagan A, Dutta A (2022) Characterization of ultrasonic-treated corn crop biomass using imaging, spectral and thermal techniques: a review. Biomass Convers Biorefinery 12(4):1393–1408. https://doi.org/10.1007/s13399-020-00748-4
78. Sharma V, Tsai ML, Sun PP, Chen CW, Nargotra P, Dong CD (2023) Sequential ultrasound assisted deep eutectic solvent-based protein extraction from Sacha inchi meal biomass: towards circular bioeconomy. J Food Sci Technol Mysore 60(4):1425–1434. https://doi.org/10.1007/s13197-023-05689-0
79. Shen P, Yin P, Zou Y, Li M, Zhang N, Tan D, Zhao H, Li Q, Yang R, Zou B, Liu B (2023) Ultra-fast piezocatalysts enabled by interfacial interaction of reduced graphene oxide/MoS_2 heterostructures. Adv Mater 35(18). https://doi.org/10.1002/adma.202212172
80. Shen XJ, Wen JL, Mei QQ, Chen X, Sun D, Yuan TQ, Sun RC (2019) Facile fractionation of lignocelluloses by biomass-derived deep eutectic solvent (DES) pretreatment for cellulose enzymatic hydrolysis and lignin valorization. Green Chem 21(2):275–283. https://doi.org/10.1039/c8gc03064b

81. Siddiqui SA, Redha AA, Salauddin M, Harahap IA, Rupasinghe HPV (2023) Factors affecting the extraction of (poly)phenols from natural resources using deep eutectic solvents combined with ultrasound-assisted extraction. Crit Rev Anal Chem. https://doi.org/10.1080/10408347.2023.2266846
82. Smith EL, Abbott AP, Ryder KS (2014) Deep Eutectic solvents (DESs) and their applications. Chem Rev 114(21):11060–11082. https://doi.org/10.1021/cr300162p
83. Struhs E, Hansen S, Mirkouei A, Ramirez-Corredores MM, Sharma K, Spiers R, Kalivas JH (2021) Ultrasonic-assisted catalytic transfer hydrogenation for upgrading pyrolysis-oil. Ultrason Sonochem 73, Article 105502. https://doi.org/10.1016/j.ultsonch.2021.105502
84. Sukor NF, Selvam VP, Jusoh R, Kamarudin NS, Abd Rahim S (2021) Intensified DES mediated ultrasound extraction of tannic acid from onion peel. J Food Eng 296, 11, Article 110437. https://doi.org/10.1016/j.jfoodeng.2020.110437
85. Suslick KS, Flannigan DJ (2008) Inside a collapsing bubble: Sonoluminescence and the conditions during cavitation. Annu Rev Phys Chem 59:659–683. https://doi.org/10.1146/annurev.physchem.59.032607.093739
86. Vassilev SV, Baxter D, Andersen LK, Vassileva CG, Morgan TJ (2012) An overview of the organic and inorganic phase composition of biomass. Fuel 94:1–33. https://doi.org/10.1016/j.fuel.2011.09.030
87. Vogel HJ, Schibli DJ, Jing WG, Lohmeier-Vogel EM, Epand RF, Epand RM (2002) Towards a structure-function analysis of bovine lactoferricin and related tryptophan- and arginine-containing peptides. Biochem Cell Biol 80(1):49–63. https://doi.org/10.1139/o01-213
88. Wang C, Song W, Jiang LZ, Du M (2014) Purification and identification of an ACE-inhibitory peptide from walnut protein hydrolysate. Eur Food Res Technol 239(2):333–338. https://doi.org/10.1007/s00217-014-2227-7
89. Wang XH, Wang JP (2019) Effective extraction with deep eutectic solvents and enrichment by macroporous adsorption resin of flavonoids from *Carthamus tinctorius* L. J Pharm Biomed Anal 176:9, Article 112804. https://doi.org/10.1016/j.jpba.2019.112804
90. Wen C, Zhang J, Zhang H, Duan Y, Ma H (2019) Effects of divergent ultrasound pretreatment on the structure of watermelon seed protein and the antioxidant activity of its hydrolysates. Food Chem 299, 125165. https://doi.org/10.1016/j.foodchem.2019.125165
91. Wen C, Zhang J, Zhang H, Dzah CS, Zandile M, Duan Y, Ma H, Luo X (2018) Advances in ultrasound assisted extraction of bioactive compounds from cash crops—A review. Ultrason Sonochem 48:538–549. https://doi.org/10.1016/j.ultsonch.2018.07.018
92. Wu D, Tu ML, Wang ZY, Wu C, Yu CP, Battino M, El-Seedi HR, Du M (2020) Biological and conventional food processing modifications on food proteins: Structure, functionality, and bioactivity. Biotechnol Adv 40, Article 107491. https://doi.org/10.1016/j.biotechadv.2019.107491
93. Wu LF, Li L, Chen SJ, Wang L, Lin X (2020) Deep eutectic solvent-based ultrasonic-assisted extraction of phenolic compounds from *Moringa oleifera* L. leaves: Optimization, comparison and antioxidant activity. Sep Purif Technol 247:11, Article 117014. https://doi.org/10.1016/j.seppur.2020.117014
94. Xia B, Liu Q, Sun D, Wang Y, Wang WJ, Liu DH (2023) Ultrasound-assisted deep eutectic solvent extraction of polysaccharides from Anji white tea: characterization and comparison with the conventional method. Foods 12(3):14, Article 588. https://doi.org/10.3390/foods12030588
95. Xia QQ, Liu YZ, Meng J, Cheng WK, Chen WS, Liu SX, Liu YX, Li J, Yu HP (2018) Multiple hydrogen bond coordination in three-constituent deep eutectic solvents enhances lignin fractionation from biomass. Green Chem 20(12):2711–2721. https://doi.org/10.1039/c8gc00900g
96. Xu F, Sun J, Wehrs M, Kim KH, Rau SS, Chan AM, Simmons BA, Mukhopadhyay A, Singh S (2018) Biocompatible choline-based deep eutectic solvents enable one-pot production of cellulosic ethanol. ACS Sustain Chem Eng 6(7):8914–8919. https://doi.org/10.1021/acssuschemeng.8b01271

97. Xu JY, Yang YY, Liu BY, Kong Y, Du BY, Guo YZ, Zhou JH, Wang X (2022) Ultrasonic assisted enhanced catalytic effect of perovskite to promote depolymerization of lignin. Int J Biol Macromol 218:431–438. https://doi.org/10.1016/j.ijbiomac.2022.07.120
98. Yang ZY, Yang Z, Yu F, Tao ZC (2020) Ultrasound-assisted heat pump intermittent drying of adzuki bean seeds: drying characteristics and parameter optimization. J Food Process Eng 43(10), Article e13501. https://doi.org/10.1111/jfpe.13501
99. Yoo DE, Jeong KM, Han SY, Kim EM, Jin Y, Lee J (2018) Deep eutectic solvent-based valorization of spent coffee grounds. Food Chem 255:357–364. https://doi.org/10.1016/j.foodchem.2018.02.096
100. Yu Q, Zhang AP, Wang W, Chen L, Bai RX, Zhuang XS, Wang Q, Wang ZM, Yuan ZH (2018) Deep eutectic solvents from hemicellulose-derived acids for the cellulosic ethanol refining of Akebia' herbal residues. Biores Technol 247:705–710. https://doi.org/10.1016/j.biortech.2017.09.159
101. Zakani B, Grecov D (2022) Effect of ultrasonic treatment on yield stress of highly concentrated cellulose nano-crystalline (CNC) aqueous suspensions. Carbohy Polym 291, Article 119651. https://doi.org/10.1016/j.carbpol.2022.119651
102. Zhang CW, Xia SQ, Ma PS (2016) Facile pretreatment of lignocellulosic biomass using deep eutectic solvents. Biores Technol 219:1–5. https://doi.org/10.1016/j.biortech.2016.07.026
103. Zhang Y, Chang S, Chen Y, Bai Y, Zhao C, Wang X, Xue JM, Wang H (2021) Low-temperature superplasticity of beta-stabilized Ti-43Al-9V-Y alloy sheet with bimodal gamma-grain-size distribution. J Mater Sci Technol 95:225–236. https://doi.org/10.1016/j.jmst.2021.03.077
104. Zhou W, Sarpong F, Zhou C (2022) Use of ultrasonic cleaning technology in the whole process of fruit and vegetable processing. Foods 11(18), Article 2874. https://doi.org/10.3390/foods11182874
105. Zhu KX, Su CY, Guo XN, Peng W, Zhou HM (2011) Influence of ultrasound during wheat gluten hydrolysis on the antioxidant activities of the resulting hydrolysate. Int J Food Sci Technol 46(5):1053–1059. https://doi.org/10.1111/j.1365-2621.2011.02585.x
106. Zulkefli S, Abdulmalek E, Rahman MBA (2017) Pretreatment of oil palm trunk in deep eutectic solvent and optimization of enzymatic hydrolysis of pretreated oil palm trunk. Renewable Energy 107:36–41. https://doi.org/10.1016/j.renene.2017.01.037

Chapter 10
Gamma Radiation-Based Technologies for Waste Biomass Valorization: Mechanism, Instrument Design, Recent Technical Advances, Limitations, and Future Aspect

Shruti S. Raut and Abha Mishra

Abstract Waste valorization involves converting waste residue into valuable products like chemicals, materials, and biofuels, capitalizing on untapped polymeric substances within the residual waste. This concept is vital in addressing rapid natural resource depletion, driving the need for sustainable waste management and renewable energy sources. Biomass waste transformation, encompassing steps such as generation, pretreatment, hydrolysis, and fermentation, plays a pivotal role in this process. Effective pretreatment technology selection is critical, given its substantial impact on production costs. Enzyme and chemical-based pretreatment methods produce toxic chemical waste. In contrast, radiation pretreatment, particularly alpha and gamma irradiation, offers eco-friendly alternatives for biomass pretreatment, presenting a promising solution to this issue. Previous research indicates that gamma irradiation enhances fruit smoothness and promotes the formation of secondary metabolites through stress-related reactions. The potential of gamma irradiation to alter the composition of lignocellulosic biomass underscores its significance for future adoption. This overview discusses gamma irradiation pretreatment technologies and their mechanisms, providing insights into their role in enhancing lignocellulosic biomass pretreatment for diverse product formations.

Keywords Radiation · Gamma radiation · Alpha radiation · Waste · Valorization · Biomass · Pretreatment

S. S. Raut · A. Mishra (✉)
Biomolecular Engineering Laboratory, School of Biochemical Engineering, Indian Institute of Technology (BHU), Varanasi, India
e-mail: abham.bce@itbhu.ac.in

© The Author(s), under exclusive license to Springer Nature Switzerland AG 2024

P. Verma and B. Kumar (eds.), *Sustainable Radiation Technologies in Waste-biomass Valorization*, Green Chemistry and Sustainable Technology,
https://doi.org/10.1007/978-3-031-63941-8_10

10.1 Introduction

Environmentally friendly and renewable alternative energy resources are being actively sought after due to the scarcity of fossil fuels and the associated environmental risks. Due to lignocellulosic biomass's numerous benefits, including their high yield and affordability, they have been an essential component for the sustainable development of commercially viable biofuels like bioethanol substitutes for fossil fuels [38, 61]. Lignocellulosic biomass (LCB) consists of cellulose, hemicelluloses, and lignin. When subjected to enzymatic hydrolysis, lignocellulosic biomass only yield approximately 20% of their hypothetical glucose yield because lignin and hemicelluloses coexist with cellulose [19]. Thus, pretreatment of lignocellulosic biomass is required to produce biofuel and valuable products. This is necessary to change the biomass's structure and produce hydrolyzable fractions that can be fermented to produce biofuel after being transformed into sugars [54]. A range of pretreatment techniques (Fig. 10.1), including chemical hydrolysis, biological conversion, steam or fiber explosion, radiation, and mechanical grinding, have been developed to increase the fermentability and digestibility of lignocellulosic biomass under various circumstances [54]. Among these, gamma radiation has gained a lot of interest lately because of its capacity to increase the enzymatic digestibility of materials that have been processed [41]. Pretreatment using γ-irradiation has several benefits, such as inexpensive capital investment, minimal influence on the environment, quick response time, mild temperature, and few harmful operations. A few hundred kGy of radiation is needed to achieve good performance, as the irradiation dosage is a crucial factor affecting pretreatment performance [39, 40]. However, it has been discovered that subjecting cellulose, a primary component of LCB, to extreme concentrations of ionizing radiation in an environment with oxygen produces certain by-products, including Hydroxymethylfurfural, CO, CO_2, carboxyl, and carbonyl chemicals, and its organic acid product, all of which are harmful to the process of fermentation in downstream processing of biofuels [58]. Wang et al. [63] discussed that rice straw treated with radiation did not contain glucuronide acid, a hazardous substance produced during the steam explosion pretreatment. Moreover, it was discovered that an appropriate dose of radiation (less than 1000 kGy) did not appear to have a detrimental impact on the bioconversion of LCB into biofuel [40]. To find out how the pretreatment with γ-irradiation affected the creation mechanism of different inhibitory substances, both qualitative and quantitative experiments were required. Khan et al. [33] demonstrated that a 100 kGy gamma irradiation reduced the total crystallinity of irradiated jute fibers from 42 to 31%. Research on pure cellulose showed that the crystallinity dropped from 87 to 45% as the dose increased by an electron beam from 0 to 1000 kGy [21]. Takács et al. [56] demonstrated that following 10 kGy of gamma irradiation, the degree of polymerization of cotton cellulose decreased from approximately 1600 to 300. This book chapter will focus on utilizing gamma radiation as a pretreatment technique for the valorization of lignocellulosic biomass.

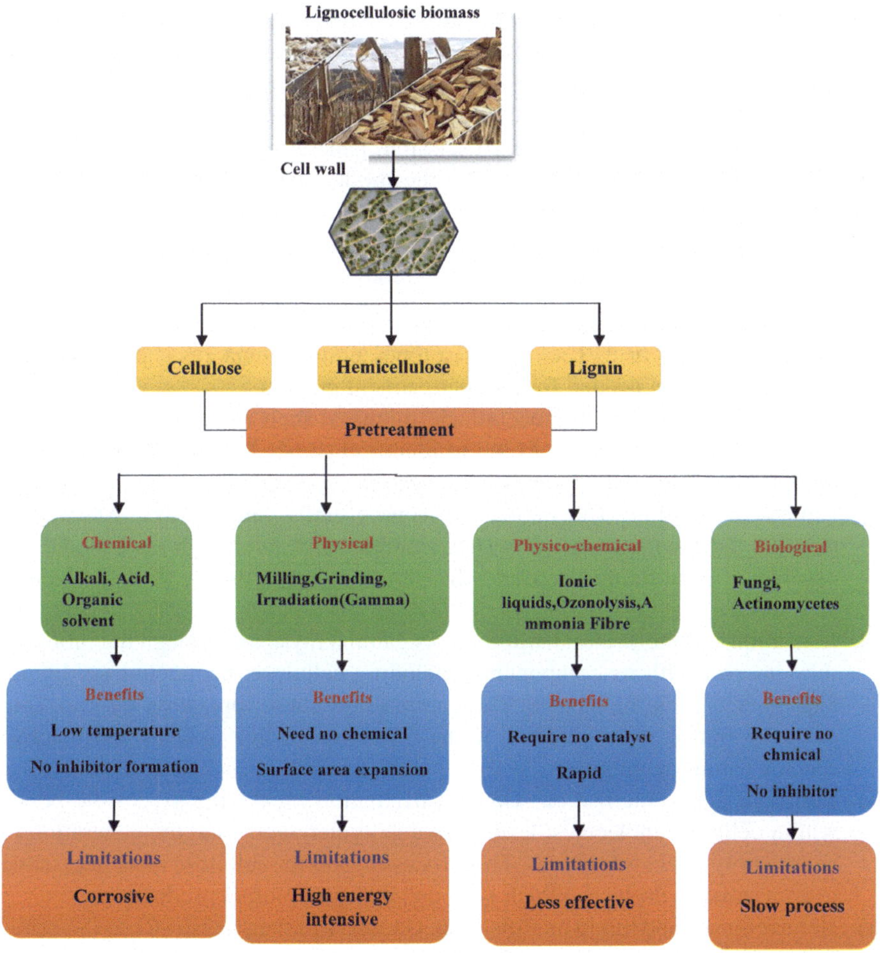

Fig. 10.1 Different pretreatment methods of lignocellulosic biomass

10.2 Mechanism of Action of Gamma Irradiation on LCB Pretreatment

High-energy ionizing radiation including gamma rays are found in the electromagnetic spectrum, with the ability to penetrate most materials easily. This form of irradiation consists of large, high-frequency radiation and varies depending on the source of radiation. Recently, there has been a significant interest in utilizing gamma radiation technology for pretreating biomass and biofibers in the production of liquid biofuels. Radioactive nuclides like ^{60}co and ^{137}ce are frequently employed in this pretreatment process [40]. The primary objective of this kind of radiation is to disrupt the arrangement of molecules inside and between cellulose by breaking down the hydrogen bonding between molecules. During the process, gamma radiation emitted from the radiation source reaches the biomass and biofibers, transferring energy to their components through collisions, which leads to ionization by causing the loss of electrons from atoms. As a result, the biomass components, particularly macromolecules made of cellulose, undergo scission, generating different radicals, both short- and long-lived [40]. Additionally, the production of fragments with lower degrees of polymerization gradually increases, altering the structure of the biomass and facilitating subsequent processes such as enzymatic saccharification.

Here is how a gamma irradiator works to pretreat biomass.

(1) Radiation source: The gamma irradiator contains a radioactive source, most commonly ^{60}Co or ^{137}Cs. These isotopes emit gamma rays, providing the ionizing radiation needed for the pretreatment process.
(2) Biomass loading: The lignocellulosic biomass, such as agricultural residues or wood chips, is loaded into the irradiator chamber. It's important to ensure uniform exposure of the biomass to the gamma rays for consistent pretreatment.
(3) Gamma irradiation: The radioactive source emits gamma rays, which pass through the biomass material. These gamma rays have high energy and are capable of penetrating deep into the biomass, interacting with its molecular structure.
(4) Ionization and molecular disruption: As gamma rays pass through the biomass, they interact with its components, including cellulose, hemicellulose, lignin, and other organic molecules. The gamma rays cause ionization by knocking electrons out of atoms, leading to the formation of free radicals and reactive species within the biomass.
(5) Cellulose depolymerization: Gamma irradiation breaks the glycosidic bonds within cellulose molecules, leading to depolymerization and the generation of shorter polysaccharide chains. This process makes the cellulose more accessible to enzymatic hydrolysis, which is a crucial step in biofuel production.
(6) Hemicellulose and lignin modification: Gamma irradiation also affects the hemicellulose and lignin components of the biomass, causing structural modifications and fragmentation. These changes reduce the recalcitrance of the biomass, making it simpler to degrade and transform into simpler sugars.
(7) Optimization and control: The pretreatment conditions, including radiation dose, exposure time, and biomass moisture content, need to be carefully optimized and controlled to achieve the desired outcomes while minimizing any negative effects on biomass quality.

10.3 Application of Gamma Irradiation in Waste Biomass Valorization

10.3.1 Utilizing Gamma Radiation to Convert Biomass Waste into Biohydrogen

Over 86% of the global energy supply is derived from finite, non-renewable resources, predominantly fossil fuels [30]. They are overutilized due to their low cost and widespread availability, resulting in detrimental environmental consequences, primarily through the emission of greenhouse gases that contribute to the phenomenon of climate change [11]. As a precautionary measure against this worldwide predicament, the scientific community is proactively investigating fuels that are environmentally friendly and free from toxins. When hydrogen gas is consumed in the air, water is produced rather than greenhouse gases; a thin, nearly invisible flame results [42]. The concept of employing biological fuel generation as a substitute for conventional hydrogen production has been discussed in recent years on account of the benign environmental impact, low cost, and mild nature of these reactions [9, 65]. Energy generation from sustainable resources contributes to environmental conservation through two key mechanisms: reduced organic waste and minimal greenhouse gas emissions. Bioelectrohydrogenesis, photofermentation, biophotolysis, and dark fermentation are all techniques utilized in the production of biohydrogen [42, 43]. When comparing Dark Fermentation Reactors (DFRs) and Microbial Electrolysis Cells (MECs), the latter exhibits a capacity of 8–9 mol H_2/mol glucose, whereas the former is limited to 4 mol H_2/mol glucose [12]. Considerable environmental issues have arisen regarding the management and dispersal of the substantial volumes of generated waste from activated sludge produced throughout the biological treatment process for wastewater. Despite this, waste-activated sludge is becoming increasingly popular due to the wide variety of applications which can be attributed to its high organic substance content. These consist of biological denitrification, anaerobic fermentation substrate for the manufacturing of sustainable fuels, and fish and crop fertilizer, among others [20, 28, 60], J. Y. [34]. Biological cultures derived from decomposed matter, soil, or waste resulting from anaerobic fermentation from treatment facilities are commonly utilized as the inoculum for biohydrogen production. Manufacturing procedures in the textile industry produce an extensive range of industrial discharge through the utilization of colorants and non-degradable harmful substances [22]. Non-biodegradable chemicals remain outside the cells of microorganisms as extracellular polymeric molecules, whereas biodegradable organic compounds comprise the majority of the organic compounds in the sediment [60]. Ionizing irradiation has been applied to activated sludge, which may cause the decomposition of cell structures as a result of the numerous radical species produced during water radiolysis. Microbial cells are extremely reactive toward these species [17, 62]. Scientists initiated an inquiry into the potential application of ionizing radiation to purify sewage sediment during the 1970s. However, early investigations were predominately focused on the disinfection of sewage [63]. Because of advancements in radiation technology, ionizing radiation is now more widely used to remediate sediment. Potential benefits include enhanced solubility of sewage sludge, removal of harmful elements, prevention of excess sludge formation, and improved anaerobic treatment feasibility [3, 31, 44, 72]. Yin and Wang proposed that the concentrations of SCOD, polysaccharides, and protein were observed to be 90%, 70%, and 190% higher in acidic conditions, respectively, following a 20 kGy pretreatment dosage, as compared to neutral conditions. T. H. [35] examined the feasibility of employing solubilized activated sludge as a carbon source for biological denitrification. Additionally, alkaline hydrolysis and radiation treatment were utilized to extract organic carbon from this source. The results indicate that gamma irradiation exhibits greater efficacy in dissolving sediment in alkaline conditions. This is supported by the observation that the solubility of soluble chemical oxygen demand (SCOD) increased with the pH level from 9 to 11, with a dosage of 20 kGy. Kim et al. [35] treated sediment with a combination of gamma irradiation (0–50 kGy) and alkali (pH = 7–11). The SCOD dissolution of the sludge treated with 20 kGy of radiation at a pH of 11 was approximately twelve times that of untreated sludge. Thus, the combined treatment of alkali and gamma radiation may have a substantial effect on the destruction of the cell walls and membranes of microorganisms, producing sediment that is more readily biodegradable. In a research conducted by Chu et al. [18], gamma irradiation was employed for sludge treatment, and there was an enhancement in solubilization from 1.7 to 6.5 as the dosage escalated from 5 to 25 kGy [18]. Chung et al. indicated that the production of the percentage of reducing sugars rose from 35.4% to 51.5% when polar bark was exposed to a range of 25–1000 kGy. Wang et al. reported rice straw treated with 400–2000 kGy gamma radiation exhibited a notable increase in glucose content (from 6.58% to 47.44%), while cellulose, hemicellulose, and lignin were rendered substantially destroyed. Chen et al. [13] studied gamma irradiation at concentrations of 10 and 20 kGy significantly enhanced the production of hydrogen from pretreated macroalgal biomass. The findings indicated that the production of hydrogen from macroalgae can be significantly enhanced through the application of appropriate doses of gamma irradiation. Specifically, 20 kGy irradiation resulted in the highest potential for hydrogen generation (34,000 mL), rate of hydrogen production (6,003 mL/h), and yield of hydrogen (14.17% of g-TSadded), all of which were 71.4%, 19.7%, and 70.0% greater than the control group, respectively. An increase in gamma irradiation dosage to 30 kGy leads to a decrease in the production of biohydrogen. Yin and Wang [71] furthermore observed that the production of hydrogen from waste-activated sludge exhibited a decrease after gamma irradiation with 30 kGy as opposed to 20 kGy. In addition, they deduced that this event transpired due to the high dose of gamma radiation destroying the complex polymer substances, which is essential for the growth of hydrogen-producing microbes. However, the concentrations of soluble carbohydrates, protein, and SCOD were significantly increased after 30 kGy irradiation. Figure 10.1 represents the application of gamma radiation in LCB pretreatment for biohydrogen production. Table 10.1 represents the application of gamma radiation in LCB pretreatment for biohydrogen production.

Table 10.1 Biohydrogen production using LCB pretreated by gamma radiation

Biomass used	Biohydrogen yield	Doses of gamma radiations (kGy)	References
Mud sludge	39.8 ± 13.1	0	[47]
Seed sludge	1.81 molH_2/mol glucose	5	[71]
Mud sludge	51.2 ± 10.7	0.1	[47]
Seed sludge	17.8 ml/g VS	5	[68]
Mud sludge	51.7 ± 17.3	0.35	[47]
Macroalgae (*Laminaria japonica*)	14.2 ml/g VS	20	[13]
Mud sludge	37.7 ± 2.40	0.6	[47]
Sewage sludge	529.4 ml/g VS	5	[70]
Grass waste	68 ml/g grass	150	[67]
Seed sludge	1.58 mol H_2/ mol glucose	5	[69]
Lolium perenne L	64.4 ml/g	5	[66]

10.3.2 Utilizing Gamma Radiation to Produce Bioethanol from Lignocellulosic Biomass

Lignocellulosic biomass possesses the capacity to serve as an economical and environmentally sustainable feedstock in the bioconversion process, wherein it is converted into fermentable carbohydrates that facilitate the production of biofuel. In addition to the presence of hemicelluloses and lignin, the crystalline structure of cellulose makes unmodified lignocelluloses difficult to hydrolyze. Thus, lignin and hemicelluloses must be degraded or altered before the initiation of the bioconversion process. Additionally, to facilitate enzymatic access to cellulose during hydrolysis, its crystalline structure must be perturbed [4]. As a result of its lignocellulosic composition, it is susceptible to ionizing radiation, producing free radicals that can change the lignin structure and cause cellulose crystal regions to break down. Radicals within crystalline and semi-crystalline regions of cellulose decompose more slowly and continue to degrade the lignocellulosic biomass after irradiation ceases, while those in amorphous regions of cellulose disappear immediately. Gamma rays or other forms of ionizing radiation may be employed in favor of enzymatic hydrolysis to penetrate the cellulose and liberate the derivative enzymes. Previous studies have demonstrated that generating sucrose from lignocellulosic biomass is possible through the irradiation-induced degradation of various lignocellulosic materials [64]. Multiple studies have reported that to observe a noticeable effect on the production of sugar via enzymatic hydrolysis from sugarcane bagasse and rice straw, radiation concentrations exceeding 1000 kGy were required [64]. The investigation of lignocellulosic biomass enzymatic hydrolysis via the combination of gamma rays and dilute acid is infrequent. One study found that a combined effect of gamma radiation as well as sulfuric acid treatment enhanced the in vitro digestion in the rumen of sugarcane bagasse [29]. Following pretreatment with diluted acid, the findings revealed a sugar yield of 56.1% (w/w). Furthermore, as the irradiation dosage increased, there was a slight improvement in the liberation of sugars with reducing properties, including glucose and xylose, which ranged from 51.5% (w/w) to 35.4% (w/w). An observed enhancement in the production of fermentable sugars (83.0%, w/w) was attributed to the integration of pretreatment which consist of 1000 kGy gamma irradiation and 3% (w/w) H_2SO_4. Increasing the dosage of irradiation resulted in a reduction in xylose content from 23.8% to 14.3%. Furthermore, the combined pretreatment of 3% H_2SO_4 and 1000 kGy irradiation further enhanced this decline to 3.4%. Due to the reduction in xylose concentration, the combination pretreatment led to an increase in glucose concentration. A significant indicator of hemicellulose removal is the depletion of xylose. Theoretically, this facilitates enhanced cellulose surface accessibility, thereby potentially enabling more efficient cellulose hydrolysis. When lignocellulosic biomass is exposed to gamma irradiation, carbohydrates are typically depolymerized and lignocellulosic biomass becomes less crystalline and more digestible [29]. The degree of crystalline structure found in cellulose derived from poplar decreased from 43.1% to 38.2% as gamma-ray exposure was increased. As the dosage of gamma irradiation increased, the level of crystalline structure within poplar bark following acid treatment was considerably diminished, at 1000 kGy, it had decreased to a minimum of 4.5%. Despite an increase in lignin content in both samples, the recovery of alkaline nitrobenzene oxidation (NBO) products decreased from 41.1% to 31.8% in gamma-irradiated samples and from 35.1% to 26.1% in combination-treated samples. [1] investigated the impact of gamma radiation on microorganisms and bioethanol synthesis. To increase the production of the bioethanol production process, this study conducted an extensive investigation into the effects of gamma irradiation on the organisms capable of producing bioethanol. Bioethanol was produced through the batch culture of *Saccharomyces cerevisiae* ATCC 7754 and *Zymomonas mobilis* ATCC 29,191 using two readily available and economical agricultural by-products, namely potato husk and sugarcane bagasse. The initial objective was to investigate the effects of different concentrations of gamma irradiation on the development of these two species. The assessed dosages varied between 0 and 3000 Gy. At 3000 Gy, both *Saccharomyces cerevisiae* ATCC 7754 and *Z. mobilis* ATCC 29,191 were rendered lethal. The microbial proliferation, bioethanol generation, and sucrose consumption were quantified over the following energy levels: 0–100, 300, 500, 1000, and 1500 Gy. The objective was to establish a correlation between the dosage of irradiation and the production of bioethanol so that the optimal irradiation treatment could be determined. Increasing irradiation doses inhibited the development of cells in both species, irrespective of the feedstock utilized. Therefore, the non-irradiated *Saccharomyces cerevisiae* ATCC 7754 exhibited a maximal cell count of 33.8 × 104 CFU/ml, whereas the *Z. mobilis* cells maintained a count of 29.2 × 104 CFU/ml. The final bioethanol concentration for potato peels underwent a significant reduction after gamma irradiation of *Z. mobilis* ATCC 29,191, decreasing from 3 gL-1 (non-irradiated) to 1.8 gL^{-1} (at 150 Gy) and from 2 gL^{-1} (non-irradiated) to 1 gL^{-1} (at 1500 Gy), respectively. *S cerevisiae* ATCC 7754 exhibited a discernible response to irradiation in comparison to its non-irradiated state. Following 300 Gy of radiation, its productivity from sugarcane bagasse increased from 4.2 gL^{-1} to 4.9 gL^{-1}, subsequently declining to 3 g^{L-1} after 1500 Gy of radiation. The production of bioethanol using potato peels increased marginally from 2.2 g L^{-1} at 0 Gy to 2.4 g L^{-1} at 100 Gy, before declining to 1.2 gL^{-1} at 1500 Gy. The fermentation process of *Saccharomyces cerevisiae* ATCC 7754 on sugar cane bagasse under 300 Gy of radiation produced the highest concentration of bioethanol (4.9 gL^{-1}), the yield of bioethanol (34.5% w/w), conversion coefficient (45.7% w/w), and sugar utilization efficiency (75.4% w/w) [1]. Short-lived and long-lived radicals induce secondary degradation in cellulosic materials through chain scission and crosslinking, respectively, based on the polymeric structure and radiation exposure. The economic feasibility and degradation rate of the lignocellulosic material degradation process are both enhanced. Lignocellulosic materials degrade quickly at modest irradiation dosages because of alkali swelling and feeble chemical bonds among lignin, hemicellulose, and cellulose units [55]. On the contrary to alternative radioactive isotopes, ^{60}Co is frequently employed in industrial irradiation processes due to its half-life of 5.26 years. Gamma particles are produced when the ^{60}Co and ^{137}Cs isotopes decay through radioactive processes. The utilization of a gamma chamber within an irradiation system could potentially exemplify a batch-oriented closed system process. This limits the capacity for treating biomass. The amount of irradiated biomass could potentially be utilized in a batch-continuous process for subsequent large-scale processing. Such systems are not new to the market and have been in existence for some time. As a result of absorbing the energy of the gamma radiation, the biomass atoms generate radicals. Destruction of cell walls and degradation of polysaccharides are both consequences of glycosidic bond scission [6, 10].

10.3.3 Utilizing Gamma Radiation to Produce Biodiesel from Biomass Waste

Unsustainable and excessive energy consumption has resulted from the progress made in all spheres of human endeavor and the exponential growth of the global population. Fossil fuels are regarded as non-renewable and environmentally unsustainable sources of energy due to their contribution to global warming, instability in prices, and depletion of resources [49]. Recent years have witnessed an international push for the development of renewable and sustainable energy alternatives to fossil fuels [5]. Microalgae possess lipid contents and rapid growth rates, which render them viable feedstock for biofuels, which are indispensable in numerous domains [25, 46]. Algae modify their metabolic pathways in response to nutritional and environmental stresses in order to produce hydrocarbons (lipids and/or carbohydrates). Biodiesel synthesis has significantly leveraged this protective mechanism [26, 71]. Gamma radiation substantially altered the metabolic phytoplankton constituents, although in a species-specific manner [24]. A growth density increase of 53.1% was observed in *Chlorella pyrenoidosa* variants that were exposed to 500 Gy of ^{60}Co irradiation [15]. The lipid content of *Nitzschia* sp. decreased to 9.9% following 900 Gy of ^{60}Co irradiation, despite an increase in lipid excretion to 19.76 mg L^{-1} [14]. Abo-state et al. evaluated the effect of gamma radiation on biodiesel production-related lipid buildup and *Chlorella vulgaris* growth. In this study, cultures exposed to a 1000 Gy irradiation dose produced biomass yields that decreased substantially (90.29%), from 1.03 ± 0.06 g L^{-1} to 0.10 ± 0.01 g L^{-1}. The results indicated that the proliferation of microalgae was significantly inhibited when exposed to large concentrations of gamma radiation (γ-rays). Consequently, the lipid content underwent a substantial change ($p < 0.05$) as a result of the radiation dosage increase from zero to one thousand Gy. The maximal lipid concentration (18.26 ± 0.21%), lipid output (7.05 ± 0.47 mg $L^{-1}d^{-1}$), and yield (0.187 ± 0.012 g L^{-1}) were documented at zero kGy. The reduction in lipid content (34.67%) was negligible in comparison to the biomass decline of 90.29 percent. Gamma radiation exerts an impact on the biological components of cells, specifically water molecules. The generation of free radicals upon the interaction between gamma rays and growth medium has the potential to alter the composition of cells [36]. Nevertheless, microalgal cells subjected to exposure to elevated concentrations of γ-rays dissolved or fragmented if they were incapable of self-repair and failed to fully recover [2, 36]. However, when exposed to low quantities of radiation, cells experienced only minor damage and rapidly reverted to their original state [23]. Protein synthesis was found to decrease in response to increases in irradiation dosage, indicating that photosynthetic photoinhibition could potentially be amplified [2]. Conversely, elevated levels of irradiation had an impact on the regulatory mechanism governing cellular metabolism [2]. Lipid accumulation was initially mitigated in microalgal cells through the inhibition of lipid synthesis that transpired during the restoration of cellular metabolism. Furthermore, elevated levels of irradiation induce lipid peroxidation in the biomembrane as a result of the disruption of double bonds in polyunsaturated fatty acids [2, 15]. In contrast to UV mutagenesis, gamma rays, heavy metals, plasma generated at atmospheric and room temperature, aeronautics, and ionizing radiation are suitable for producing stable mutants due to their ability to induce chromosomal aberration and inability to induce photoreactivation [16, 73]. The utilization of gamma radiation in the production of microalgal variants is one of their many industrial applications. Cheng et al. [16] reported that the gamma radiation of up to 900 Gy was deemed safe for *Chlorella* species. The vitality of *Chlorella vulgaris* was found to be diminished when exposed to gamma irradiation at concentrations exceeding 1 kGy [57]. The degradation of cellular DNA can be attributed to reactive oxygen species (ROS), which are generated via the radiolysis of water using gamma radiation. superoxide radical, hydrogen peroxide, hydroxyl radical, and singlet oxygen are all ROS. Schroeder et al. [51] studied that by subjecting *Chlorella sp.* KM504965, oleaginous microalgae, to gamma radiation, it is possible to enhance its biomass and biodiesel production. In this study, log phase cultures of *Chlorella sp.* KM504965 were exposed to a range of ^{60}Co ray concentrations (100, 200, 300, 400, 500, 700, 800, 900, 1000, and 1100 Gy) within the Gamma Chamber. The purpose of these exposures was to induce mutations in the cells. The findings indicated that the viability of cells was enhanced when exposed to gamma irradiation intensities ranging from 100 to 600 Gy. Plates containing cultures exposed to 1000 and 1100 Gy exhibited no growth. The application of energies of 700, 800, and 900 Gy to plates resulted in the manifestation of distinct colonies. The results obtained from the development of the naturally occurring form (plates containing cultures that were not exposed to radiant radiation) indicated that the viability of *Chlorella* sp. was below 10% and 1% at energies of 800 and 900 Gy, respectively. Following gamma irradiation, fluctuations in total chlorophyll and carotenoids were detected. During the initial stationary phase of Cl801, the chlorophyll concentration peaked at 12.82 mg L^{-1}, representing an increase of 1.12 times. The chlorophyll content of the mutant Cl802 was 1.04 times that of the wild type. Senthamilselvi et al. developed three superior mutants that were identified through the utilization of gamma radiations: Cl801, which demonstrated enhanced growth rate, carbohydrate percentage, and biomass for green ethanol production; Cl805, which exhibited a larger amount of lipid and biomass content for the synthesis of biodiesel; and Cl803, which demonstrated elevated protein content for the production of single-cell proteins. Cl805 exhibited a 1.4-fold increase in lipid accumulation, a 1.3-fold higher production of biomass, and a 1.4-fold higher rate of growth. Furthermore, the fuel parameters ascertained using standards of national and international biodiesel were met by the biodiesel obtained from the Cl805 mutant (CN: 67.48, IV: 42.22 mg I2100 g^{-1}, KV: 4.87 mm^2 s^{-1}). Cl805 is a mutant strain that exhibits possible utility in the production of biofuels. Figure 10.2 depicts an overview of the physical pretreatment of waste using gamma radiation.

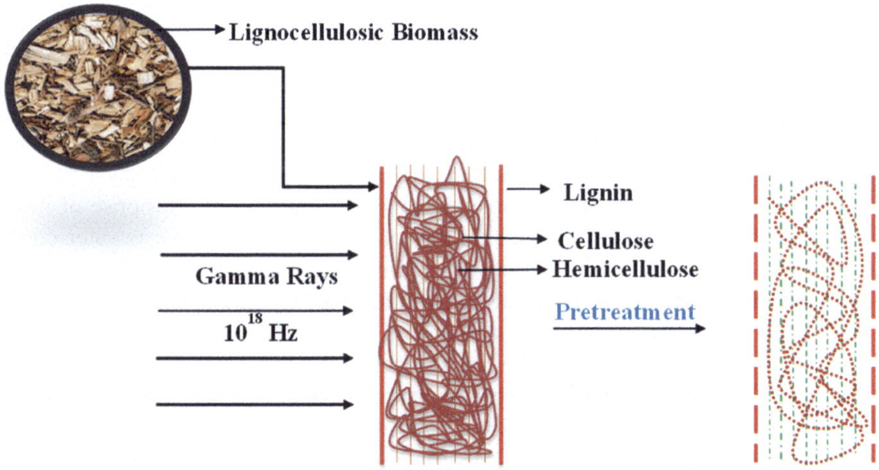

Fig. 10.2 Schematic representation of lignocellulosic waste pretreatment

10.3.4 Utilization of Gamma Radiation in the Treatment of Wastewater

An increasing number of contaminants are entering water supplies worldwide via industrial, municipal, and agricultural discharge. Possibly extinction or a hazard to human health could result from this. Individuals must take steps to reduce the presence of these contaminants to safeguard present and future generations [25]. Furthermore, many countries face two significant challenges when attempting to satisfy the growing need for potable water: ensuring the accessibility of potable water that is safe to consume and effectively managing and disposing of the escalating quantities of wastewater. Reusing wastewater represents a potential solution to these challenges. Several publications by the International Atomic Energy Agency (IAEA) have underscored the immense capacity of radiation technology to remediate waste emissions originating from diverse industrial and municipal processes [59]. Irradiation can be an effective treatment method for purifying water, wastewater, sediment, food, and medication by eliminating any pathogenic microorganisms that may be present [8, 37, 45, 62]. Gamma radiation is utilized to decompose hazardous organisms and organic substances while preserving vital nutrients like phosphorus and nitrogen. These nutrients are essential for the utilization of wastewater in agricultural applications [45, 53]. The microbicidal effect of ionizing radiation can be realized through two mechanisms: firstly, by directly disrupting nucleic acids (single- and double-strand breaks); and secondly, by utilizing indirect interaction between intracellular or extracellular molecules and radicals generated by radiolysis of water, specifically OH free radicals, H atoms, and solvated electrons e^- aqu. Bicarbonate and other scavengers have the potential to reduce the indirect impact of radicals, while oxygen can enhance their efficacy. The type of radiation, the dose rate, the temperature, the quantity of water within the cell, the age of the microbes, the energy source (e.g., gamma or neutron radiation), and the composition of the irradiation medium (e.g., air, nitrogen, vacuum) are all potential determinants that could impact how microbes respond to ionizing radiation [52]. Verde et al. [60] investigated the effects of various attributes, including substrate composition and dose rate, on the response of inactivation of indigenous microorganisms present in the sample of wastewater could provide valuable insights into identifying optimal treatment parameters for gamma radiation. The present study observed that the increased resistance power of microorganisms to gamma radiation was correlated with decreased dose rates. This result could potentially be attributed to the extended time available for microorganisms to repair radiation-damaged cells, thereby facilitating growth, and delaying inactivation. Even after extensive radiation exposures, the other common bacteria were found to be inactive at the time the pigmented bacteria with radiation resistance were isolated, according to this study. The results of this study provide reliability to the hypothesis that bioremediation could potentially draw advantages from implementing reduced dosage rates. By accelerating the degradation of wastewater substrates, ionizing radiation facilitates bacterial metabolism of the molecules, thereby reducing chemical wastewater pollution. Thus, it becomes a more efficacious therapeutic approach.

Furthermore, due to the extended duration of irradiation required, the genetic repair process of bacteria operates more efficiently when low dose rates are applied. Higher dosage rates were demonstrated to be more effective at disinfecting; therefore, they could potentially be utilized in tertiary treatment [48, 48].

10.3.5 Utilizing Gamma Radiation to Convert Lignocellulosic Biomass to Proteins

Several scientists have explored the feasibility of bioconverting lignocellulosic biomass into protein-rich materials via radiation pretreatment, to identify environmentally sustainable applications for this material. To isolate protein-rich organic compounds from biomass maize stover, Awafo et al. examined the effects of pretreatment with radiation [7]. The bioconversion of 100 kGy gamma irradiation into biomass maize stover into a protein-rich organic substance was unaffected. There was no noticeable difference observed in the irradiated material when compared to the unirradiated biomass. As stated previously, increased irradiation enriches the protein profile due to depolymerization, a rise in the specific surface area available for bioconversion, and a reduction in crystallinity [7].

10.3.6 Other Applications of Gamma Radiation to Valorize Waste

Orozco et al. investigated the effect of gamma radiation on the degradation of orange peels. When reducing sugars were produced, irradiation induced a color change from yellow to brown. High-performance liquid chromatography (HPLC) was employed to quantify sucrose, glucose, and fructose in the orange rind biomass after irradiation. After subjecting orange peels to additional irradiation at 1800 kGy, the concentrations of fructose and glucose in the sugar solution decreased due to glycolytic breakdown. With only slight variations in functional groups, the FTIR spectra of orange peels that have been exposed to irradiation or not exposed to radiation are remarkably similar. The hypothesized origin of these modifications was a decrease in chemical bonds induced by gamma irradiation. Hemicellulose and cellulose undergo chain cleavage at gamma irradiation doses of 600 kGy or higher, according to thermal gravimetric data; lignin is less significantly affected. The biomass structure is shielded from radiation by lignin due to its increased resistance to it. The process by which cellulose is transformed into compounds with reduced molecular weight is enhanced when oxygen is present and carbonyl groups form on the cellulose structure.

Kim et al. [34] reported that in the response of radiation on lignocellulose modification when addressing *Brachypodium* stems, a ^{60}Co gamma source is utilized.

Alterations in the cell walls, a reduction in cell size, and an augmentation in the quantity of fractured portions on the cell surfaces were observed as the level of irradiation escalated. This phenomenon transpired because gamma irradiation generates reactive oxygen species, the effects of which augment enzymatic hydrolysis. The achievement of objectives such as the development of bioethanol feedstock was significantly influenced by the reduction of lignin concentration. Kim et al. discovered that the concentration of lignin in irradiated cell walls fell from 2.4% to 18.4% [32]. To know the yield resulting from gamma radiation of agriculture solid waste conversion to simpler sugars, the degree of structural alteration in glucose resulting from the enzymatic hydrolysis of *Brachypodium* biomasses was quantified. Gamma radiation doses of 0, 50, 100, 150, 200, 250, and 300 kGy were applied to the samples, respectively. It was demonstrated that doses of irradiation from 200 to 300 kGy increased the glucose yield. At the irradiation doses of 50 and 100 kGy, the glucose yield was decreased in comparison to the plant samples at control (0 kGy, no exposure). The diminished glucose yield was due to inadequate pretreatment that failed to induce surface cracking of the lignocellulose or induce other by-product conversions, as earlier reported by [64]. Enzymatic hydrolysis of lignocellulose may be affected by crystallinity, and gamma rays pretreatment of *Brachypodium* can modify the surface crystalline nature of complex carbohydrates. When lignocellulosic biomass subjected to extremely low exposure to gamma rays, which requires an extended duration of lower energy irradiation, depolymerization, further degradation, and increases in the surface area of lignocellulose yield of hydrolysis with the help of enzyme [50]. Yang et al. investigated the enzymatic hydrolysis of wheat straw as influenced by gamma irradiation before treatment. A greater degree of weight loss and a more consistent particle size distribution were both outcomes of augmenting the irradiation dosage for the chopped biomass. Chain scission, by-product gas emission (CO, CO_2, and H_2), and leakage of bound water all contributed to the 3.70% reduction in weight of the biomass which is treated by gamma rays at 500 kGy of wheat straw. With an increase in irradiation dosage, there was a corresponding rise in the yield of fine powder below a 200 mesh screen size. At 500 kGy, the yield increased to 61.3%, from 15.2% in the control condition. Based on the findings, it can be deduced that the radiolytic degradation of wheat straw biomass resulted in the formation of a fragile structure. Additionally, the combination of pulverization and irradiation degradation effectively reduced in size of wheat straw particles. [64] reported that an upward trend in the yield of glucose was observed in response to an increase in irradiation dosage. At doses ranging from 50 to 300 kGy, the yield was moderate; however, it reached 13.40% between 300 and 500 kGy, after which it began to decrease with increasing doses. The reduced amount of by-products potentially hindered the enzymatic degradation of cellulose, resulting in a diminished conversion to glucose. Furthermore, an investigation was conducted by the researchers to assess the impact of biomass wheat straw particle sizes on radiation pretreatment [64]. Powder fractions of 20, 60, 100, 140, and 180 mesh were examined,. The enzymatic hydrolysis of biomass samples that had been irradiated to a dosage of 500 kGy was subsequently examined. The correlation between granularity and glucose yield was significantly stronger for unirradiated biomass pulverized wheat straw ranging in mesh size from 20 to 180 particles than for biomass measuring 1 cm in diameter. By increasing the area of the surface of complex polymer substances through pulverization or grinding, its crystallinity is reduced and enzymatic hydrolysis is enhanced. A glucose yield of 10.24% was observed as a result of the synergistic effect between grinding and irradiation of biomass straw with a particular particle size of 140 mesh that was subjected to gamma rays of 500 kGy. The effects of radiation therapy on the biomass of wheat straw were also examined by Yang et al. [64]. When compared to the primary effects at the same irradiation dosage, the effect after the experiment resulted as a change in glucose (or decreasing sugar) sample yield for variable quantities of time after irradiation. The impact of subjecting biomass powder with a lattice size of 140 to 400 kGy of irradiation was investigated during a variety of storage periods (0, 1, 4, 9, 15, 22, and 30 days). The yield of glucose increased until the 22nd day after gamma rays treatment, peaking at 10.21%, and then decreased on the 30th day. Lowering the yield of glucose on the 30th day may be attributed to the buildup of by-products that impede the degradation by the enzyme. Chung et al. [19] examined the method by which bioethanol is generated from poplar bark via gamma-ray enzymatic hydrolysis with diluted sulfuric acid. By subjecting biomass poplar bark to a pretreatment involving gamma rays of 1000 kGy and a diluted sulfuric acid procedure containing 3% w/w H_2SO_4, the yield of fermentable sugars was considerably increased. The lignin content of poplar bark is 20.2% by dry weight. Despite irradiation, the xylose content of biomass poplar bark exhibited a mere reduction from 23.8% to 14.3%. Through the combination of 1000 kGy gamma irradiation and a 3% diluted sulfuric acid solution, a substantial reduction in xylose concentration to 3.4% was achieved. Additionally, the combined pretreatment technique produced cellulose-rich solids. When xylose is lost, the surface area of the biomass residue becomes larger and becomes accessible for enzymatic degradation; this occurs due to the removal of hemicelluloses. Table 10.2 represents biofuel production using waste biomass pretreated by gamma radiation.

10.4 Future Perspectives

In the context of blood irradiation, ^{137}Cs and ^{60}Co are occasionally employed as radiation sources. The energy levels and half-lives of these gamma rays are comparatively high, with ^{137}Cs having a half-life of 30.10 years and ^{60}Co of 5.27 years. These particular gamma sources are the most frequently employed in materials processing utilizing radiation. Present-day commercial gamma-ray irradiators utilize ^{60}Co as their gamma source. Manufacturers of these sources provide a large range of irradiator configurations. The processed material is subjected to radiation in a shielded containment, whereas the source is generally maintained in either a dry or wet environment. In newly constructed industrial irradiators, the second type of solution has become increasingly prevalent in recent years. A variety of packaged goods may be supported by their product handling system per the condition that the product units meet the mass and dimension specifications for the transportation network.

Table 10.2 Gamma radiation-induced waste biomass valorization

Feedstock used	Product yield	Gamma radiation doses	Reference
Wheat straw	69.7% w/w increase in glucose yield	1000 kGy	[31]
Undaria sp	Sugar concentration increased by 0.048 g/l	0–500 kGy	[72]
Poplar leaves	83% increase in sugar yield	25–1000 kGy	[19]
Chlorella sp. KM504965	1.4-fold lipid accumulation	100–1100 kGy	[52]
Microalgae *Chlamydomonas reinhardtii*	20% more lipid accumulation	400 kGy	[34]
Sewage sludge	16–17 ml	20 kGy	[71]
Mud Sludge	0.029 ± 0.018 ppm methane	0.6 kGy	[47]
Anaerobic sludge	30 ml/g biogas	8.28 kGy	[22]

The duration of product irradiation may be altered in a variety of ways to meet the dosage requirements of individual products, contingent upon the activity of the source. According to global market data, the installed activity of ^{60}Co is expanding at a yearly rate of 6%. Its trends in the retail sale and manufacturing of single-use medical supplies, which fluctuate between 5.6% and 6.6%, are monitored in some ways. Within the United States territory are situated fifty of the over two hundred large irradiators in the globe. Other locations span across forty additional countries. Nordion constructed over 120 systems and 200 irradiators in total. This organization provides an intriguing option known as Gamma FIT, which enables nations to begin with technology deployment and subsequently enhance it in order to accommodate growing service demand as their businesses expand. Research and development facilities that are converting their Small-Scale Panoramic Systems to two-pass automatic tube irradiators will find this to be an exceptional offer. The Parallel Row Pallet Irradiator, another product offered by Nordion, is utilized for the irradiation of transport pallets containing intact materials, thereby lowering the costs associated with offloading and refilling unit shipments. As one of the numerous gamma irradiators provided by Symec Engineers (India) Pvt. Ltd., the Tote Box continuous processing facility is designed for the irradiation of agricultural and medicinal products. Significant quantities of potent ^{60}Co radioactive sources discovered in gamma irradiators present a novel and complex security challenge [35]. India and China, two countries that jointly manufacture ^{60}Co, serve as notable illustrations of locations where gamma irradiators operate significantly, albeit subject to certain limitations. Despite the implementation of stricter safety regulations, North American manufacturers of gamma irradiators may still encounter competition from EtO systems, the development of which is occasionally prohibited by local governments. In this era of sustainable economies, chemical usage should also be restricted.

References

1. Abdelhafez AA, Moussa M, El-kheima S (2014) Effect of gama irradiation of bioethanol producing microorganisms on bioethanol formation from sugarcane bagasse and potato peels. Egypt J Microbiol 49(1):55–79. https://doi.org/10.21608/ejm.2014.242
2. Agarwal R, Rane SS, Sainis JK (2008) Effects of 60Co γ radiation on thylakoid membrane functions in Anacystis nidulans. J Photochem Photobiol B 91(1):9–19. https://doi.org/10.1016/j.jphotobiol.2008.01.006
3. Al-Bachir M, Al-Adawi MA, Shamma M (2003) Synergetic effect of gamma irradiation and moisture content on decontamination of sewage sludge. Biores Technol 90(2):139–143. https://doi.org/10.1016/S0960-8524(03)00124-X
4. Alvira P, Tomás-Pejó E, Ballesteros M, Negro MJ (2010) Pretreatment technologies for an efficient bioethanol production process based on enzymatic hydrolysis: a review. Biores Technol 101(13):4851–4861. https://doi.org/10.1016/j.biortech.2009.11.093
5. Ameen M, Azizan MT, Ramli A, Yusup S, Alnarabiji MS (2019) Catalytic hydrodeoxygenation of rubber seed oil over sonochemically synthesized Ni-Mo/γ-Al_2O_3 catalyst for green diesel production. Ultrason Sonochem 51(July 2018):90–102. https://doi.org/10.1016/j.ultsonch.2018.10.011
6. Arshad R, Bokhari TH, Khosa KK, Bhatti IA, Munir M, Iqbal M, Iqbal DN, Khan MI, Iqbal M, Nazir A (2020) Gamma radiation induced degradation of anthraquinone Reactive Blue-19 dye using hydrogen peroxide as oxidizing agent. Radiat Phys Chem 168(September 2019), 108637. https://doi.org/10.1016/j.radphyschem.2019.108637
7. Awafo VA, Chahal DS, Charbonneau R (1995) Effect of irradiation, as a pretreatment, on bioconversion of corn stover into protein-rich mycelial biomass of Pleurotus sajor-caju. Radiat Phys Chem 46(4–6 PART 2):1299–1302. https://doi.org/10.1016/0969-806X(95)00374-7
8. Basfar AA, Abdel Rehim F (2002) Disinfection of wastewater from a Riyadh wastewater treatment plant with ionizing radiation. Radiat Phys Chem 65(4–5):527–532. https://doi.org/10.1016/S0969-806X(02)00346-8
9. Bedoya A, Castrillón JC, Ramírez JE, Vásquez JE, Zabala MA (2008) Biological production of hydrogen: a literature survey. In: Producción biológica de hidrógeno: Una aproximación al estado del arte. DYNA (Colombia) 75(154):137–157. http://www.scielo.org.co/pdf/dyna/v75n154/a14v75n154.pdf
10. Betiku E, Adetunji OA, Ojumu TV, Solomon BO (2009) A Comparative study of the hydrolysis of gamma irradiated lignocelluloses. Braz J Chem Eng 26(2):251–255. https://doi.org/10.1590/S0104-66322009000200002
11. Bhardwaj N, Kumar B, Verma P (2020) Microwave-assisted pretreatment using alkali metal salt in combination with orthophosphoric acid for generation of enhanced sugar and bioethanol. Biomass Convers Biorefinery. https://doi.org/10.1007/s13399-020-00640-1
12. Call D, Logan BE (2008) Hydrogen production in a single chamber microbial electrolysis cell lacking a membrane. Environ Sci Technol 42(9):3401–3406. https://doi.org/10.1021/es8001822
13. Chen Y, Yin Y, Wang J (2022) Biohydrogen production using macroalgal biomass of Laminaria japonica pretreated by gamma irradiation as substrate. Fuel 309(October 2021):122179. https://doi.org/10.1016/j.fuel.2021.122179
14. Cheng J, Chu W, Liu S, Dong P, Wu Z (2014) An overview on the research of iron-based high-T c superconductors probed by x-ray absorption spectroscopy. J Supercond Novel Magn 27(1):9–15. https://doi.org/10.1007/s10948-013-2405-z
15. Cheng J, Huang Y, Feng J, Sun J, Zhou J, Cen K (2013) Mutate Chlorella sp. by nuclear irradiation to fix high concentrations of CO_2. Biores Technol 136:496–501. https://doi.org/10.1016/j.biortech.2013.03.072
16. Cheng J, Lu H, Huang Y, Li K, Huang R, Zhou J, Cen K (2016) Enhancing growth rate and lipid yield of Chlorella with nuclear irradiation under high salt and CO_2 stress. Biores Technol 203:220–227. https://doi.org/10.1016/j.biortech.2015.12.032

17. Chu L, Wang J, Liu Y (2015) Degradation of sulfamethazine in sewage sludge mixture by gamma irradiation. Radiat Phys Chem 108:102–105. https://doi.org/10.1016/j.radphyschem.2014.12.002
18. Chu L, Wang J, Wang B (2011) Effect of gamma irradiation on activities and physicochemical characteristics of sewage sludge. Biochem Eng J 54(1):34–39. https://doi.org/10.1016/j.bej.2011.01.004
19. Chung BY, Lee JT, Bai HW, Kim UJ, Bae HJ, Gon Wi S, Cho JY (2012) Enhanced enzymatic hydrolysis of poplar bark by combined use of gamma ray and dilute acid for bioethanol production. Radiat Phys Chem 81(8):1003–1007. https://doi.org/10.1016/j.radphyschem.2012.01.001
20. De Gioannis G, Muntoni A, Polettini A, Pomi R (2013) A review of dark fermentative hydrogen production from biodegradable municipal waste fractions. Waste Manage 33(6):1345–1361. https://doi.org/10.1016/j.wasman.2013.02.019
21. Driscoll M, Stipanovic A, Winter W, Cheng K, Manning M, Spiese J, Galloway RA, Cleland MR (2009) Electron beam irradiation of cellulose. Radiat Phys Chem 78(7–8):539–542. https://doi.org/10.1016/j.radphyschem.2009.03.080
22. Fei X, Chen T, Jia W, Shan Q, Hei D, Ling Y, Feng J, Feng H (2020) Enhancement effect of ionizing radiation pretreatment on biogas production from anaerobic fermentation of food waste. Radiat Phys Chem 168(January 2019):108534. https://doi.org/10.1016/j.radphyschem.2019.108534
23. Fuma S, Ishii N, Takeda H, Miyamoto K, Yanagisawa K, Doi K, Kawaguchi I, Tanaka N, Inamori Y, Polikarpov GG (2009) Effects of acute γ-irradiation on the aquatic microbial microcosm in comparison with chemicals. J Environ Radioact 100(12):1027–1033. https://doi.org/10.1016/j.jenvrad.2009.06.007
24. Golz AL, Bradshaw C (2019) Gamma radiation induced changes in the biochemical composition of aquatic primary producers and their effect on grazers. Front Environ Sci 7(JUL):3389. https://doi.org/10.3389/fenvs.2019.00100
25. Goswami RK, Agrawal K, Verma P (2021b) Bioremediation of heavy metals from wastewater: a current perspective on microalgae-based future. Lett Appl Microbiol 1–21
26. Goswami RK, Agrawal K, Verma P (2022) Microalgae Dunaliella as biofuel feedstock and β-carotene production: an influential step towards environmental sustainability. Energy Convers Manage X 1–16:100154
27. Goswami RK, Mehariya S, Karthikeyan OPK, Verma P (2021a) Advanced microalgae-based renewable biohydrogen production systems: a review. Biores Technol 320:A:124301. https://doi.org/10.1016/j.biortech.2020.124301
28. Guo X, Liu J, Xiao B (2013) Bioelectrochemical enhancement of hydrogen and methane production from the anaerobic digestion of sewage sludge in single-chamber membrane-free microbial electrolysis cells. Int J Hydrogen Energy 38(3):1342–1347. https://doi.org/10.1016/j.ijhydene.2012.11.087
29. Han YW, Catalano EA, Ciegler A (1983) Chemical and physical properties of sugarcane bagasse irradiated with 7 rays. J Agric Food Chem 31(1):34–38. https://doi.org/10.1021/jf00115a009
30. He M, he, & Wei, C. hai. (2010) Performance of membrane bioreactor (MBR) system with sludge Fenton oxidation process for minimization of excess sludge production. J Hazard Mater 176(1–3):597–601. https://doi.org/10.1016/j.jhazmat.2009.11.071
31. Hyun Hong S, Taek Lee J, Lee S, Gon Wi S, Ju Cho E, Singh S, Sik Lee S, Yeoup Chung B (2014) Improved enzymatic hydrolysis of wheat straw by combined use of gamma ray and dilute acid for bioethanol production. Radiat Phys Chem 94(1):231–235. https://doi.org/10.1016/j.radphyschem.2013.05.056
32. Kadier A, Kalil MS, Abdeshahian P, Chandrasekhar K, Mohamed A, Azman NF, Logroño W, Simayi Y, Hamid AA (2016) Recent advances and emerging challenges in microbial electrolysis cells (MECs) for microbial production of hydrogen and value-added chemicals. Renew Sustain Energy Rev 61:501–525. https://doi.org/10.1016/j.rser.2016.04.017
33. Khan F, Ahmad SR, Kronfli E (2006) Γ-radiation induced changes in the physical and chemical properties of lignocellulose. Biomacromol 7(8):2303–2309. https://doi.org/10.1021/bm060168y

34. Kim JY, Na CS, Kim DS, Kim JB, Seo YW (2015) The effect of chronic gamma ray irradiation on lignocellulose of *Brachypodium distachyon*. Cellulose 22(4):2419–2430. https://doi.org/10.1007/s10570-015-0687-y
35. Kim TH, Nam YK, Park C, Lee M (2009) Carbon source recovery from waste activated sludge by alkaline hydrolysis and gamma-ray irradiation for biological denitrification. Biores Technol 100(23):5694–5699. https://doi.org/10.1016/j.biortech.2009.06.049
36. Kohli AK (2018) Gamma irradiator technology: challenges and future prospects. Univ. J. Phys. Appl. 12(3):47–51. https://doi.org/10.13189/ujpa.2018.120303
37. Kovács E, Keresztes. (2002) Effect of gamma and UV-B/C radiation on plant cells. Micron 33(2):199–210. https://doi.org/10.1016/S0968-4328(01)00012-9
38. Kumar B, Bhardwaj N, Verma P (2019) Pretreatment of rice straw using microwave assisted $FeCl_3$-H_3PO_4 system for ethanol and oligosaccharides generation. Biores Technol Reports 7(2019):100295. https://doi.org/10.1016/j.biteb.2019.100295
39. Kumar B, Bhardwaj N, Agrawal K, Chaturvedi V, Verma P (2020) Current perspective on pretreatment technologies using lignocellulosic biomass: an emerging biorefinery concept. Fuel Process Technol 199:106244. https://doi.org/10.1016/j.fuproc.2019.106244
40. Liu Y, Chen J, Wu X, Wang K, Su X, Chen L, Zhou H, Xiong X (2015) Insights into the effects of γ-irradiation on the microstructure, thermal stability and irradiation-derived degradation components of microcrystalline cellulose (MCC). RSC Adv 5(43):34353–34363. https://doi.org/10.1039/c5ra03300d
41. Liu Y, Guo L, Wang L, Zhan W, Zhou H (2017) Irradiation pretreatment facilitates the achievement of high total sugars concentration from lignocellulose biomass. Biores Technol 232:270–277. https://doi.org/10.1016/j.biortech.2017.01.061
42. Malik SN, Ghosh PC, Vaidya AN, Waindeskar V, Das S, Mudliar SN (2017) Comparison of coagulation, ozone and ferrate treatment processes for color, COD and toxicity removal from complex textile wastewater. Water Sci Technol 76(5):1001–1010. https://doi.org/10.2166/wst.2017.062
43. Nair LG, Agrawal K, Verma P (2022) An overview of sustainable approaches for bioenergy production from agro-industrial wastes. Energy Nexus 100086
44. Orozco RS, Hernández PB, Ramírez NF, Morales GR, Luna JS, Montoya AJC (2012) Gamma irradiation induced degradation of orange peels. Energies 5(8):3051–3063. https://doi.org/10.3390/en5083051
45. Park W, Hwang MH, Kim TH, Lee MJ, Kim IS (2009) Enhancement in characteristics of sewage sludge and anaerobic treatability by electron beam pre-treatment. Radiat Phys Chem 78(2):124–129. https://doi.org/10.1016/j.radphyschem.2008.09.010
46. Rawat KP, Sharma A, Rao SM (1998) Microbiological and physicochemical analysis of radiation disinfected municipal sewage. Water Res 32(3):737–740. https://doi.org/10.1016/S0043-1354(97)00220-0
47. Recalde C, Aguay D, Pérez M, Fiallos F, Lopez D, Jaramillo D, García VJ (2018) Biohydrogen from activated sludge from textile wastewater treated with low doses of gamma radiation. Radiat Phys Chem 153(August):234–238. https://doi.org/10.1016/j.radphyschem.2018.08.026
48. Rittmann BE (2008) Opportunities for renewable bioenergy using microorganisms. Biotechnol Bioeng 100(2):203–212. https://doi.org/10.1002/bit.21875
49. Sabbagh S, ASEM, YYA-D (2014) Waste water sterilization by cobalt Co-60 for the agricultural irrigation: a case study. Int J Water Res Arid Environ 3(1):11–18, 3(January), 47–54. https://www.academia.edu/download/35983025/2e.pdf
50. Schenk PM, Thomas-Hall SR, Stephens E, Marx UC, Mussgnug JH, Posten C, Kruse O, Hankamer B (2008) Second generation biofuels: high-efficiency microalgae for biodiesel production. BioEnergy Research 1(1):20–43. https://doi.org/10.1007/s12155-008-9008-8
51. Schroeder LR, Gentile VM, Atalla RH (1986) Nondegradative preparation of amorphous cellulose. J Wood Chem Technol 6(1):1–14. https://doi.org/10.1080/02773818608085213
52. Senthamilselvi D, Kalaiselvi T (2022) Gamma ray mutants of oleaginous microalga Chlorella sp. KM504965 with enhanced biomass and lipid for biofuel production. Biomass Convers Biorefinery 13(17):15501–15517. https://doi.org/10.1007/s13399-022-02400-9

53. Swinwood JF, Waite TD, Kruger P, Rao SM (1994) Radiation technologies for waste treatment: a global perspective. IAEA Bull 36(1):11–15. https://www.iaea.org/sites/default/files/publications/magazines/bulletin/bull36-1/36102681115.pdf
54. Taherzadeh MJ, Karimi K (2008) Pretreatment of lignocellulosic wastes to improve ethanol and biogas production: a review. In: International Journal of Molecular Sciences, Vol. 9, Issue 9. https://doi.org/10.3390/ijms9091621
55. Tahri L, Elgarrouj D, Zantar S, Mouhib M, Azmani A, Sayah F (2010) Wastewater treatment using gamma irradiation: Tétouan pilot station, Morocco. Radiat Phys Chem 79(4):424–428. https://doi.org/10.1016/j.radphyschem.2009.11.008
56. Takács E, Wojnárovits L, Borsa J, Földváry C, Hargittai P, Zöld O (1999) Effect of γ-irradiation on cotton-cellulose. Radiat Phys Chem 55(5–6):663–666. https://doi.org/10.1016/S0969-806X(99)00245-5
57. Tan J, Li Y, Tan X, Wu H, Li H, Yang S (2021) Advances in pretreatment of straw biomass for sugar production. Front Chem 9(June):1–28. https://doi.org/10.3389/fchem.2021.696030
58. Tissot C, Grdanovska S, Barkatt A, Silverman J, Al-Sheikhly M (2013) On the mechanisms of the radiation-induced degradation of cellulosic substances. Radiat Phys Chem 84:185–190. https://doi.org/10.1016/j.radphyschem.2012.06.020
59. Toghyani MA, Karimi F, Hosseini Tafreshi SA, Talei D (2020) Two distinct time dependent strategic mechanisms used by *Chlorella vulgaris* in response to gamma radiation. J Appl Phycol 32(3):1677–1695. https://doi.org/10.1007/s10811-020-02106-3
60. Verde SC, Silva T, Matos P (2016) Effects of gamma radiation on wastewater microbiota. Radiat Environ Biophys 55(1):125–131. https://doi.org/10.1007/s00411-015-0617-2
61. Viikari L, Vehmaanperä J, Koivula A (2012) Lignocellulosic ethanol: From science to industry. Biomass Bioenerg 46:13–24. https://doi.org/10.1016/j.biombioe.2012.05.008
62. Wang CC, Chang CW, Chu CP, Lee DJ, Chang BV, Liao CS (2003) Producing hydrogen from wastewater sludge by *Clostridium bifermentans*. J Biotechnol 102(1):83–92. https://doi.org/10.1016/S0168-1656(03)00007-5
63. Wang K, qin, Xiong, X. yao, Chen, J. ping, Chen, L., Su, X., & Liu, Y. (2012) Comparison of gamma irradiation and steam explosion pretreatment for ethanol production from agricultural residues. Biomass Bioenerg 46:301–308. https://doi.org/10.1016/j.biombioe.2012.08.013
64. Wen G, Voroney RP, Winter JP, Bates TE (1997) Effects of irradiation on sludge organic carbon and nitrogen mineralization. Soil Biol Biochem 29(9–10):1363–1370. https://doi.org/10.1016/S0038-0717(97)00057-6
65. Yang C, Shen Z, Yu G, Wang J (2008) Effect and aftereffect of γ radiation pretreatment on enzymatic hydrolysis of wheat straw. Biores Technol 99(14):6240–6245. https://doi.org/10.1016/j.biortech.2007.12.008
66. Yang G, Wang J (2018a) Kinetics and microbial community analysis for hydrogen production using raw grass inoculated with different pretreated mixed culture. Biores Technol 247(July 2017):954–962. https://doi.org/10.1016/j.biortech.2017.09.041
67. Yang G, Wang J (2018b) Pretreatment of grass waste using combined ionizing radiation-acid treatment for enhancing fermentative hydrogen production. Biores Technol, 255(December 2017):7–15. https://doi.org/10.1016/j.biortech.2018.01.105
68. Yang G, Wang J, Shen Y (2019) Antibiotic fermentation residue for biohydrogen production using different pretreated cultures: performance evaluation and microbial community analysis. Biores Technol 292(August):122012. https://doi.org/10.1016/j.biortech.2019.122012
69. Yang G, Yin Y, Wang J (2019) Microbial community diversity during fermentative hydrogen production inoculating various pretreated cultures. Int J Hydrogen Energy 44(26):13147–13156. https://doi.org/10.1016/j.ijhydene.2019.03.216
70. Yin Y, Hu J, Wang J (2014) Gamma irradiation as a pretreatment method for enriching hydrogen-producing bacteria from digested sludge. Int J Hydrogen Energy 39(25):13543–13549. https://doi.org/10.1016/j.ijhydene.2014.01.147
71. Yin Y, Wang J (2016) Gamma irradiation induced disintegration of waste activated sludge for biological hydrogen production. Radiat Phys Chem 121:110–114. https://doi.org/10.1016/j.radphyschem.2016.01.009

72. Yoon M, Choi J il, Lee JW, Park DH (2012). Improvement of saccharification process for bioethanol production from Undaria sp. by gamma irradiation. Radiat Phys Chem 81(8):999–1002. https://doi.org/10.1016/j.radphyschem.2011.11.035
73. Yu SJ, Shen XF, Ge HQ, Zheng H, Chu FF, Hu H, Zeng RJ (2016) Role of sufficient phosphorus in biodiesel production from diatom *Phaeodactylum tricornutum*. Appl Microbiol Biotechnol 100(15):6927–6934. https://doi.org/10.1007/s00253-016-7641-2

Chapter 11
Electric Field and Pulsed Electricity-Based Technologies for Agricultural Waste Biomass Valorization: Mechanism, Recent Technical Advances, Limitations, and Future Aspect

Cunshan Zhou, Man Zhou, Haoxin Li, and Xingyu Fan

Abstract This chapter presents an overview of recent advances in electro-technologies used in the valorization of agri-food waste biomass. Pulsed electric fields (PEF), high voltage electrical discharges (HVED), ohmic heating (OH), and moderate electric fields (MEF) as extraction technologies for the valorization of agri-food waste biomass are comprehensively described. As a class of green and safe media, deep eutectic solvent (DES) is considered generally recognized as safe (GRAS) and can assist these electro-technologies for the valorization of agricultural waste biomass. Particularly, instrumental design, critical parameters, and recent advances in electro-technologies in waste biomass valorization are highlighted. Numerous high value compounds that are interesting for their technological functions, nutritional properties, or health-promoting attributes have been valorized from various agri-food waste biomass. Furthermore, current limitations involved in the equipment and infrastructure, energy consumption, scalability, process optimization, product selectivity and diversity, and environmental impact assessment remain to be addressed. Finally, future perspectives in the context of agricultural biomass valorization via electric-based techniques are also included.

Keywords Agri-food waste biomass · Valorization · Pulsed electric fields · High voltage electrical discharges · Ohmic heating · Deep eutectic solvents

C. Zhou (✉) · M. Zhou · H. Li · X. Fan
School of Food and Biological Engineering, Jiangsu University, Zhenjiang, China
e-mail: cunshanzhou@163.com

© The Author(s), under exclusive license to Springer Nature Switzerland AG 2024
P. Verma and B. Kumar (eds.), *Sustainable Radiation Technologies in Waste-biomass Valorization*, Green Chemistry and Sustainable Technology,
https://doi.org/10.1007/978-3-031-63941-8_11

11.1 Introduction

Each year, 1.3 billion tons of food are lost or wasted globally and account for one-third of all food produced for human use, according to the Food and Agriculture Organization (FAO). As a result, agri-food waste biomass is one of the most important stakeholders in waste and by-product valorization. The disposal of these wastes is expensive for the food sector and harmful to the environment. The 2030 Agenda for Sustainable Development Goals 12—Responsible Consumption and Production—indicates growing worldwide concern about this issue.

There has been increasing interest in electric fields and pulsed electricity-based technologies, including pulsed electric fields (PEF), high voltage electrical discharges (HVED), ohmic heating (OH), and moderate electric fields (MEF), as non-thermal, green assistant techniques for the valorization of agri-food waste biomass in recent years. Agri-food waste biomass can be converted into a variety of valuable compounds. These electro-technologies can be used with or after solid(liquid)–liquid extraction to increase selectivity and yield in the recovery of important chemicals. Recently, deep eutectic solvent (DES) has received increasing attention as the extraction media for the valorization of agri-food waste biomass [62]. DES has many merits including the availability of raw materials, ease of synthesis, tunable attributes, as well as good biodegradability and recyclability. These characteristics are instrumental in the valorization of agri-food waste biomass [61].

This chapter provides an in-depth discussion of the basics, instrumental design, and other recent breakthroughs in PEF, HVED, OH, MEF, and DES as a valorization method for agri-food waste biomass. Furthermore, covered are the existing constraints and anticipated prospects for the utilization of electric-based methods to value agricultural biomass.

11.2 Fundamentals of Electric Field and Pulsed Electricity-Based Technologies

11.2.1 Basic Principles

Electric fields and pulsed electricity-based technologies, as new and developing bioprocess technologies, rely on using electric current in biomass to achieve technological objectives. Electro-technologies, namely PEF, HVED, OH, and MEF, have flourished and demonstrated great promise for the biomass derived from agri-food waste [47]. Electro-technologies can improve the extraction and diffusion of chemicals, help with separation, and stabilize biomaterials, all of which are important benefits in the various processes of agri-food waste biomass valorization [49].

When applying an electric field on biomaterials, there are two effects: electroporation (non-thermal) and Joule heating (thermal). Electroporation is an increasingly prevalent electric field effect. When a cell is located in an electric field, it develops a transmembrane potential in its membrane. The effectiveness of the electric field is determined by the field strength applied and cell radius. When the transmembrane potential surpasses a certain threshold (i.e., 0.2–1.0 V), the electric field causes the membrane to electro-permeabilize. The creation of pores in biological cell membranes occurs upon application of the necessary critical transmembrane potential. When an electric field is applied, the pores that have formed with a radius of around 0.5 nm may grow and destroy cells, causing irreversible electroporation (Fig. 11.1a). After removing the applied voltage, the cell cannot return to its original location in irreversible electroporation. This effect may promote the flux of small molecules and cellular materials through the membrane, as well as the damage and destruction of cells [26]. Figure 11.1b illustrates the electroporation phenomena of membrane permeabilization in an electric field. Compression from the enlarged vacuole brought on by electromechanical compression causes elongated cell architectures in response to MEF treatment [24]. Figure 11.1c shows that compressive pressures during the MEF treatment can improve membrane permeability, leading to preferable extraction performances than those of traditional methods. Electro-permeabilization, on the other hand, is more likely at moderate temperature (<50 °C), while high temperature can cause thermal denaturation, cell expansion, and membrane damage.

The Joule effect causes heat to be generated in the medium as another impact of being exposed to electric fields. Because biological or bio-based systems function as semiconductors, applying electric fields to them will cause heat dissipation. Ohmic heating (OH) is the term for this phenomenon, which the Joule effect explains. It delivers a rapid and uniform heating rate as well as high energetic efficiency. Heating originates from the friction that occurs during the molecular agitation caused by the movement of charged molecules. Interests in the sterile and extraction processes have increased due to the Joule effect's rapid, uniform, and energy-efficient (>90%) heating [26]. Cell membranes may become thermally and electrically permeable due to the joule effect, resulting in modifications in permeability and structure that contribute to the release of greater quantities of target molecules [11].

Table 11.1 lists the main mechanism and key process parameters of PEF, HVED, OH, and MEF. The nature of the electric flow including the electric field strength, electrical pulse, pulse width, number of pulses, current, voltage, frequency, and energy consumption is one of the criteria used to categorize various approaches. These parameters determine the induced effects and are critical in defining the equipment design and operational requirements. In the upcoming section, the main electro-technologies applied to the valorization of agricultural waste biomass will be provided.

11.2.2 Pulsed Electric Fields

Pulsed electric fields (PEF) employ electrical pulses to a product positioned between two electrodes at high voltages (0.1–80 kV/cm) and short durations. PEF is typically

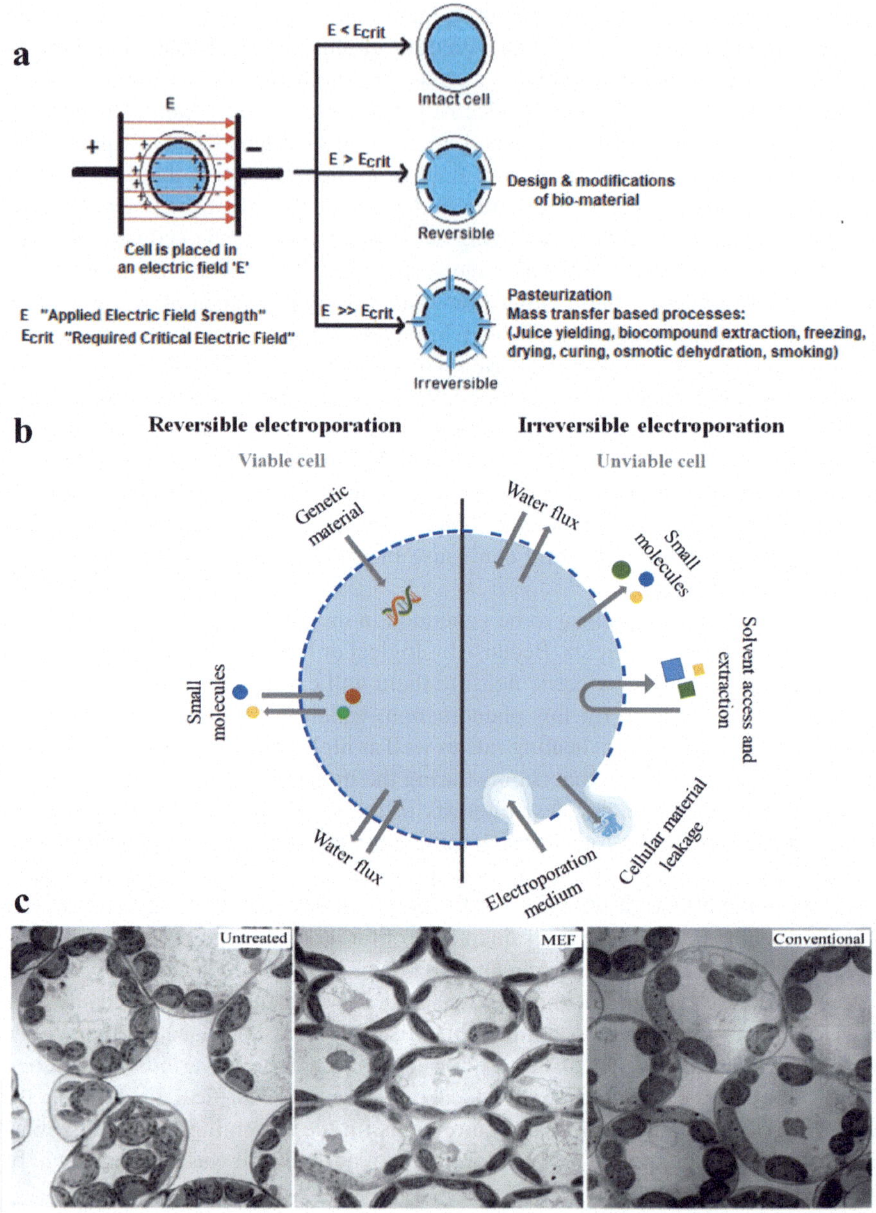

Fig. 11.1 Permeabilization of the cell membrane exposed in electrical fields (**a**), different levels and consequences of electroporation (**b**), and cell membrane permeabilization of fresh mint leaves upon MEF and conventional method (**c**). *Source* [6, 26, 24]

Table 11.1 Main mechanism and key process parameters of PEF, HVED, OH, and MEF

	PEF	HVED	OH	MEF
Main mechanisms involved	Increased mass transfer and the creation of holes in the cell membrane	Shock waves, mechanical forces from bubbles and turbulence, UV radiations, and active radicals	Volumetric heating, which is quick and uniform	Enhanced mass transfer from electroporation
Electric field strength (kV/cm)	0.1–80	20–80	0.1–0.8	<0.1
Other key process parameters	Pulsation; Frequency	Time (discharge duration time t_i plus number of pulses n); Energy per pulse	Frequency; Electrical conductivity	Frequency; Process temperature

performed at room temperature or a little bit warmer. Therefore, PEF is considered a nonthermal treatment technique. Electroporation or electro-permeabilization is the fundamental principle of PEF. When biological cells are exposed to an applied voltage and accompanying electric field that exceeds the critical transmembrane potential during PEF treatment, electroporation takes place.

Pore formation happens in the cell membrane after providing the requisite critical transmembrane voltage. Electroporation is the process of forming reversible or irreversible pores. These qualities enable PEF to be extensively applied in the biomass value-adding process of agricultural waste [6]. PEF has demonstrated significant promise in the extraction of useful compounds from a variety of biomass derived from agri-food waste. Valuable proteins, polyphenols, isothiocyanates, carbohydrates, and pigments have been retrieved from flaxseed hulls [9], grape pomace [7, 55], papaya seeds [43], rapeseed waste [60], pear peels and pulps [34], coffee by-products [8], waste chicken meat [27], Brewer's spent grains [35, 39], tomato waste [4, 45], olive waste [5], fish residue [22], orange peel [21], noni-processing wastes [36], shrimp waste [15], and almond hull [55].

11.2.3 High Voltage Electrical Discharges

High voltage electrical discharges (HVED) are based on dielectric breakdown. The liquid medium becomes ionized due to the application of a high voltage (30–40 kV) and intense pulse (~10 kA) between electrodes for a brief period (μs–ms). During the streamer discharge process, relatively modest shock waves, tiny bubble fragments, UV radiations, and active radicals are generated in the solvent. Shock waves gain significant strength after transitioning to arc discharge because of the amplified electrohydraulic action. Intense turbulence, UV radiation, a significant number of bubbles with plasma inside, and highly concentrated free radicals are all produced by the arc channel [37]. These occurrences lead to the disintegration of intricate structures and a rise in mass transfer, hence enhancing extraction efficiency. HVED has successfully extracted various high-value compounds from inclusive agri-food waste biomass, such as grape pomace [7], papaya seeds [43], sugar beet pulp [3], pomegranate waste [30], coffee by-products [17], peanut shells [59], and mustard seed meal [10].

11.2.4 Ohmic Heating

By applying an electric current to materials and inducing ions to migrate toward electrodes with opposing charges, a process known as ohmic heating is used to heat materials. Ions encounter opposition to their motion and gain kinetic energy as a result of collisions with one another. When the applied electric field is moderate, the sample has enough conductivity, and the process lasts long enough, considerable heat deposition happens via the Joule effect, and therefore OH takes place. The amount of heat produced is determined by the current, voltage, electric field, and electrical conductivity of biomass. Ohm's law governs how much electrical current can flow through a substance when it turns into electrical resistance.

OH is possibly the most successful example of electro-technology with practical applications, as it is widely used and marketable in the food business as a pasteurization strategy. OH has wide applications because of the combination of the below-described two mechanisms: (1) increased permeability as a result of the applied electric field, and (2) volumetric heating. Firstly, electro-permeabilization causes molecular leakage due to increased transmembrane potential. Secondly, volumetric heating in OH raises temperatures significantly quicker than traditional conduction and convection heating, and an increase in applied voltage raises the heating rate [48]. Although OH combines electrical and thermal effects, it is mainly referred to as thermal processing.

OH is an effective method to extract valuable compounds from agri-food waste biomass. Many valuable compounds, including polyphenols, pectin, carotenoids, oils, chitosan, and amino acids, have been recovered from grape pomace [18], orange and citrus waste [51, 58], tomato by-products [11, 12], wheat bran [2], carp viscera [1], pineapple core [23], crab shells [31], pomegranate peel [57], poultry slaughterhouse waste [32], and winemaking by-products [13].

11.2.5 Moderate Electric Fields

Moderate electric field (MEF) is a technique for regulated and perhaps reversible permeabilization that employs relatively low (typically below 1 kV/cm) and non-pulsed electric fields with arbitrary waveform (typically sinusoidal or square) electric fields [49]. Despite the applied field strength being relatively low, MEF treatment promotes at least reversible electroporation of the cell membranes, enhancing permeability. Due to the exposure of the food samples to the electric field, electrical effects and thermal effects are observed. The Joule effect is rather modest in comparison to OH, and the electrical effect becomes more apparent. As a result, the sample is subjected to both thermal and electric fields. MEF facilitates the diffusion of intracellular compounds from the intracellular matrix by promoting cell membrane permeabilization [46]. MEF has shown great potential in the extraction of pectin from yellow passion fruit peel [16] and oleuropein from olive leaves [14]. Besides, MEF can aid fermentation processes through thermal and electrical effects, namely fast temperature adjustment, micronutrient release from substrates, and increased cell permeability [25].

The acronyms MEF and OH are usually used interchangeably, but they have some differences. Generally, OH refers to thermal treatment, whereas MEF emphasizes the electrical effect when a thermal aspect is attenuated or compared on a similar basis. Although PEF remains the most commonly-mentioned extraction technology, MEF requires lower operational conditions and related heating characteristics, electric generators, electrode materials, and control systems, compared to more difficult high-energy pulsed applications. These features may promote the implementation of MEF.

11.3 Instrumental Design of Electric Field and Pulsed Electricity-Based Technologies

11.3.1 Instrumental Design of Pulsed Electric Field

PEF equipment comprises a pulse generator, a treatment chamber, two or more electrodes, and a control system [28]. Figure 11.2a schematically depicts these fundamental components. The electric generator produces a pulse voltage not exceeding 50 kV. One electrode is linked to the electric generator, a grounded electrode is present in the treatment chamber, and the waste biomass is placed between the electrodes (Fig. 11.2b). The stored electrical energy in the storage capacitor is released through the treatment chamber and the potential difference induces an electric field in the biomass material.

The configurations of electrodes can be parallel, co-axial, or co-linear, as shown in Fig. 11.2c. The most commonly used configuration is the collinear type since it greatly minimizes the electrode surface and the required amperage. The widely used pulses are unipolar or bipolar, with waveforms of square or exponential waves (Fig. 11.2d). The square waveform is most suitable for optimizing PEF application performance since the intensity of power attenuation remains unchanged throughout the pulse width.

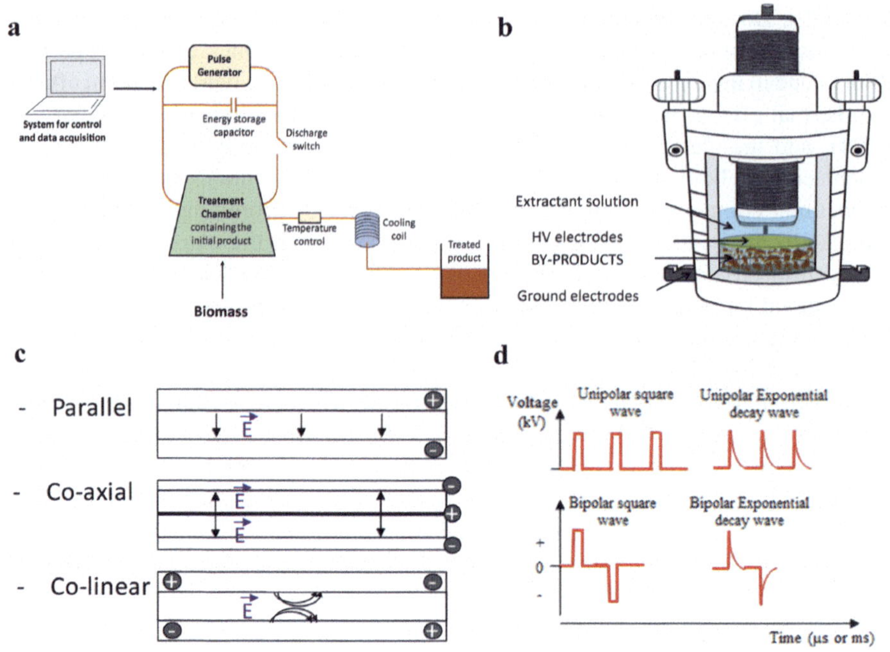

Fig. 11.2 Schematic of fundamental PEF systems (**a**), PEF batch device (**b**), electrode configurations (**c**), and pulse protocols (**d**). *Source* [28, 46, 47]

Several process parameters are critical when designing a PEF process, including electric field strengths E, number of pulses n, pulse duration t_i, pulse shape, and frequency [47]. These variables should be considered to evaluate the efficiency. The specific energy input (W) of the process (depending on the input voltage, the ohmic resistance of processed products, and the processing time) is intimately related to the processing cost and eco-footprint [28]. Generally, W increases with the increase of PEF strength.

11.3.2 Instrumental Design of High Voltage Electrical Discharges

An HVED apparatus consists of two main basic components: (i) a high-voltage pulse generator (as in PEF) and (ii) a treatment chamber. The generator is comprised of a high-voltage direct current generator, a capacitor, a high-voltage switch, a chamber, and an automatic control system. Figure 11.3a shows a cylindrical treatment chamber. It contains two electrodes: one connected to the high-voltage generator (needle geometry) and one connected to the ground electrode (plane geometry), where the pulses are discharged. The electrical energy is stored in a low-inductance capacitor that has already been charged using the high-voltage generator. An oscilloscope is attached to the voltage and current measurements. The automatic control system provides the capability to control treatment, pulse frequency, and mixing speed [47]. As shown in Fig. 11.3, three types of HVED systems have been developed, namely batch, continuous, and circulating operations [37].

Several parameters are critical in the application of HVED, including the pulse amplitude U, number of pulses n, discharge energy input W, total treatment duration t, and the inter-electrodes gap distance. The energy at electrode breakdown varies with the variation of electrode clearance. If the clearance is less than 5 mm, some points of the electrode needle may experience arc discharge, while the intensity of corona discharge at other locations will rapidly decrease. The gap distance is adjusted based on the conductivity of the sample [37].

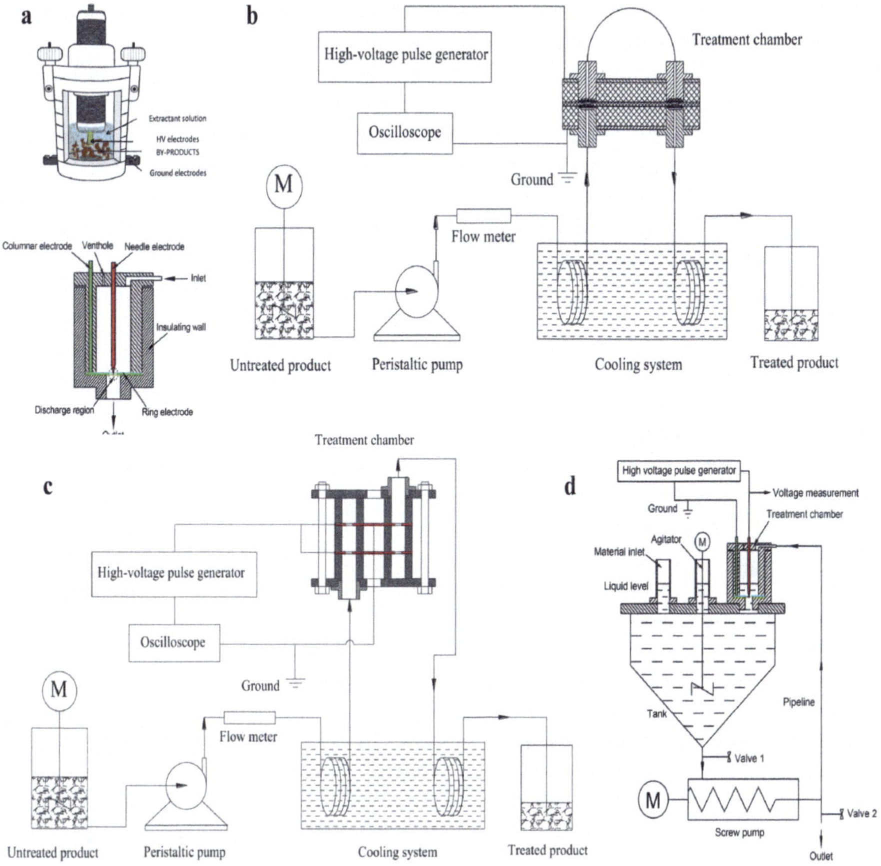

Fig. 11.3 Schematic diagrams of batch (**a**), continuous (**b** and **c**), and circulating (**d**) systems of HVED. *Source* [37]

11.3.3 Instrumental Design of Ohmic Heating

The OH system includes no less than two electrodes to apply current to the fluid. The electrodes are designed according to their locations and positions. This design can be configured as either a static system in the container or a continuous flow through it. Generally, OH systems come in two broad designs. The open geometries are easy to clean and can protect products from fouling and damage.

Figure 11.4a illustrates an OH system setup for the recovery of valuable compounds from wheat bran [44]. Generally, an OH system is made up of four parts: power supply (voltage generator), control system (thermocouple, digital temperature gauge, pressure measuring device, and control unit), electrodes, and extraction chamber. OH electrodes are usually arranged in four different arrangements (Fig. 11.4b): (1) parallel plate, (2) parallel rod, (3) collinear, and (4) staggered rod [54]. Which configuration is adopted depends on the conductivity of the heating material, applied voltage, current distribution, and capital cost. The electrode gap affects the amount of heat generated and electric field strength. Generally, the electrode gap is set between 9 and 14 cm. Various conductive electrode materials including stainless steel, titanium, platinized-titanium, and graphite have been developed for electrode design. Electrodes are commonly selected according to price and correction resistance. Usually, stainless steel is preferred.

The volume and density of the material to be treated determine the size of the OH chamber. PVC pipes, acrylic sheets, perspex sheets, wood, glass, and other materials are commonly employed in the construction of OH chambers. Meanwhile, a high

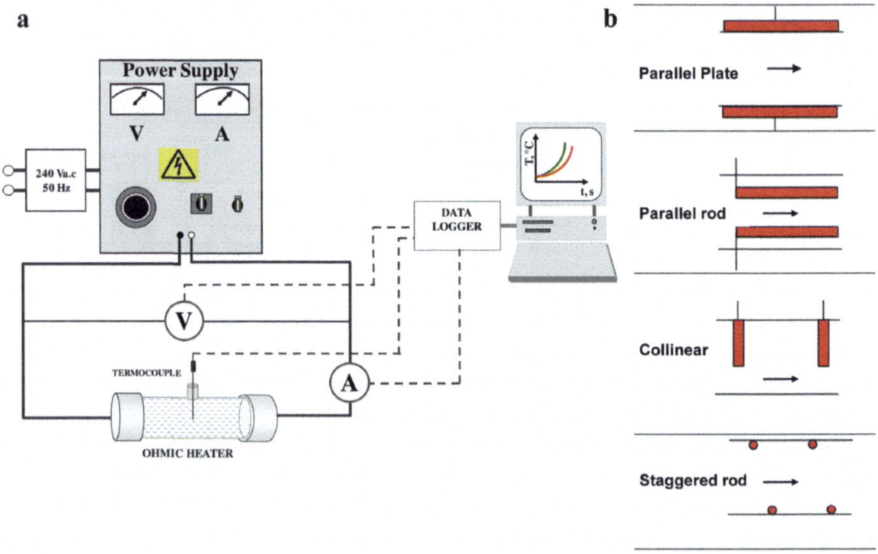

Fig. 11.4 Schematic diagram of the OH setup (**a**) and typical electrode configurations in flow through OH (**b**). *Source* [44, 54]

frequency of power supply (commonly 50 Hz and 60 Hz) significantly prevents corrosion and visible metal dissolving. Low frequency will result in the electroporation of cell membranes, leading to a rapid rise in electrical conductivity and altering the texture of end products. Therefore, using a high frequency reduces electroporation, while increasing the time required to reach the set temperature [33].

Designing an applicable OH system necessitates multiple considerations (for example, selecting proper electrodes and modifying electrical parameters) as well as engineering calculations. There are several important parameters for the OH system: electric conductivity, electrical field strength, current, applied voltage, temperature, frequency, and waveform. In summary, the most significant equations to consider when constructing OH systems are the power equation and Ohm's law, which combined derive Joule's first law [25].

11.3.4 Instrumental Design of Moderate Electric Fields

Figure 11.5a shows a schematic illustration of the MEF system setup for the valorization of olive leaves [14]. The MEF apparatus generally consists of a transformer/wave moderator, a microprocessor (which records temperature, current, and voltage data), an oscilloscope, and a polytetrafluoroethylene extraction cell. Two different waveforms, namely sine and square waveforms (as shown in Fig. 11.5b) are usually adopted in MEF. Electric field strength, applied frequency, wave type, and temperature are key parameters in the application of MEF [14]. High frequency (>17 kHz) can greatly prevent electrochemical reactions which induce the degradation and corrosion of electrodes as well as the contamination of products. Although low frequencies originate undesired electrochemical effects, MEF with low frequencies usually has a better efficiency on cell permeabilization. Besides, the influence of sample direction on the electric field was an effective parameter (Fig. 11.5c) [24].

11.4 Recent Advances in Retrieving Valuable Products from Agri-Food Waste Biomass Using Electric Fields and Pulsed Electricity-Based Technologies

Agri-food waste biomass is abundant in bioactive molecules and can be valorized into various valuable compounds and products. Figure 11.6 illustrates the application of four electric fields and pulsed electricity-based technologies in retrieving various valuable compounds from agri-food waste biomass. They are appealing for their (i) nutritional properties (e.g., proteins, carbohydrates, and phenolic compounds), (ii) technological functions (e.g., pigments), or (iii) significant health benefits (e.g., antioxidant effect) [47].

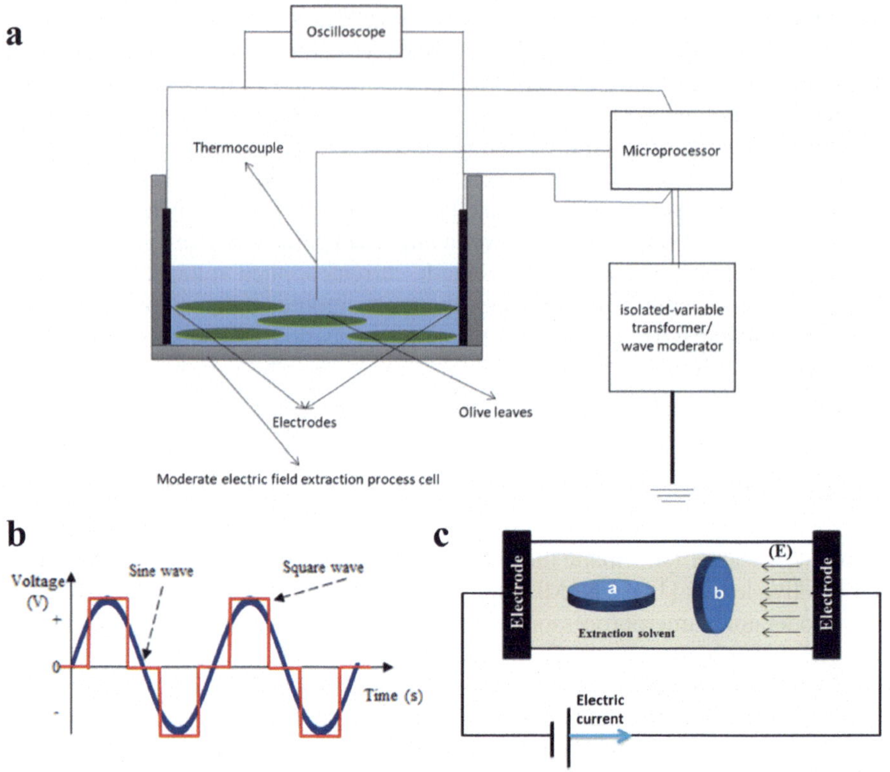

Fig. 11.5 Schematic illustration of MEF system setup (**a**), sine and square waveforms (**b**), and sample orientation (perpendicular and parallel) concerning the electric field (**c**). *Source* [14, 24]

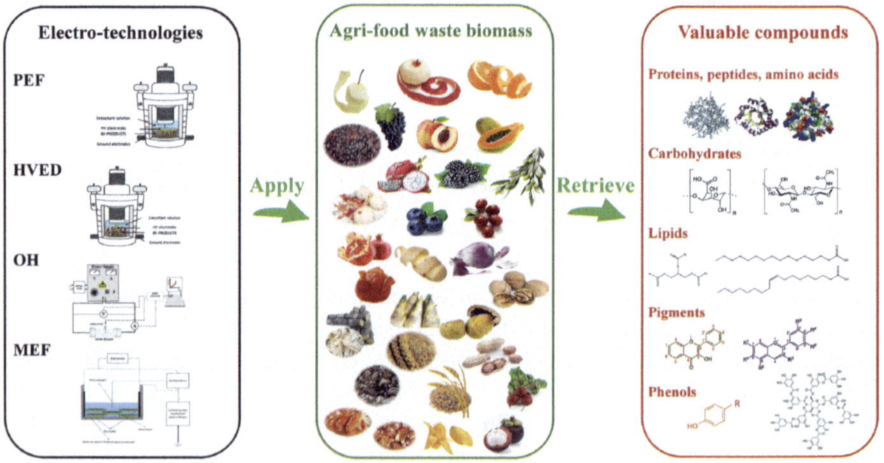

Fig. 11.6 Retrieving various valuable compounds from agri-food waste biomass using four electric fields and pulsed electricity-based technologies

11.4.1 Retrieving Proteins, Peptides, and Amino Acids from Agri-Food Waste Biomass

Proteins, peptides, and amino acids are the fundamental macromolecules of life. They are deeply involved in a series of biochemical reactions and life activities. The extraction of protein from papaya peels using conventional methods and methods assisted by PEF or HVED was studied. The HVED-assisted technique had a better extraction performance of proteins than PEF-assisted extraction [41]. Similar findings were reported from the olive kernel [50] and mango peels [42]. HVED treatment was demonstrated to be more efficacious than PEF from the perspectives of treatment time and energy input. The efficiency of HVED-assisted extraction is high, which can reflect the suspension turbulence and pill fragmentation caused by electrical discharges. Nevertheless, PEF treatment obtained more stable and clearer extracts than HVED, which was beneficial for the follow-up steps. Besides, an HVED pretreatment which was followed by a solid-DES extraction improved the selectivity and the yield of proteins [30]. All tested DESs favored the extraction of proteins. In particular, ChCl:gluc extracted more than 94% protein. The combined use of HVED and DES treatment demonstrated an efficient and green strategy for valorizing pomegranate seeds. Comparing HVED with ultrasounds, it was observed that HVED had a better performance in releasing 70% proteins.

The extraction yields of proteins from rapeseed stems and leaves were highly increased by PEF [60]. Under moderate PEF treatment, the protein extraction yield did not significantly improve. Whereas high PEF treatment significantly promoted protein yield (~80%). Franco and co-workers used PEF to extract antioxidants from commercial fish residues [22]. PEF-assisted water extracted more antioxidants than water alone, suggesting the great potential of PEF in retrieving antioxidative proteins from fish processing by-products. A PEF-based process combining high voltage, short pulses followed by low voltage, long pulses has been used to extract protein from waste chicken breast muscle [27]. This two-step protocol enabled the extraction high content of protein (78 $\pm$ 8 mg/mL) with antioxidant properties. PEF pre-treatment remarkably improved the extraction yields of protein from dark Brewer's spent grains [35]. After PEF treatment, the recovered protein from olive pomace was twice that of the control group [4]. The amounts of total amino acids retrieved by OH-subcritical water were 70.48% higher than that of the control. Kadem and co-workers found that OH-subcritical water extracted much more (79.40%) amino acids than the control treatment (15.48%) from poultry slaughterhouse waste [32]. PEF is an efficient and green technology that contributes to Sustainable Development Goals achievement via decreasing power costs, improving process greenness, and increasing resource efficiency through waste valorization.

11.4.2 Retrieving Carbohydrates from Agri-Food Waste Biomass

PEF, HVED, OH, and MEF are prime extraction strategies for recovering carbohydrates from diverse agri-food waste biomass. The yield, physicochemical characteristics, and structure of orange peel soluble dietary fiber (SDF) were examined in relation to the impacts of PEF-assisted extraction [21]. PEF-extracted SDF had a lower molecular weight and higher antioxidant activity than the control sample. PEF can selectively extract soluble carbohydrates from almond hulls, primarily glucose, fructose, and sucrose [55, 56]. HVED preparation increased the pectin extraction rate from sugar beet pulp while preserving chemical composition and structure [3]. Many researchers have used OH to extract pectin with high quality from orange waste [52, 53], citrus waste [51], lemon waste [58], and pomegranate peel [57]. Moreover, the determination of OH rate and system performance coefficients at different conditions (such as voltage gradient) is crucial for designing industrial OH extraction. de Oliveira and co-workers found that MEF extraction was more efficient and environment-benign than conventional heating extraction to extract pectin from passion fruit peel [16].

11.4.3 Retrieving Lipids from Agri-Food Waste Biomass

Some lipids are considered bioactive ingredients in humans. Pare and co-workers investigated the oil recovery from soybean seed under the effect of OH parameters [40]. The maximum oil recovery rate (73%) was achieved when the sample was heated using a 600 V electric field strength and held at 90°C for 10 min. Asaad [1] used OH to extract oil from *carp viscera* processing by-products. OH had greater heating rate values, lower specific energy consumption, and less extraction time than conventional heating. Salgado-Ramos and co-workers first reported the combined sequential extraction by PEF and supercritical (SC) fluid extraction (SFE) with CO_2 (SC-CO_2) for the valorization of almond hull biomass [56]. PEF + SFE selectively extracted the lipids fractions. Different from conventional methods, PEF does require boiling and evaporating oil and water before condensation. In addition, electrical field strength helps oil cells break down to release oil. These two factors contribute to the enhancement of oil extraction.

11.4.4 Retrieving Pigments from Agri-Food Waste Biomass

Commonly used pigments in the food industry include anthocyanins, lycopene, betanin, chlorophyll, astaxanthin, etc. These pigments can be recovered from the low-cost and abundant agri-food waste biomass. For example, anthocyanins can be isolated from winemaking by-products [12]; lycopene can be isolated from tomato processing by-products [11, 13, 45]. Barba and co-workers compared the alternative treatments (PEF and HVED) on solvent-free extraction of anthocyanins from fermented grape pomace. They found that the PEF-assisted method was more effective than the HVED-assisted method in terms of extracting anthocyanins [7]. The extraction yields of anthocyanins were enhanced by up to 55% compared to extraction PEF treatment resulting in a high recovery of betanin from red prickly pear [34]. The high recovery was ascribed to the increased cell permeability induced by PEF, which allows betaine molecules to move from intracellular to extracellular. PEF has recently been used to recover lycopene from tomato processing by-products [4, 45]. In a more recent study, PEF extracted 585.90 μg/g of astaxanthin from shrimp side streams [15].

11.4.5 Retrieving Phenolic Compounds from Agri-Food Waste Biomass

PEF, HVED, OH, and MEF are cost-effective and environmentally benign methods for retrieving phenolic compounds. PEF treatment increased polyphenol extraction from rapeseed hulls by 80% [9]. The resulting polyphenols were less turbid and had larger particle sizes than milling samples. This feature simplified the subsequent purifying procedure. Increased recovery of polyphenols from PEF-treated mango peel [42], citrus waste [19], coffee industry by-products [8], and spent grains [35, 39] were also reported. The HVED-assisted technique outperformed the PEF-assisted technique in terms of total phenolic compound extraction efficiency [41, 42, 50]. Additionally, OH is an efficient technique to retrieve phenolic compounds. For example, when the grape peel was subjected to an OH treatment, the extraction yield of polyphenolic compounds increased by 36% [18], OH treatment of tomato peel increased the recovery rates of rutin by 77% when compared to the control [11]. More recently, researchers have developed several more efficient and environment-benign methods based on HVED or PEF combined with DESs to retrieve phenolic compounds [20, 20]. HVED treatment followed by DES extracted more flavonoids from grapefruit peels and reduced the energy input of HVED [20]. PEF combined with DESs recovered more target flavonoids from noni pomace waste than conventional organic extraction [36].

11.4.6 Retrieving Other Valuable Products from Agri-Food Waste Biomass

Recently, economically valuable bioactive compounds have been extracted from various agri-food waste biomass. The advantages of the PEF technique have been proved for miscellaneous bioactive substances like calcium malate [38], chondroitin sulfate [29], methylxanthine, and essential oils [58]. PEF is an efficacious and eco-friendly method to prepare calcium malate from eggshells, and PEF-treated calcium malate was beneficial for calcium absorption in vitro [38]. PEF has demonstrated its advantages as a cost-effective method to extract chondroitin sulfate [29]. Under the optimal PEF conditions, the supreme yield of chondroitin sulfate was quickly obtained (5 min). Moreover, the quality of obtained chondroitin sulfate was quite good, with no impurities such as glycosaminoglycans. PEF-assisted extraction yielded greater (~20%) methylxanthines from cocoa bean shells and coffee silver skin than the conventional method [8]. More recently, the ohmic heating-assisted extraction/hydro distillation process has tremendously extracted more essential oil (limonene) when compared to conventional heating extraction/hydro distillation [58].

11.5 Future Prospects

Electric field and pulsed electricity-based technologies have shown great potential in the valorization of agri-food waste biomass. However, there are certain limitations and future perspectives that need to be considered. Table 11.2 summarizes the pros, cons, and prospects of four electro-technologies.

11.5.1 Limitations

Although PEF, HVED, OH, and MEF are progressing toward broader industrial applications, some limitations still impede their commercialization into the valorization of agri-food waste biomass. Some technical challenges, economic disadvantages, market adoption, and regulatory ramifications remain to be tackled. The following are the primary constraints to the adaptation of these technologies:

11.5.1.1 Equipment and Infrastructure

The requirement for more reliable electrical systems is widely perceived as a major barrier to the commercialization of these electro-technologies. Therefore, there is a tremendous potential to optimize processes through energy efficiency. The treatment chamber of PEF causes fouling, corrosion, and chemical changes in electrodes, as well as water electrolysis, interfering with the product during the treatment. Implementing electric field-based technologies requires specialized equipment and infrastructure. The initial investment for setting up these systems can be high, especially for small-scale farmers or developing regions. Additionally, the availability of skilled operators to handle and maintain the equipment may be limited.

Table 11.2 Applications, advantages, disadvantages, and prospect of electric field and pulsed electricity-based technologies

Methods	Advantages	Disadvantages	Prospects
PEF	✓ High selectivity ✓ Low operating temperature and temperature rise ✓ Low energy input	× Low efficiency of cell destruction	• Scientific studies to improve industrial-level operations • Model electroporation-related phenomena • Combination with DES
HVED	✓ High cell disruption ✓ Reduced solvent consumption ✓ Low operating temperature	× Production of free radicals × Less selectivity × Requirement of high-energy input	• Mechanisms and kinetics • Prevention of oxidation by free radicals • Increasement of processing capacity and determination of key factors in a pilot experiment • Combination with DES
OH	✓ Uniformity heating ✓ Higher energy efficiency ✓ Nonthermal addition effects (electroporation) ✓ Reduction of surface fouling ✓ No residual heat transfer after the current cutoff	× Short of general information × The necessity of using electroconductive solutions × Narrow frequency band × Difficult to monitor and control	• Identification and measurement of cold spots and hot spots • Developing advanced control devices to ensure the heating rate • Analyzing the relationship between OH parameters and physical–chemical properties of materials
MEF	✓ Improved cell permeability and microstructural changes ✓ Alteration of fermentation substrate and microbial metabolism	× High costs of electrodes × Generation of electrochemical reactions	• Process optimization and engineering calculations • Reduction of capital costs • Development of several fermented products using MEF

11.5.1.2 Energy Consumption

Electro-assisted technologies often require a significant amount of electrical energy to operate. This can increase the overall operational costs and dependency on the power grid. Due to high electricity consumption leading to massive carbon dioxide emissions, the use of renewable energy is a major choice. Finding sustainable and cost-effective energy sources for these technologies will be crucial in the future.

11.5.1.3 Scalability

While electro-assisted technologies have shown promise in laboratory-scale experiments, scaling them up for commercial applications can be challenging. The limited protocols also limit the comparison of laboratory scale data to commercial scale data. Ensuring consistent and efficient treatment of large volumes of agricultural waste biomass remains a hurdle that needs to be addressed.

11.5.1.4 Process Optimization

Sequential studies are needed to optimize the process parameters and design of electro-assisted technologies. Additionally, a portion of the input electrical energy is transformed into thermal energy, influencing the stability of thermo-sensitive compounds. Understanding the effects of different variables such as electric field strength, pulse frequency, and treatment duration on biomass valorization efficiency is essential for maximizing yields and product quality. For instance, during the OH process, there are still many challenges in controlling the heat rate because of the change in the electric conductivity of the heating material.

11.5.1.5 Product Selectivity and Diversity

The current focus of electro-assisted technologies has primarily been on biofuel production or extraction of value-added compounds. There is a need to explore other potential applications and products derived from agricultural waste biomass, such as bioplastics, biochar, and bio-based chemicals. Diversifying the range of products can enhance the economic viability of these technologies.

11.5.1.6 Environmental Impact Assessment

While electro-assisted technologies offer advantages in terms of waste reduction and resource recovery, their environmental impact needs a thorough evaluation. Life cycle assessments should be conducted to understand potential emissions, waste generation, and overall sustainability of these processes.

11.5.2 Future Prospects

The future prospects of these electro-technologies are as follows.

11.5.2.1 Mechanisms and Kinetics

It is of great importance to investigate the mechanisms and kinetics of these technologies. To model electroporation-related phenomena, interdisciplinary research from multiple scientific and engineering fields is needed. Another under-investigated problem of PEF is the electrochemical reactions occurring on the electrode surface. Besides, an in-depth investigation of the mechanisms of HVED is required. Although intense liquid turbulence and shock waves generated by HVED are known to improve mass transfer, specificity impacts, and kinetic models need continued study. The oxidation mechanisms of HVED by free radicals need to be studied and possible preventive methods should be developed.

11.5.2.2 Process Optimization and Parameter Control

Further research is required to investigate the effects of different process parameters on biomass valorization outcomes. Understanding the optimum electric field strength, pulse frequency, treatment time, and other variables can lead to improved yields, better product quality, and reduced energy consumption. The development of electrodes with novel materials is necessary to prevent fouling and corrosion. Additional research should be performed to optimize HVED extraction systems, for example, the development of continuous HVED systems to increase the processing capacity. To guarantee accurate and efficient treatment, real-time monitoring and control systems should also be created. During OH, it is crucial to precisely identify and measure the hot and cold regions.

11.5.2.3 Scale-Up and Commercialization

One important area for the future is the development of electro-assisted technologies for industrial use. This entails creating dependable and strong systems that can manage substantial amounts of biomass from agri-food waste while maintaining a steady and high-quality output. The cost-effectiveness of the technology should be taken into account during commercialization efforts to make it available to a wider variety of stakeholders. Moreover, the goal of research should be to make these technologies more efficient overall. This can involve maximizing biomass conversion through improved treatment, minimizing energy losses during operation, and optimizing the energy input–output ratio. These technologies can become much more economically viable and sustainable by increasing their efficiency.

11.5.2.4 Integration with Existing Systems

It is crucial to incorporate electro-assisted technology into the current agricultural waste management systems. For future applications, combining these electro-technologies with faster solvent extraction or enzyme-assisted extraction is a promising path. It is important to investigate compatibility with additional valorization processes including pyrolysis, anaerobic digestion, and fermentation to develop comprehensive and effective waste management strategies.

11.5.2.5 Co-Product Development

Exploring the potential for co-product development, in addition to biofuels and value-added compounds, is an essential future element. This entails broadening the breadth of agricultural waste biomass-derived goods, such as bio-based materials, biofertilizers, and animal feed additives. Increasing biomass usage can improve the overall sustainability and economic viability of electro-assisted technologies.

11.5.2.6 Environmental Sustainability

Future research should focus on evaluating and reducing the environmental impact of these technologies. This involves employing in conjunction with DES, analyzing greenhouse gas emissions, conducting life cycle assessments, and implementing waste generation and resource consumption reduction initiatives. Developing environmentally friendly and sustainable methods will be critical for these technologies' long-term acceptance and adoption.

11.6 Conclusions

With the rapid consumption of fossil fuels and environmental deterioration, the transition to a more sustainable bioenergy-based economy is crucial. The viability of replacing fossil fuels with waste biomass is high due to the readily available, abundant, and renewable nature. Agri-food waste biomass represents a huge potential resource for retrieving valuable compounds. The application of these energy-efficient electro-technologies, namely PEF, HVED, OH, and PEF can be a good choice for the valorization of agri-food waste biomass into high value compounds. These electro-technologies employ less solvent and can be jointly used with DES, improving their environmental friendliness and sustainability. Besides, the development of industrial-scale equipment is highly realizable. Overcoming the limitations and exploring future aspects will contribute to their wider adoption and commercial viability in the agricultural sector.

Acknowledgements The authors would like to thank the following funding in gathering the information and assessing different electro-technologies: National Natural Science Foundation of China (22378175 and 22308130).

The authors declare non-usage of AI tools during writing.

References

1. Al-Hilphy AR, Al-Mtury AAA, Al-Shatty SM, Hussain QN, Gavahian M (2022) Ohmic heating as a by-product valorization platform to extract oil from carp (*Cyprinus carpio*) viscera. Food Bioprocess Technol 15(11):2515–2530. https://doi.org/10.1007/s11947-022-02897-y
2. Al-Hilphy AR, Al-Musafer AM, Gavahian M (2020) Pilot-scale ohmic heating-assisted extraction of wheat bran bioactive compounds: effects of the extract on corn oil stability. Food Res Int 137. https://doi.org/10.1016/j.foodres.2020.109649
3. Almohammed F, Koubaa M, Khelfa A, Nakaya M, Mhemdi H, Vorobiev E (2017) Pectin recovery from sugar beet pulp enhanced by high-voltage electrical discharges. Food Bioprod Process 103:95–103. https://doi.org/10.1016/j.fbp.2017.03.005
4. Andreou V, Dimopoulos G, Dermesonlouoglou E, Taoukis P (2020) Application of pulsed electric fields to improve product yield and waste valorization in industrial tomato processing. J Food Eng 270. https://doi.org/10.1016/j.jfoodeng.2019.109778
5. Andreou V, Psarianos M, Dimopoulos G, Tsimogiannis D, Taoukis P (2020) Effect of pulsed electric fields and high pressure on improved recovery of high-added-value compounds from olive pomace. J Food Sci 85(5):1500–1512. https://doi.org/10.1111/1750-3841.15122
6. Arshad RN, Abdul-Malek Z, Roobab U, Qureshi MI, Khan N, Ahmad MH, Aadil RM (2020) Effective valorization of food wastes and by-products through pulsed electric field: A systematic review. J Food Process Eng 44(3). https://doi.org/10.1111/jfpe.13629
7. Barba FJ, Brianceau S, Turk M, Boussetta N, Vorobiev E (2015) Effect of alternative physical treatments (ultrasounds, pulsed electric fields, and high-voltage electrical discharges) on selective recovery of bio-compounds from fermented grape pomace. Food Bioprocess Technol 8(5):1139–1148. https://doi.org/10.1007/s11947-015-1482-3
8. Barbosa-Pereira L, Guglielmetti A, Zeppa G (2018) Pulsed electric field assisted extraction of bioactive compounds from cocoa bean shell and coffee silverskin. Food Bioprocess Technol 11(4):818–835. https://doi.org/10.1007/s11947-017-2045-6
9. Boussetta N, Soichi E, Lanoisellé JL, Vorobiev E (2014) Valorization of oilseed residues: Extraction of polyphenols from flaxseed hulls by pulsed electric fields. Ind Crops Prod 52:347–353. https://doi.org/10.1016/j.indcrop.2013.10.048
10. Chadni M, Boussetta N, Guerin C, Lagalle F, Zoghlami A, Perre P, Ioannou I (2023) Improvement of sinapine extraction from mustard seed meal by application of emerging technologies. Foods 12(3). https://doi.org/10.3390/foods12030520
11. Coelho M, Pereira R, Rodrigues AS, Teixeira JA, Pintado ME (2019) Extraction of tomato by-products' bioactive compounds using ohmic technology. Food Bioprod Process 117:329–339. https://doi.org/10.1016/j.fbp.2019.08.005
12. Coelho MC, Ghalamara S, Campos D, Ribeiro TB, Pereira R, Rodrigues AS, Pintado M (2023) Tomato processing by-products valorisation through ohmic heating approach. Foods 12(4). https://doi.org/10.3390/foods12040818
13. Coelho MC, Ghalamara S, Pereira R, Rodrigues AS, Teixeira JA, Pintado ME (2023) Innovation and winemaking by-product valorization: an ohmic heating approach. Processes 11(2). https://doi.org/10.3390/pr11020495
14. Cokgezme OF, Icier F (2022) Frequency and wave type effects on extractability of oleuropein from olive leaves by moderate electric field assisted extraction. Innov Food Sci Emerg Technol 77. https://doi.org/10.1016/j.ifset.2022.102985

15. De Aguiar Saldanha Pinheiro AC, Marti-Quijal FJ, Barba FJ, Benitez-Gonzalez AM, Melendez-Martinez AJ, Castagnini JM, Rocculi P (2023) Pulsed electric fields (PEF) and accelerated solvent extraction (ASE) for valorization of red (*Aristeus antennatus*) and Camarote (*Melicertus kerathurus*) shrimp side streams: antioxidant and HPLC evaluation of the carotenoid astaxanthin recovery. Antioxidants 12(2). https://doi.org/10.3390/antiox12020406
16. de Oliveira CF, Giordani D, Gurak PD, Cladera-Olivera F, Marczak LDF (2015) Extraction of pectin from passion fruit peel using moderate electric field and conventional heating extraction methods. Innov Food Sci Emerg Technol 29:201–208. https://doi.org/10.1016/j.ifset.2015.02.005
17. Deng Y, Ju T, Xi J (2018) Circulating polyphenols extraction system with high-voltage electrical discharge: design and performance evaluation. ACS Sustain Chem Eng 6(11):15402–15410. https://doi.org/10.1021/acssuschemeng.8b03827
18. El Darra N, Grimi N, Vorobiev E, Louka N, Maroun R (2013) Extraction of polyphenols from red grape pomace assisted by pulsed ohmic heating. Food Bioprocess Technol 6(5):1281–1289. https://doi.org/10.1007/s11947-012-0869-7
19. El Kantar S, Boussetta N, Lebovka N, Foucart F, Rajha HN, Maroun RG, Vorobiev E (2018) Pulsed electric field treatment of citrus fruits: improvement of juice and polyphenols extraction. Innov Food Sci Emerg Technol 46:153–161. https://doi.org/10.1016/j.ifset.2017.09.024
20. El Kantar S, Rajha HN, Boussetta N, Vorobiev E, Maroun RG, Louka N (2019) Green extraction of polyphenols from grapefruit peels using high voltage electrical discharges, deep eutectic solvents and aqueous glycerol. Food Chem 295:165–171. https://doi.org/10.1016/j.foodchem.2019.05.111
21. Fan R, Wang L, Fan J, Sun W, Dong H (2022) The pulsed electric field assisted-extraction enhanced the yield and the physicochemical properties of soluble dietary fiber from orange peel. Front Nutr 9:925642. https://doi.org/10.3389/fnut.2022.925642
22. Franco D, Munekata PES, Agregan R, Bermudez R, Lopez-Pedrouso M, Pateiro M, Lorenzo JM (2020) Application of pulsed electric fields for obtaining antioxidant extracts from fish residues. Antioxidants 9(2). https://doi.org/10.3390/antiox9020090
23. Gavahian M, Chu R (2022) Ohmic heating extraction at different times, temperatures, voltages, and frequencies: a new energy-saving technique for pineapple core valorization. Foods 11(14). https://doi.org/10.3390/foods11142015
24. Gavahian M, Chu YH, Sastry S (2018) Extraction from food and natural products by moderate electric field: mechanisms, benefits, and potential industrial applications. Compr Rev Food Sci Food Saf 17(4):1040–1052. https://doi.org/10.1111/1541-4337.12362
25. Gavahian M, Tiwari BK (2020) Moderate electric fields and ohmic heating as promising fermentation tools. Innov Food Sci Emerg Technol 64. https://doi.org/10.1016/j.ifset.2020.102422
26. Geada P, Rodrigues R, Loureiro L, Pereira R, Fernandes B, Teixeira JA, Vicente AA (2018) Electrotechnologies applied to microalgal biotechnology—Applications, techniques and future trends. Renew Sustain Energy Rev 94:656–668. https://doi.org/10.1016/j.rser.2018.06.059
27. Ghosh S, Gillis A, Sheviryov J, Levkov K, Golberg A (2019) Towards waste meat biorefinery: extraction of proteins from waste chicken meat with non-thermal pulsed electric fields and mechanical pressing. J Clean Prod 208:220–231. https://doi.org/10.1016/j.jclepro.2018.10.037
28. Gomez B, Munekata PES, Gavahian M, Barba FJ, Marti-Quijal FJ, Bolumar T, Lorenzo JM (2019) Application of pulsed electric fields in meat and fish processing industries: an overview. Food Res Int 123:95–105. https://doi.org/10.1016/j.foodres.2019.04.047
29. He G, Yin Y, Yan X, Yu Q (2014) Optimisation extraction of chondroitin sulfate from fish bone by high intensity pulsed electric fields. Food Chem 164:205–210. https://doi.org/10.1016/j.foodchem.2014.05.032
30. Hernández-Corroto E, Boussetta N, Marina ML, García MC, Vorobiev E (2022) High voltage electrical discharges followed by deep eutectic solvents extraction for the valorization of pomegranate seeds (*Punica granatum* L.). Innov Food Sci Emerg Technol 79. https://doi.org/10.1016/j.ifset.2022.103055

31. Hu L, Qiu W, Feng Y, Jin Y, Deng S, Tao N, Jin Y (2022) Effect of recycling ohmic heating on the preparation of chitosan from the *Portunus trituberculatus* crab shells. Food Bioprocess Technol 15(12):2762–2775. https://doi.org/10.1007/s11947-022-02913-1
32. Kadem ZA, Al-Hilphy AR, Alasadi MH, Gavahian M (2023) Combination of ohmic heating and subcritical water to recover amino acids from poultry slaughterhouse waste at a pilot-scale: new valorization technique. J Food Sci Technol 60(1):24–34. https://doi.org/10.1007/s13197-022-05556-4
33. Kaur M, Kumar S, Samota MK, Lalremmawii. (2023) Ohmic heating technology systems, factors governing efficiency and its application to inactivation of pathogenic microbial, enzyme inactivation, and extraction of juice, oil, and bioactive compounds in the food sector. Food Bioprocess Technol. https://doi.org/10.1007/s11947-023-03126-w
34. Koubaa M, Barba FJ, Grimi N, Mhemdi H, Koubaa W, Boussetta N, Vorobiev E (2016) Recovery of colorants from red prickly pear peels and pulps enhanced by pulsed electric field and ultrasound. Innov Food Sci Emerg Technol 37:336–344. https://doi.org/10.1016/j.ifset.2016.04.015
35. Kumari B, Tiwari BK, Walsh D, Griffin TP, Islam N, Lyng JG, Rai DK (2019) Impact of pulsed electric field pre-treatment on nutritional and polyphenolic contents and bioactivities of light and dark brewer's spent grains. Innov Food Sci Emerg Technol 54:200–210. https://doi.org/10.1016/j.ifset.2019.04.012
36. Li J, Chen W, Niu D, Wang R, Xu F-Y, Chen B-R, Zeng X-A (2022) Efficient and green strategy based on pulsed electric field coupled with deep eutectic solvents for recovering flavonoids and preparing flavonoid aglycones from noni-processing wastes. J Clean Prod 368. https://doi.org/10.1016/j.jclepro.2022.133019
37. Li Z, Fan Y, Xi J (2019) Recent advances in high voltage electric discharge extraction of bioactive ingredients from plant materials. Food Chem 277:246–260. https://doi.org/10.1016/j.foodchem.2018.10.119
38. Lin S, Wang L, Jones G, Trang H, Yin Y, Liu J (2012) Optimized extraction of calcium malate from eggshell treated by PEF and an absorption assessment in vitro. Int J Biol Macromol 50(5):1327–1333. https://doi.org/10.1016/j.ijbiomac.2012.03.013
39. Martín-García B, Tylewicz U, Verardo V, Pasini F, Gómez-Caravaca AM, Caboni MF, Dalla Rosa M (2020) Pulsed electric field (PEF) as pre-treatment to improve the phenolic compounds recovery from brewers' spent grains. Innov Food Sci Emerg Technol 64. https://doi.org/10.1016/j.ifset.2020.102402
40. Pare A, Nema A, Singh VK, Mandhyan BL (2014) Combined effect of ohmic heating and enzyme assisted aqueous extraction process on soy oil recovery. J Food Sci Technol 51(8):1606–1611. https://doi.org/10.1007/s13197-012-0685-0
41. Parniakov O, Barba FJ, Grimi N, Lebovka N, Vorobiev E (2014) Impact of pulsed electric fields and high voltage electrical discharges on extraction of high-added value compounds from papaya peels. Food Res Int 65:337–343. https://doi.org/10.1016/j.foodres.2014.09.015
42. Parniakov O, Barba FJ, Grimi N, Lebovka N, Vorobiev E (2016) Extraction assisted by pulsed electric energy as a potential tool for green and sustainable recovery of nutritionally valuable compounds from mango peels. Food Chem 192:842–848. https://doi.org/10.1016/j.foodchem.2015.07.096
43. Parniakov O, Roselló-Soto E, Barba FJ, Grimi N, Lebovka N, Vorobiev E (2015) New approaches for the effective valorization of papaya seeds: Extraction of proteins, phenolic compounds, carbohydrates, and isothiocyanates assisted by pulsed electric energy. Food Res Int 77:711–717. https://doi.org/10.1016/j.foodres.2015.03.031
44. Pataro G, Barca GMJ, Pereira RN, Vicente AA, Teixeira JA, Ferrari G (2014) Quantification of metal release from stainless steel electrodes during conventional and pulsed ohmic heating. Innov Food Sci Emerg Technol 21:66–73. https://doi.org/10.1016/j.ifset.2013.11.009
45. Pataro G, Carullo D, Falcone M, Ferrari G (2020) Recovery of lycopene from industrially derived tomato processing by-products by pulsed electric fields-assisted extraction. Innov Food Sci Emerg Technol 63. https://doi.org/10.1016/j.ifset.2020.102369

46. Poojary MM, Barba FJ, Aliakbarian B, Donsi F, Pataro G, Dias DA, Juliano P (2016) Innovative alternative technologies to extract carotenoids from microalgae and seaweeds. Mar Drugs 14(11). https://doi.org/10.3390/md14110214
47. Puértolas E, Barba FJ (2016) Electrotechnologies applied to valorization of by-products from food industry: main findings, energy and economic cost of their industrialization. Food Bioprod Process 100:172–184. https://doi.org/10.1016/j.fbp.2016.06.020
48. Rifna EJ, Misra NN, Dwivedi M (2023) Recent advances in extraction technologies for recovery of bioactive compounds derived from fruit and vegetable waste peels: a review. Crit Rev Food Sci Nutr 63(6):719–752. https://doi.org/10.1080/10408398.2021.1952923
49. Rocha CMR, Genisheva Z, Ferreira-Santos P, Rodrigues R, Vicente AA, Teixeira JA, Pereira RN (2018) Electric field-based technologies for valorization of bioresources. Bioresour Technol 254:325–339. https://doi.org/10.1016/j.biortech.2018.01.068
50. Roselló-Soto E, Barba FJ, Parniakov O, Galanakis CM, Lebovka N, Grimi N, Vorobiev E (2015) High voltage electrical discharges, pulsed electric field, and ultrasound assisted extraction of protein and phenolic compounds from olive kernel. Food Bioprocess Technol 8(4):885–894. https://doi.org/10.1007/s11947-014-1456-x
51. Sabanci S, Çevik M, Göksu A (2021) Investigation of time effect on pectin production from citrus wastes with ohmic heating assisted extraction process. J Food Process Eng 44(6). https://doi.org/10.1111/jfpe.13689
52. Saberian H, Hamidi-Esfahani Z, Ahmadi Gavlighi H, Barzegar M (2017) Optimization of pectin extraction from orange juice waste assisted by ohmic heating. Chem Eng Process 117:154–161. https://doi.org/10.1016/j.cep.2017.03.025
53. Saberian, H., Hamidi-Esfahani, Z., Ahmadi Gavlighi, H., Banakar, A., & Barzegar, M. (2017). The potential of ohmic heating for pectin extraction from orange waste. *Journal of Food Processing and Preservation, 42*(2). https://doi.org/10.1111/jfpp.13458
54. Sakr M, Liu S (2014) A comprehensive review on applications of ohmic heating (OH). Renew Sustain Energy Rev 39:262–269. https://doi.org/10.1016/j.rser.2014.07.061
55. Salgado-Ramos M, Martí-Quijal FJ, Huertas-Alonso AJ, Sánchez-Verdú MP, Cravotto G, Moreno A, Barba FJ (2023) Sequential extraction of almond hull biomass with pulsed electric fields (PEF) and supercritical CO_2 for the recovery of lipids, carbohydrates and antioxidants. Food Bioprod Process 139:216–226. https://doi.org/10.1016/j.fbp.2023.04.003
56. Salgado-Ramos M, Martí-Quijal FJ, Huertas-Alonso AJ, Sánchez-Verdú MP, Moreno A, Barba FJ (2023) A preliminary multistep combination of pulsed electric fields and supercritical fluid extraction to recover bioactive glycosylated and lipidic compounds from exhausted grape marc. Lwt 180. https://doi.org/10.1016/j.lwt.2023.114725
57. Sharifi A, Hamidi-Esfahani Z, Ahmadi Gavlighi H, Saberian H (2022) Assisted ohmic heating extraction of pectin from pomegranate peel. Chem Eng Process Process Intensif 172. https://doi.org/10.1016/j.cep.2021.108760
58. Tunç MT, Odabaş Hİ (2021) Single-step recovery of pectin and essential oil from lemon waste by ohmic heating assisted extraction/hydrodistillation: a multi-response optimization study. Innov Food Sci Emerg Technol 74. https://doi.org/10.1016/j.ifset.2021.102850
59. Yan LG, Deng Y, Ju T, Wu K, Xi J (2018) Continuous high voltage electrical discharge extraction of flavonoids from peanut shells based on "annular gap type" treatment chamber. Food Chem 256:350–357. https://doi.org/10.1016/j.foodchem.2018.02.129
60. Yu X, Bals O, Grimi N, Vorobiev E (2015) A new way for the oil plant biomass valorization: polyphenols and proteins extraction from rapeseed stems and leaves assisted by pulsed electric fields. Ind Crops Prod 74:309–318. https://doi.org/10.1016/j.indcrop.2015.03.045
61. Zhou M, Fakayode OA, Ahmed Yagoub AE, Ji Q, Zhou C (2022) Lignin fractionation from lignocellulosic biomass using deep eutectic solvents and its valorization. Renew Sustain Energy Rev 156. https://doi.org/10.1016/j.rser.2021.111986
62. Zhou M, Fakayode OA, Ren M, Li H, Liang J, Zhou C (2023) Green and sustainable extraction of lignin by deep eutectic solvent, its antioxidant activity, and applications in the food industry. Crit Rev Food Sci Nutr 1–19. https://doi.org/10.1080/10408398.2023.2181762

Chapter 12
Innovative Laser Technologies: A Sustainable Approach to Waste Biomass Utilizations

Husam Talib Hamzah, Sridevi Veluru, Thamer Adnan Abdullah, Venkata Rao Poiba, and Jampana Vishnu Varma

Abstract This book chapter explores the transformative role of innovative laser technologies in addressing the global challenge of waste and biomass utilization. Laser-based processes have emerged as powerful tools for sustainable waste management and biomass conversion into valuable products. This chapter provides a comprehensive overview of laser-assisted techniques, their environmental benefits, and their applications in waste-to-energy conversion, biomass valorization, and the mitigation of environmental pollution. By examining cutting-edge research and practical examples, readers will gain insight into the promising prospects of laser technologies for a greener and more sustainable future.

Keywords Laser technologies · Waste biomass · Sustainable waste management · Laser-assisted processes · Waste-to-energy conversion · Green technology

12.1 Introduction

Besides the significant growth in population, lifestyle changes and luxurious exposure enhance demand for everything quantitatively and qualitatively. Most of these demands are met by petroleum-based businesses. For instance, fossil fuels provide 85% of world energy [65]. Consequently, CO_2 emissions from energy sources increased from 32.3 BMT in 2012 to 35.6 BMT in 2020 and 43.2 BMT in 2040

H. T. Hamzah (✉) · S. Veluru · V. R. Poiba · J. V. Varma
Department of Chemical Engineering, College of Engineering, Andhra University, Visakhapatnam, India
e-mail: alwatanaliraqi@gmail.com

T. A. Abdullah
Applied Sciences Department, University of Technology, Baghdad, Iraq

© The Author(s), under exclusive license to Springer Nature Switzerland AG 2024

P. Verma and B. Kumar (eds.), *Sustainable Radiation Technologies in Waste-biomass Valorization*, Green Chemistry and Sustainable Technology,
https://doi.org/10.1007/978-3-031-63941-8_12

[15], causing global warming [35]. Alternative resources are chosen due to environmental, geopolitical, socioeconomic, sustainability, and other concerns about fossil-based resources. To date, biomass is the most sustainable and greener option. Waste biomass is more appealing to these projects due to its non-competitiveness with food, quantity, and poor disposal. The latest World Bank analysis predicts that global waste will rise by 70% by 2050 without rapid action [31]. Fossil fuels and waste stress our ecology. Using waste biomass to make high-value products reduces environmental impact, increases income, and reduces petroleum dependence. Not new: biomass utilization. Similar to fossil fuels, biomass can be co-produced into products. Like a petroleum refinery, a biorefinery that produces many products from one feedstock has excellent potential [27, 34]. Biomass can be used to synthesize all fossil-based materials in the future. It's not straightforward to switch from petroleum to renewable carbon sources for energy, chemicals, materials, etc. [11]. Thus, multi-sector collaboration is essential to create cutting-edge biomass utilization technology. Converting biomass waste into profitable products and energy is essential to sustainable development [50]. By improving resource efficiency and reducing dependency on finite resources, this procedure reduces the environmental impact of waste disposal and promotes a circular economy [18].

Many new laser applications research are underway. Thermal treatments utilizing lasers are a new crucial field where heat is produced by random movement of particles in a material (atomic or molecular particles), and the 'photons' that comprise the laser beam, the so-called 'light particles' have no mass and no temperature [46]. Current interest is the use of biomass waste in producing grenewable energy and valuable products. Biofuels, chemicals, fertilizers, and materials can be made from municipal solid waste, agricultural residues, industrial by-products, forestry, and agriculture biomass [48]. Laser biomass graphitization can pattern supercapacitor electrodes from renewable materials. On bamboo, a renewable biomass resource, laser-induced graphene (LIG) was microstructured via laser-based graphitization. Bamboo graphitization and microstructuring by femtosecond laser pulses created conductive, high-surface-area structures [47]. Laser-based graphitization is a promising single-step graphitic carbon patterning method. Lin et al. discovered in 2014 that laser-induced graphene (LIG) was made from graphitic carbon on polyimide, resurrecting the laser-based approach [39]. Natural resources and manufactured polymers, including silk [38], coconut [14], and wood [21], are both sources of LIG. The utilization of biomass-derived LIG as an electrode material for supercapacitors has also been the subject of multiple reports [47]. Ye et al. [74], utilized wood-derived LIG for supercapacitor electrodes. Because capacitance depends on electrode area, surface form matters. Bioenergy production from biomass sources has garnered global renewable energy studies to understand its progress in most emerging and developed countries due to the energy crisis. Gasification turns biomass or fossil fuels into CO, H_2, CO_2, and possibly hydrocarbon compounds like methane (CH_4). Additionally, pyrolysis is the heat breakdown of macromolecular molecules without oxygen [33]. They are practical and analytical. Improved quick pyrolysis and gasification techniques for bioenergy (biofuel, biogas, and biochar) production have received interest in the last decade since they enable biomass conversion into bio-yield and value-added products to be easy and flexible [52]. Electron microscopy, diffraction, and Raman scattering have all shown that the laser-irradiated product's structure differs greatly from Bernal stacked graphitic carbon's. It has never been seen before that an organic chemical can be used to grow a turbostratic phase with the use of a laser [5]. Laser breakdown and carbonization to transform synthetic polymers or natural chemicals into graphene-like compounds in huge quantities appears promising. Using a CO_2 laser pulsing at 10.6 μm for 14 μs, investigators have considered carbon-containing substances potential building blocks for graphene-like materials [39]. A second irradiation can convert amorphous carbon, produced by breaking organic materials into carbon with a low-fluence CO_2 laser, into structures similar to graphene. When carbons with varying sp2/sp3 fractions are exposed to laser light, they can transform into turbostratic graphene, with interlayer spacing ranging from 3.5 to 4.0 Å [4]. Laser photoactivated nanomaterials improve biomass biohydrogen production. Hydrogen output and photo-fermentation efficiency depend on light saturation. Attia et al. [7] used laser-photoactivated nanomaterials to bio-stimulate hydrogen-producing purple non-sulfur (PNS) bacteria to improve their endurance to uneven light irradiation and overcome light saturation. Recently, patterning laser-induced graphene (LIG) into commercial polymers with low-cost and easily available infrared lasers has garnered attention because of its simplicity and ability to construct electrical and electrochemical devices [25]. Multiflora cuttings treated with laser beams increased fresh and dry mass yield and energy value. In 2009–2013, Mydlniki near Cracow hosted a one-factor field experiment. [32], examined the effects of two laser beam dosages on multiflora rose cutting yield, chemical composition, and energy value. Laser radiation did not affect multiflora rose fresh yield, but larger laser doses increased dry mass concentration and yield in the above-ground sections. Laser-driven waste biomass technologies, which offer efficiency and sustainability, have been studied. Laser technologies are popular due to their unique characteristics. Lasers produce coherent light beams by stimulated emission. High energy densities and localized heating are possible with laser beam control and focus. Lasers are ideal for many applications, including waste biomass conversion [73]. Laser-driven calorimetry evaluates sample-environment energy exchange during thermal and chemical reactions [51, 55]. Algal thermochemical characteristics are estimated using energy conservation to estimate sample-specific energy release rate, total specific energy release, and temporal mass loss. Again, laser heating offers advantages compared to thermal-analysis methods: smaller sample sizes and higher heating rates than TGA, heating samples in minutes instead of hours, and heating rates suitable, lowering energy generated from preferred vaporization of lower-temperature organics [16]. This chapter reviews laser-driven waste biomass solutions and examines their potential to solve laser waste biomass conversion. Laser operation concepts, laser parameters, Laser-Driven Pyrolysis, application, and products from pyrolysis will be covered in the chapter.

12.2 Principles of Laser Technologies

Laser technologies have transformed several industries, and their waste biomass potential is increasingly being recognized. Laser-driven waste biomass conversion provides an efficient and environmentally beneficial alternative to older methods [66]. Laser operation must be understood before applying lasers to waste biomass utilization. Laser indicates "Light Amplification by Stimulated Emission of Radiation." By stimulating emission, coherent and monochromatic light is emitted. An external energy source excites a solid, liquid, or gas in a laser system. Excitation raises medium atoms and molecules' energy levels. Excited particles produce photons with particular wavelengths upon returning to their ground state. Stimulated emission amplifies the emitted photons by engaging them with other excited particles, releasing more photons with the same wavelength and phase. An essential part of biomass valorization is stimulated emission, which is useful for procedures like laser-induced breakdown spectroscopy (LIBS) and laser-induced fluorescence (LIF). The process of laser-induced biological sample separation (LIBS) involves vaporizing and exciting biomass with laser radiation, which causes it to release photons that may be used to analyze its constituent composition [13, 66].

12.3 Laser-Driven Pyrolysis of Agricultural Residues

Agricultural waste like crop stalks, husks, and straws is common biomass, burning or decomposing these residues pollutes the air and wastes resources. Laser-driven pyrolysis turns agricultural leftovers into biochar and bio-oil sustainably [59]. Laser-driven pyrolysis device converts biomass into biochar and bio-oil. Laser technology controls the pyrolysis process, yielding more charcoal and bio-oil than older approaches [67]. Biochar improved soil fertility and carbon sequestration, while bio-oil generated heat and energy. Basu [10], described pyrolysis as the thermochemical degradation of biomass into various valuable products in the non-existence of oxidizing agents that prevent gasification (Fig. 12.1). It's one of the numerous gasifier reaction zones. Pyrolysis breaks biomass's complex hydrocarbon molecules into more straightforward gas and liquid components. Pyrolysis is similar to thermolysis, carbonization, dry distillation, devolatilization, and cracking [24]. The temperature range for biomass pyrolysis is 300–650 °C, while gasification usually takes place at 800–1000 °C. The heating rate divides pyrolysis into slow and fast processes. Carbonization, a slow pyrolysis process, produces charcoal or char, while fast pyrolysis maximizes liquid or bio-oil production [63].

Pyrolysis thermally decomposes organic molecules into beneficial compounds without oxygen [49]. Laser-driven pyrolysis uses lasers to improve waste biomass utilization. The results demonstrate that changes in reaction chamber pressure, laser beam focus, and carrier gas selection significantly affect the average particle size.

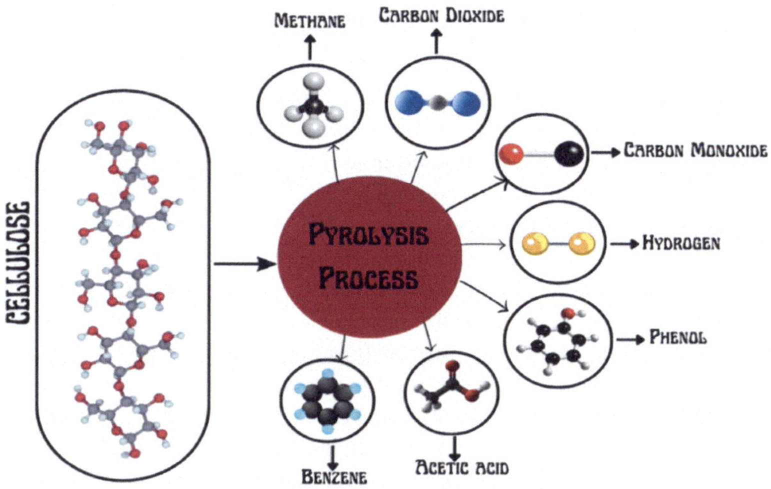

Fig. 12.1 Pyrolysis divides giant hydrocarbon molecules into smaller ones

Particle size is unaffected by changes in laser power and carrier gas flow rate (Fig. 12.2) [46].

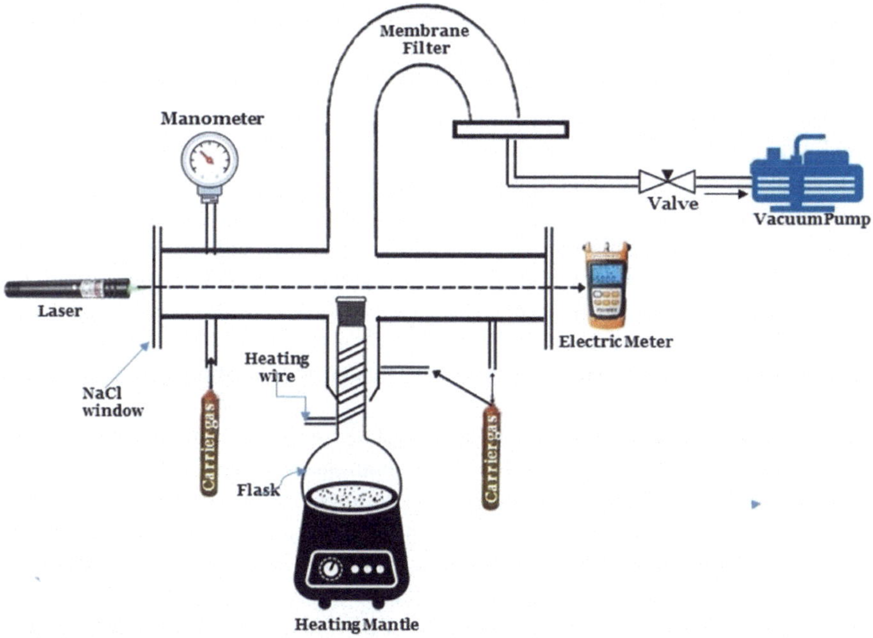

Fig. 12.2 Diagram of the laser pyrolysis system

12.4 Insight into Products of Thermal Pyrolysis

From a scientific and economic perspective, thermal pyrolysis offers more significant advantages than other alternatives for treating municipal solid waste, according to long-term study experiences. Organic materials undergo heat decomposition in an oxygen-free atmosphere during pyrolysis. While gasification does not necessitate an external heat source, pyrolysis must because it is thermally self-sustaining [68]. In a simultaneous gasification and pyrolysis process, the heat required to carry out the reactions is supplied by the combustion reaction that occurs during gasification. This process requires no additional heat source for pyrolysis, it is all done within the gasification process. As previously stated by [70], these conditions break down biomass into carbon oxides, solids (char), water, hydrogen, hydrocarbons, and organic molecules. There are aerosols and vapors in the gaseous product stream, some of which can condense and some of which cannot. Bio-oil condenses from the gaseous exhaust stream [58] state that woody biomass contains mainly cellulose and lignin polymers, extractives, and a few inorganic materials. Thermal decomposition of biomass produces products based on component proportions and decomposition temperature. A thermal gravimetric study reveals that lignin, hemicelluloses, and cellulose exhibit very different weight-loss profiles (i.e., different temperatures of maximum weight loss) when subjected to the same heating rate. Pyrolysis involves many serial and parallel processes with different activation energies. Bark, extractives, and proteins are also minor components. The range of products is affected by the rate of biomass heating and the terminal temperature. Decomposing biomass may have a different temperature than the pyrolysis tank. According to Piskorz et al. [54], carbonyl, phenolic, and carboxylic functional groups were shown to be the most abundant in bio-oil. There is a division in these categories between products made from lignin and those made from the cellulose parts of biomass. The former are more likely to dissolve in water than the latter. "Heavy tar" (or pyrolytic lignin) fraction isolation from biomass, with its primary source being the lignin component [57], typically occurs when sufficient water is added to a single-phase bio-oil. The first time that the high sensitivity of cellulose's pyrolytic decomposition routes to highly minute levels of inorganic contaminants (<0.1%) that specifically change its degradation properties was brought to light by Broido and Kilzer [12]. They encourage the breakdown of anhydrosugars into smaller pieces and the subsequent generation of char in particular.

Furthermore, char creation can be accelerated significantly by some inorganic components, especially alkaline salts. As an example, research has shown that the pyrolysis yield of char from pure crystalline cellulose can be enhanced more than ten times by adding just a few parts per million of sodium salts [58]. In subsequent work, Jeong et al. [30], showed that improving anhydrosugar synthesis by de-mineralizing biomass, by eliminating alkali cations, can significantly increase organic liquid yields. In addition to the cost of demineralization by acid treatment, sugary bio-oil has a higher viscosity, making tar recovery more complex. Laser-driven pyrolysis uses high-intensity laser beams to initiate and control the reaction. The material's organic constituents break into smaller molecules under severe heat, producing valuable products. The organic molecules in waste biomass material selectively absorb laser energy, enabling laser-driven pyrolysis. Organic molecules absorb laser radiation at different wavelengths. Target specific organic components and maximize pyrolysis efficiency by carefully adjusting the laser wavelength [69]. For several industries, laser-driven pyrolysis creates valuable products. Biochar is a carbon-rich soil additive made via laser-driven pyrolysis. Agriculture uses biochar to improve soil fertility, water retention, and nutrient availability. Besides biochar, laser-driven pyrolysis produces bio-oil and syngas. Biosolids can be turned into biofuels or used to create chemicals and products. Syngas, a carbon monoxide and hydrogen mixture, can create heat and power or manufacture chemicals. Many industries can use laser-driven pyrolysis products such as farming, gardening, and environmental cleaning employ biochar.

Biomass pyrolysis can be performed using a pulsed laser, according to the following study. The experiments were conducted in a pyrolytic system driven by pulsed lasers (Fig. 12.3). The system comprised gas feeding, quartz reactor, condensing, cleaning, and gas detection/collection. A 100-bar tablet press compressed 13 mm biomass tablets with 100 mg of yeast or co-yeast. A quartz reactor continually purged these tablets with 150 mL/min carrier gas (Ar, N_2, air, or water vapor). Laser-driven pyrolysis of a biomass tablet was performed using a pulsed laser (LSF20D, Hgtech laser) at 1064 nm and 100 ns. A computer was connected to the laser system, and EzCad2 was used to specify experimental parameters to design the laser's movement. This study used a linear scanning mode for laser-driven pyrolysis with 0.5 mJ pulse energy, 20 W output power, 500 mm/s scanning speed, and 0.05 mm scanning line spacing (Fig. 12.3) [37].

Advanced laser-driven gasification converts waste biomass into high-value products sustainably and efficiently. By using high-intensity laser beams, this technique produces syngas and other valuable by-products by heating and vaporizing waste

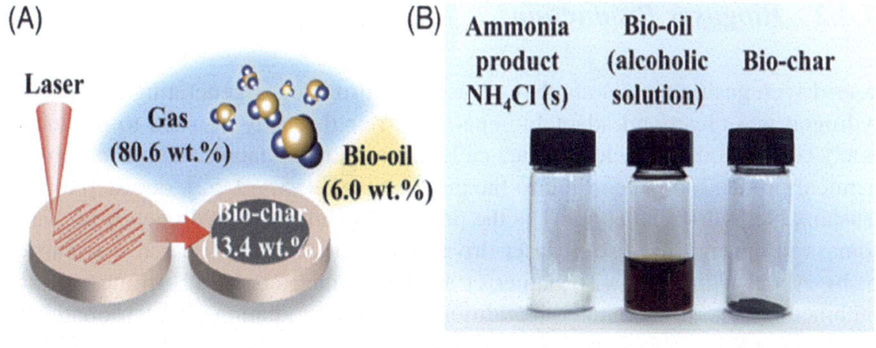

Fig. 12.3 **A** A schematic and **B** photographic evidence of yeast pyrolysis by laser [37]

biomass. Syngas can be processed and used in many industries. Laser-driven pyrolysis is expected to debut in facilitating the direct transformation of nitrogen-rich biomass into ammonia. The pulsed laser-induced local-transient thermal effect identified the transformation of biological nitrogen resources, such as cheap and plentiful yeasts, into tiny gaseous molecules. Ammonia production rates as high as 260.4 mg/h were achieved as a result of this discovery, which is a significant improvement over thermochemical ammonia synthesis. Utilizing the little heat output of a low-power laser (20W) maintains a temperature and pressure setting that is safe for the complete ammonia synthesis reaction system [56]. The reactor sphere is heated by directing the high-power laser beam into the chamber. The sample is heated as energy emanates from the sphere. Using two thermocouples, the temperature of the sample, as well as the temperature of the reactor wall, can be determined (Fig. 12.4) [56].

12.4.1 Synthesis Gas (Syngas) Production

"Syngas" or "synthesis gas" results from laser gasification. Carbon monoxide (CO) and hydrogen (H_2) are the building blocks of syngas, which have multiple useful chemical and fuel applications [53]. Syngas can be customized for individual purposes by adjusting laser parameters and feedstock mix. Syngas produces methanol, a significant ingredient in many chemicals and fuels. By using Fischer–Tropsch synthesis, it can be transformed into higher hydrocarbons like diesel and jet fuel. Syngas can make ammonia, a key fertilizer ingredient [4, 5, 39, 52]. They were used anaerobic digestion to produce biogas and methane from biomass (such as crop waste and livestock dung) as a sustainable energy source. In addition, the use of light and laser radiations to improve biogas and methane production is explored in the work of [1], a process known as "photobiostimulation" occurs when the laser light speeds up the division and growth of bacteria.

12.4.2 Biogases Production

Laser-driven gasification is another attractive hydrogen gas-generating technology. Hydrogen is a clean and adaptable energy carrier that may be used to produce a variety of compounds, including fuel cells, ammonia, methanol, and other forms of chemical production. According to the publications [4, 5, 39, 52], one step toward a sustainable hydrogen economy is the practical synthesis of hydrogen from waste biomass is through the use of laser-driven gasification. The research was carried out by Abdelsalam et al. [2], Samer et al. [61], Samer et al. [62] to measure the combined effects of nanomaterial addition and laser irradiation on microorganisms that are engaged in the production of biogas. The impact of nanomaterials and laser radiation on biomass-to-biohydrogen production was the focus of the study [6, 7];

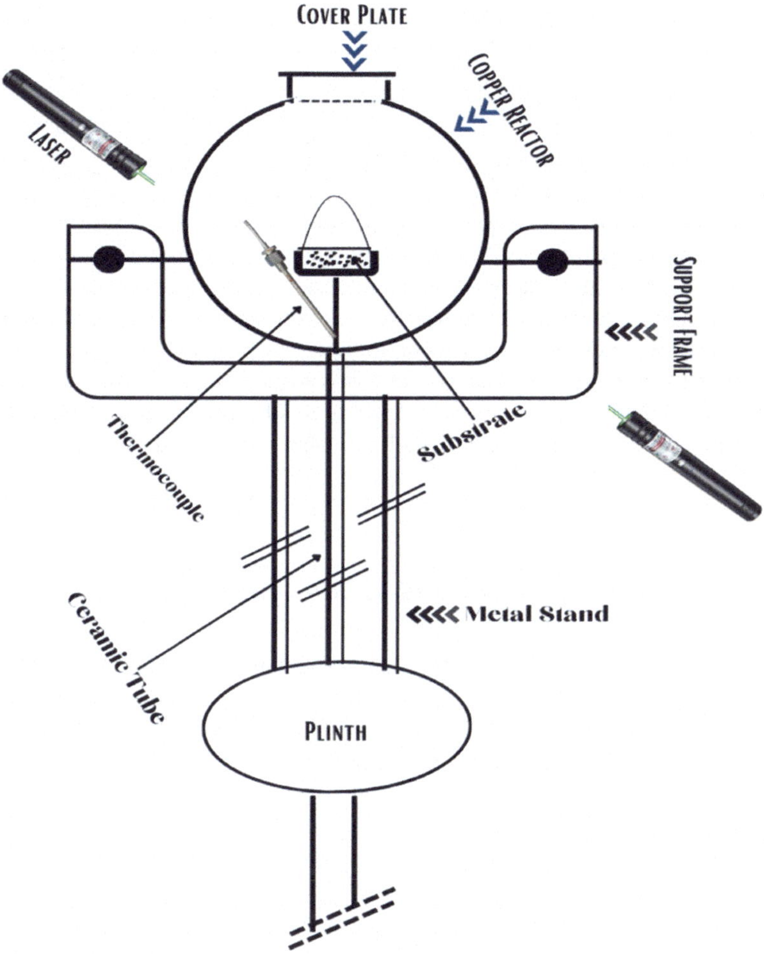

Fig. 12.4 Reactor sphere with laser-driven calorimeter experimental setup schematic

by Adding nanoparticles bio-stimulates bacterial cells, increasing their production and development rate and biohydrogen generation from biomass.

Exposing graphitic carbon nitride nanosheets and nickel nanoparticles to a helium–neon green laser source of radiation via a wavelength of 543 nm for 1 h, putting them to the bacterial inoculum, mixing them with biomass and water, and keeping them in the photobioreactor exposed to 3600 lm white light emitting diodes (LEDs) at 30 °C for 26 days was the production process. The bioenvironmental circumstances and bio-responses of bacteria can be improved, boosting biohydrogen output by 287% over the usual technique [7].

12.4.3 Biochar Production

Biochar, also known as black carbon, is created by exposing wood to a 355 nm UV pulsed laser. Jeong et al. [30] reported that, after UV laser irradiation, beech, cork oak, and ash surfaces became graphitic. Since each wood has a varied lignin content, suitable laser irradiation settings (laser fluence) were studied to produce pH sensors from these woods. Based on the amount of oil-like material produced after laser irradiation, a disposable pH sensor can be used one to three times. Wood and biochar are environmentally friendly, so the sensor can be easily discarded once used on agricultural land. It can then filter soil. Soil quality is improved, emissions of greenhouse gases are decreased, metal toxicity is decreased, and phosphorus and potassium utilization are increased by the biochar that is made from biomass pyrolysis. In water and air filtration systems, biochar is a feasible alternative to activated carbon, which is created from fossil fuels, biochar can take the place of activated carbon [4, 5, 39, 52].

12.4.4 Specialty Chemicals and Value-Added Products

Laser-driven pyrolysis can manufacture many specialty chemicals and value-added products from waste biomass [3]. Controlling process parameters allows product composition and attributes to be tailored. Several valuable chemicals, such as phenols, furans, and levoglucosan, can be produced in this way and used in food, cosmetics, and medicine. Laser-driven pyrolysis can also extract essential oils and bioactive compounds from biomass sources sustainably and effectively [41]. In a batch anaerobic system with PPW, the effects of laser exposure on bioethanol production were investigated by adding graphitic carbon nitride nanosheets (g-C_3N_4) in different quantities. At the same time, the yeast *Saccharomyces cerevisiae* was incubated. Attia et al. [6, 7] looked at how the production of hyaluronic acid was affected by the presence of nanomaterials and lasers. The use of laser photoactivated nanomaterials for bio-stimulation of hydrogen-producing purple non-sulfur (PNS) bacteria has the potential to improve their endurance to fluctuations in light intensity, potentially allowing them to overcome the problem of light saturation. When bacteria absorb CNT nanoparticles, their HA synthesis increases because the nanoparticles "serve as co-enzymes" during UV-induced mutations.

12.5 Principles of Laser-Driven Hydrothermal Conversion

Hydrothermal liquefaction of biomass in hot-compressed or sub-/supercritical water or solvent produces bio-crude oil. Hydrothermal liquefaction usually occurs at 250–450 °C and 5–20 MPa. It can use biomass feedstock [17]. Researchers are interested in hydrothermal liquefaction because of its low tar production, remarkable efficiency, and low operating temperature, unlike pyrolysis [23]. The lignin content of biomass feedstock determines biomass liquefaction product distribution. A lower conversion rate and less oil produced result from a raw material with a higher lignin content. Higher lignin content causes more substantial condensation and polymerization processes [26]. The utilization of waste biomass through the utilization of laser-driven hydrothermal conversion is a strategy that is both novel and environmentally efficient. In this laser-based hydrothermal process, waste biomass is converted into products that are of great value [45]. Lin et al. [40], used direct laser creation at normal temperature and air pressure to create Nanodiamonds (NDs) on paper created from a composite film of nano lignin and cellulose nanofibers (LCNF). Results showed that high laser power was the most critical factor in ND formation. With less laser power, they could produce acceptable products like graphene nanoribbons and carbon nano-onions. Jayan et al., 2023, stated that Laser-induced graphene (LIG) is a promising material for electrical applications. Special porosity, electrical conductivity, composition, and morphology enable flexible electronics, especially bioelectronics, sensors, and actuators, which require LIG. Due to environmental problems with synthetic polymers used to synthesize LIG, biomass-based LIG manufacturing is a new step toward sustainable development and growing research topics. Awad et al., 2020, created a novel method for converting sorghum bran ash into silica powder using laser combustion. Sorghum milled in Khartoum, Sudan, is where the bran for sorghum originated. Five grams of sorghum bran were burned for varying durations using a 60-W Nd: YAG laser, the combustion process did not produce any flames.

12.6 Production of Nanomaterials

Photocatalysis driven by a laser can also be used to make nanomaterials with unique qualities and uses [22]. Waste biomass materials can be turned into nanoparticles or nanocomposites by using specific photocatalysts and changing the settings on the laser [22]. Biomedical devices, sensors, energy storage, and chemistry are just a few of the many uses that these nanoparticles could have. For instance, cellulose nanofibers and nanocrystals can be produced from cellulose waste, these have several potential applications, such as the delivery of medicinal medications or the reinforcement of composite materials [64]. Functional carbon nanomaterials of excellent quality can be synthesized from renewable raw sources using waste heat. Because of this, the approach becomes more lucrative and long-lasting. Making

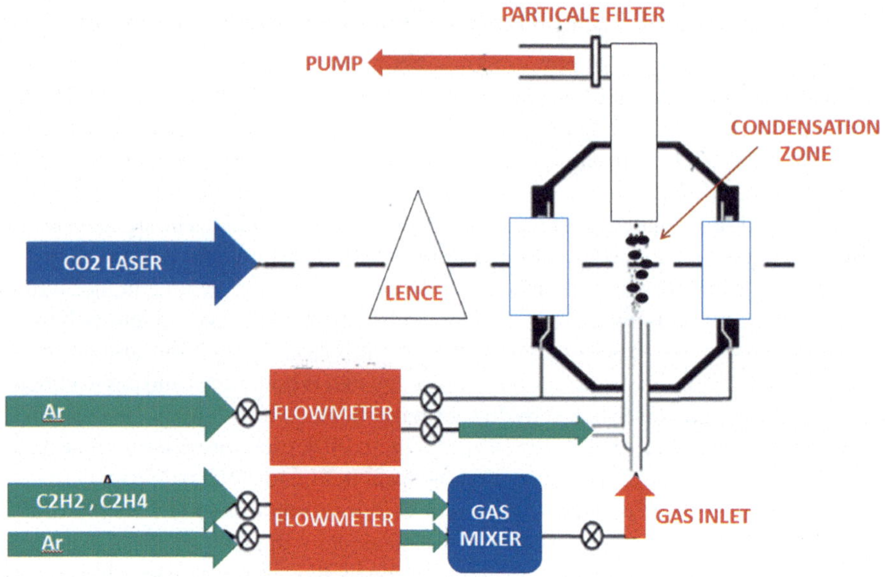

Fig. 12.5 Schematic of the laser pyrolysis setup

high-quality 3D graphene foams (3DGFs) from by-products of biomass pyrolysis can increase the process's economic benefits and sustainability in the long run. The 3DGFs that emerge from this process are excellent energy storage devices and environmental protectors [75]. Figure 12.5 shows the results of another study on the pyrolysis setup using a CO_2 laser conducted by [28], and focused on the generation of carbon nanoparticles. The significant component of the iron-carbon nanostructures containing cementite Fe_3C was produced using the CO_2-laser-induced co-pyrolysis of toluene and iron pentacarbonyl with an ethylene sensitizer.

12.7 Principles of Laser-Driven Waste Biomass Valorization

Innovative and sustainable laser-driven waste biomass valorization converts waste into valuable products. This technology converts waste biomass into valuable products using high-intensity laser beams to start chemical reactions [3]. Laser-driven waste biomass valorization increases efficiency, reduces environmental impact, and produces valuable products [44]. There are two ways to create biomass vapors: ablation with a third harmonic Nd: YAG laser (355 nm) or building a bespoke hot-stage pyrolysis reactor. The vapors are transported to an ionization region after being absorbed in a free-jet expansion of He. An improved sensitivity to detect lignin-based products from vapors can be achieved by laser ablation, which allows for the selective isolation and volatilization of particular tissues within the same plant material [20].

12.8 Laser Parameters Critical to Biomass Valorization and Other Applications

Four aspects of the laser (power, wavelength, pulse duration, and repetition rate) have a significant impact on the waste biomass conversion process [19]. Depending on the specific application and desired outcome, different laser parameters should be chosen. To efficiently thermally decompose the waste biomass feedstock, a high-power laser with a longer wavelength is typically used in laser-driven pyrolysis. In contrast, a shorter wavelength, high-pulse-repetition-rate laser might be more effective in laser-driven gasification for stimulating the right chemical reactions. Laser settings are optimized by balancing energy input and desired results. Higher laser power can speed conversions but potentially increase energy use and feedstock damage. Choosing the right wavelength maximizes waste biomass material's laser energy absorption and conversion efficiency [73]. For optimal reaction kinetics and energy savings, pulse duration and repetition rate must be tuned. The coefficient of determination (R^2) of the relationship between crop biomass density (as defined by fresh and dry matter) and the mean height of the laser reflection point was the primary metric for prospective evaluations of oilseed rape, winter wheat, and grassland crop stands [20, 42, 43].

12.9 Laser-Driven Pre-Treatment

Laser-driven conversion depends on waste biomass feedstock composition. Waste biomass chemical composition, moisture content, and physical properties affect conversion efficiency and product yield. Laser-enabled drying, size reduction, and chemical treatment improve waste biomass conversion [5]. Feedstock moisture reduction prevents laser-driven process energy losses from water evaporation. Feedstock quality and pre-treatment adjustments can boost conversion efficiency and yields [68]. Pre-treatment of cyanobacteria biomass is carried out using medium-intensity laser radiation, and the destruction of cells isolates lipids. Total lipids in the organic fraction were measured. Thin-layer chromatography. Lipid production improves with laser pre-treatment of cyanobacterial biomass and is also recommended for biodiesel fuel [71]. The measured ions showed biomass properties in the softest laser-biomass interaction conditions (with low laser intensity). The photolysis of covalent bonds produced most of the detected species from biomass subunits. Under more intense laser irradiation, ion mass spectra were produced that were associated with (i) the samples' chemical components, (ii) the gas phase recombination products that were ejected from the samples, and (iii) the compounds that were synthesized through laser pyrolysis, which was meant to mimic thermal pyrolysis [8].

12.9.1 Conclusion

Innovative technologies can turn waste biomass into valuable products while reducing environmental impact and conserving resources. Despite their obstacles, continuous research and development provide laser-driven waste biomass systems a bright future. Academia, industry, and government must invest in laser-driven technology for sustainability. Lasers can improve waste management, resource efficiency, and circularity. Laser-driven waste biomass technologies are promising for future generations. In laser-driven waste biomass valorization, innovation, and sustainability create profitable products. The technique is more efficient, ecologically friendly, and produces valuable products than traditional conversion. The laser waste biomass valorization process produces syngas, activated carbon, biofuels, and other products. Waste biomass consumption reduces waste, resource utilization, and greenhouse gas emissions. This chapter discusses laser-driven waste biomass systems for greener waste removal and biomass utilization. Discussions include biofuel production, waste biomass valorization, gasification, hydrothermal conversion, photocatalysis, and laser-driven pyrolysis. Considering laser techniques for waste biomass valorization, lasers can help in the conversion of waste biomass better than conventional procedures. Rural solid waste (MSW) and animal manure used to make syngas or methane using the above-mentioned technologies threaten the environment despite its promise. In underdeveloped countries with low waste management and recycling rates, this is crucial. Both biomass and bioenergy quality assessments are crucial to biomass use.

12.10 Conflicts of Interest

The authors declare that there is no conflict of interest.

12.11 Declaration on Non-Usage of the AI Tools

The authors clarify that they did not use artificial intelligence tools in preparing this chapter.

References

1. Abdelsalam EM, El-Hussein A, Samer M (2020) Photobiostimulation of anaerobic digestion by laser irradiation and photocatalytic effects of trace metals and nanomaterials on biogas production. Int J Energy Res 45(1):141–150. https://doi.org/10.1002/er.5817
2. Abdelsalam E, Hijazi O, Samer M, Yacoub I, Ali A, Ahmed R, Bernhardt H (2019) Life cycle assessment of the use of laser radiation in biogas production from anaerobic digestion of manure. Renew Energy 142:130–136. https://doi.org/10.1016/j.renene.2019.04.090
3. Ahuja V, Bhatt AK, Varjani S, Choi KY, Kim SH, Yang YH, Bhatia SK (2022) Quantum dot synthesis from waste biomass and its applications in energy and bioremediation. Chemosphere 293:133564. https://doi.org/10.1016/j.chemosphere.2022.133564
4. Antonelou A, Benekou V, Dracopoulos V, Kollia M, Yannopoulos SN (2018) Laser-induced transformation of graphitic materials to two-dimensional graphene-like structures at ambient conditions. Nanotechnology 29(38):384001. https://doi.org/10.1088/1361-6528/aacf85
5. Athanasiou M, Samartzis N, Sygellou L, Dracopoulos V, Ioannides T, Yannopoulos SN (2021) High-quality laser-assisted biomass-based turbostratic graphene for high-performance supercapacitors. Carbon 172:750–761. https://doi.org/10.1016/j.carbon.2020.10.042
6. Attia YA, Al Nazawi AM, Elsayed H, Sadik MW (2021) Carbon nanotubes catalyzed UV-trigger production of hyaluronic acid from *Streptococcus equi*. Saudi J Biol Sci 28(1):484–491. https://doi.org/10.1016/j.sjbs.2020.10.032
7. Attia YA, Samer M, Moselhy MA, Arisha AH, Abdelqader AA, Abdelsalam EM (2021) Influence of laser photoactivated graphitic carbon nitride nanosheets and nickel nanoparticles on purple non-sulfur bacteria for biohydrogen production from biomass. J Clean Prod 299:126898. https://doi.org/10.1016/j.jclepro.2021.126898
8. Aubriet F, Ghislain T, Hertzog J, Sonnette A, Dufour A, Mauviel G, Carré V (2018) Characterization of biomass and biochar by LDI-FTICRMS – Effect of the laser wavelength and biomass material. J Am Soc Mass Spectrom 29(10):1951–1962. https://doi.org/10.1007/s13361-018-2005-z
9. Awad AAG, Pn Gawbah MA, Orsod MU, Alsabah YA, Marouf AAS (2020) Investigation of the effects of laser-based combustion on Sorghum Bran. Albaydha Univ J 2(2):52–59. https://doi.org/10.56807/buj.v2i2.52
10. Basu P (2010) Pyrolysis and torrefaction. In: Biomass gasification design handbook, 65–96. https://doi.org/10.1016/b978-0-12-374988-8.00003-9
11. Bender TA, Dabrowski JA, Gagné MR (2018) Homogeneous catalysis for the production of low-volume, high-value chemicals from biomass. Nat Rev Chem 2(5):35–46. https://doi.org/10.1038/s41570-018-0005-y
12. Broido A, Kilzer FJ (1963) A critique of the present state of knowledge of the mechanism of cellulose pyrolysis. Fire Res Abstr Rev 5:157
13. Brown, & Sarah, (2022) Advancements in Laser Technologies for Waste-Biomass Conversion. Waste Biomass Manage 18(3):87–105
14. Chyan Y, Ye R, Li Y, Singh SP, Arnusch CJ, Tour JM (2018) Laser-induced graphene by multiple lasing: toward electronics on cloth, paper, and food. ACS Nano 12(3):2176–2183. https://doi.org/10.1021/acsnano.7b08539
15. Conti J, Holtberg P, Diefenderfer J, LaRose A, Turnure JT, Westfall L (2016) International Energy Outlook 2016 With Projections to 2040. https://doi.org/10.2172/1296780
16. de Carvalho Rocha WF, Presser C, Bernier S, Nazarian A, Sheen DA (2020) Laser-driven calorimetry and chemometric quantification of standard reference material diesel/biodiesel fuel blends. Fuel 281:118720. https://doi.org/10.1016/j.fuel.2020.118720
17. Demirbaş A (2000) Mechanisms of liquefaction and pyrolysis reactions of biomass. Energy Convers Manage 41(6):633–646. https://doi.org/10.1016/s0196-8904(99)00130-2
18. Dessie W, Luo X, Wang M, Feng L, Liao Y, Wang Z, Qin Z (2020) Current advances on waste biomass transformation into value-added products. Appl Microbiol Biotechnol 104(11):4757–4770. https://doi.org/10.1007/s00253-020-10567-2

19. Devi M, Wang H, Moon S, Sharma S, Strauss V (2023) Laser-carbonization – a powerful tool for micro-fabrication of patterned electronic carbons. Adv Mater 35(38). https://doi.org/10.1002/adma.202211054
20. Ehlert D, Horn HJ, Adamek R (2008) Measuring crop biomass density by laser triangulation. Comput Electron Agric 61(2):117–125. https://doi.org/10.1016/j.compag.2007.09.013
21. Fiodorov V, Ratautas K, Mockus Z, Trusovas R, Mikoliūnaitė L, Račiukaitis G (2022) Laser-assisted selective fabrication of copper traces on polymers by electroplating. Polymers 14(4):781. https://doi.org/10.3390/polym14040781
22. Forsythe RC, Cox CP, Wilsey MK, Müller AM (2021) Pulsed laser in liquids made nanomaterials for catalysis. Chem Rev 121(13):7568–7637. https://doi.org/10.1021/acs.chemrev.0c01069
23. Gollakota A, Kishore N, Gu S (2018) A review on hydrothermal liquefaction of biomass. Renew Sustain Energy Rev 81:1378–1392. https://doi.org/10.1016/j.rser.2017.05.178
24. Hamzah HT, Sridevi V, Surya DV, Ramesh P, Rao CS, Palla S, Abdullah TA (2024) Synergistic effects and product yields in microwave-assisted in-situ co-pyrolysis of rice straw and paraffin wax. Process Saf Environ Prot 182:45–55. https://doi.org/10.1016/j.psep.2023.11.048
25. Hawes GF, Yilman D, Noremberg BS, Pope MA (2019) Supercapacitors fabricated via laser-induced carbonization of biomass-derived poly(furfuryl alcohol)/graphene oxide composites. ACS Applied Nano Materials 2(10):6312–6324. https://doi.org/10.1021/acsanm.9b01284
26. Huang HJ, Yuan XZ (2015) Recent progress in the direct liquefaction of typical biomass. Prog Energy Combust Sci 49:59–80. https://doi.org/10.1016/j.pecs.2015.01.003
27. IEA (2007) Bioenergy Project Development & Biomass Supply, IEA, Paris. https://www.iea.org/reports/bioenergy-project-development-biomass-supply, Licence: CC BY 4.0
28. Jäger C, Mutschke H, Huisken F, Alexandrescu R, Morjan I, Dumitrache F, Schneeweiss O (2006) Iron-carbon nanoparticles prepared by CO_2 laser pyrolysis of toluene and iron pentacarbonyl. Appl Phys A 85(1):53–62. https://doi.org/10.1007/s00339-006-3665-2
29. Jayan JS, Jayan SS (2023) Biomass-derived laser-induced graphene and its advances in the electronic applications. Adv Eng Mater 25(16). https://doi.org/10.1002/adem.202300248
30. Jeong SY, Lee CW, Lee JU, Ma YW, Shin BS (2020) Laser-induced biochar formation through 355 nm pulsed laser irradiation of wood, and application to eco-friendly pH sensors. Nanomaterials 10(10):1904. https://doi.org/10.3390/nano10101904
31. Kaza S, Yao LC, Bhada-Tata P, Van Woerden F (2018) What a Waste 2.0: a global snapshot of solid waste management to 2050. Urban Development. © Washington, DC: World Bank. http://hdl.handle.net/10986/30317 License: CC BY 3.0 IGO
32. Klima K, Kieć J, Lepiarczyk A, Synowiec A (2014) Impact of laser beams treatment on the biomass yield and energy value of multiflora rose. Agric Eng 18(3):57–63
33. Kumar B, Bhardwaj N, Agrawal K, Chaturvedi V, Verma P (2020) Current perspective on pretreatment technologies using lignocellulosic biomass: an emerging biorefinery concept. Fuel Process Technol 199:106244. https://doi.org/10.1016/j.fuproc.2019.106244
34. Kumar B, Verma P (2020) Biomass-based biorefineries: An important architype towards a circular economy. Elsevier, Fuel, p 119622
35. Li M, Luo N, Lu Y (2017) Biomass energy technological paradigm (BETP): trends in this sector. Sustainability 9(4):567. https://doi.org/10.3390/su9040567
36. Li T, Miao K, Zhao Z, Li Y, Wang H, Watanabe A, Wang K (2022) Understanding cellulose pyrolysis under hydrogen atmosphere. Energy Convers Manage 254:115195. https://doi.org/10.1016/j.enconman.2021.115195
37. Li Y, Wu T, Wang Y, Li J, Yu W, Zhang G, Zhou W (2023) Green and large-scale production of ammonia: Laser-driven pyrolysis of nitrogen-enriched biomass. SusMat 3(4):533–542. https://doi.org/10.1002/sus2.146
38. Li Z, Lu L, Xie Y, Wang W, Lin Z, Tang B, Lin N (2021) Preparation of laser-induced graphene fabric from silk and its application examples for flexible sensor. Adv Eng Mater 23(9). https://doi.org/10.1002/adem.202100195
39. Lin J, Peng Z, Liu Y, Ruiz-Zepeda F, Ye R, Samuel ELG, Tour JM (2014) Laser-induced porous graphene films from commercial polymers. Nat Commun 5(1). https://doi.org/10.1038/ncomms6714

40. Lin Y, Zhang Q, Deng Y, Shen K, Xu K, Yu Y, Fang G (2021) Fabricating nanodiamonds from biomass by direct laser writing under ambient conditions. ACS Sustain Chem Eng 9(8):3112–3123. https://doi.org/10.1021/acssuschemeng.0c07607
41. Liu H, Dong M, Xiong J, Yang J, Ning J, Liang Y, Lu J (2022) Reaction characteristics and mechanism of laser-driven dry reforming of methane. SSRN Electron J. https://doi.org/10.2139/ssrn.4243124
42. Liu X, Feng X, He Y (2019) Rapid discrimination of the categories of the biomass pellets using laser-induced breakdown spectroscopy. Renewable Energy 143:176–182. https://doi.org/10.1016/j.renene.2019.04.137
43. Ma Y, Fang S, Peng Y, Gong Y, Wang D (2019) Remote estimation of biomass in winter oilseed rape (*Brassica napus* L.) using canopy hyperspectral data at different growth stages. Appl Sci 9(3):545. https://doi.org/10.3390/app9030545
44. Magallanes G, Kärkäs MD, Stephenson CRJ (2022) Depolymerization of lignin by homogeneous photocatalysis. *Springer Handbook of Inorganic Photochemistry*, 1537–1562. https://doi.org/10.1007/978-3-030-63713-2_52
45. Mas N, Hueso JL, Martinez G, Madrid A, Mallada R, Ortega-Liebana MC, Santamaria J (2020) Laser-driven direct synthesis of carbon nanodots and application as sensitizers for visible-light photocatalysis. Carbon 156:453–462. https://doi.org/10.1016/j.carbon.2019.09.073
46. Mat Junoh M, Ani FN (2016) The application of laser in thermal treatment of solid particles and gas-phase of biomass processing-a review. Jurnal Teknologi 78(3). https://doi.org/10.11113/jt.v78.7498
47. Miyakoshi R, Hayashi S, Terakawa M (2022) Simultaneous laser-based graphitization and microstructuring of bamboo for supercapacitors derived from renewable resources. RSC Adv 12(46):29647–29652. https://doi.org/10.1039/d2ra05641k
48. Nair LG, Agrawal K, Verma P (2022a) An overview of sustainable approaches for bioenergy production from agro-industrial wastes. Energy Nexus p100086
49. Nair LG, Agrawal K, Verma P (2022b) An insight into the principles of lignocellulosic biomass-based zero-waste biorefineries: a green leap towards imperishable energy-based future. Biotechnol Genetic Eng Rev 1–51
50. Nair LG, Agrawal K, Verma P (2023) Organosolv pretreatment: an in-depth purview of mechanics of the system. Biores Bioprocess Springer 10(1)
51. Nazarian A, Presser C (2008) Thermal and chemical kinetic characterization of multiphase and multicomponent substances by laser heating. Int J Heat Mass Transf 51:1365–1378
52. Tursunov O, Jan W, Dobrowolski, (2015) A brief review of application of laser biotechnology as an efficient mechanism for the increase of biomass for bio-energy production via clean thermo-technologies. Am J Renew Sustain Energy 1(2):66–71
53. Paykani A, Chehrmonavari H, Tsolakis A, Alger T, Northrop WF, Reitz RD (2022) Synthesis gas as a fuel for internal combustion engines in transportation. Prog Energy Combust Sci 90:100995. https://doi.org/10.1016/j.pecs.2022.100995
54. Piskorz J, Radlein D, Scott DS (1986) On the mechanism of the rapid pyrolysis of cellulose. J Analy Appl Pyrol 9:121–137
55. Presser C, Nazarian A (2017) Laser-driven calorimetry of single-component liquid hydrocarbons. Energy Fuels 31(7):6732–6747. https://doi.org/10.1021/acs.energyfuels.7b00534
56. Presser C, Nazarian A, Ohaion-Raz T, Lerner A, Dubkin H, Dabush B, Paz Tal O (2021) Thermochemical behavior of Chlorella sp. and *Chlamydomonas reinhardtii* algae: comparison of laser-driven calorimetry with thermogravimetric analysis. Algal Res 56:102325. https://doi.org/10.1016/j.algal.2021.102325
57. Radlein DS, Piskorz J, Scott DS (1987) Lignin derived oils from the fast pyrolysis of poplar wood. J Anal Appl Pyrol 12(1):51–59
58. Radlein D, Quignard A (2013) A short historical review of fast pyrolysis of biomass. Oil & Gas Science and Technology – Revue D'IFP Energies Nouvelles, 68(4), 765–783. https://doi.org/10.2516/ogst/2013162
59. Rex P, Mohammed Ismail K, Meenakshisundaram N, Barmavatu P, Sai Bharadwaj A (2023) Agricultural biomass waste to biochar: a review on biochar applications using machine learning

approach and circular economy. Chem Eng 7(3):50. https://doi.org/10.3390/chemengineering7030050
60. Saeed S, Samer M, Mohamed MSM, Abdelsalam E, Mohamed YMA, Abdel-Hafez SH, Attia YA (2022) Implementation of graphitic carbon nitride nanomaterials and laser irradiation for increasing bioethanol production from potato processing wastes. Environ Sci Pollut Res 29(23):34887–34897. https://doi.org/10.1007/s11356-021-18119-5
61. Samer M, Abdelsalam EM, Mohamed S, Elsayed H, Attia Y (2021) Impact of photoactivated cobalt oxide nanoparticles addition on manure and whey for biogas production through dry anaerobic co-digestion. Environ Dev Sustain 24(6):7776–7793. https://doi.org/10.1007/s10668-021-01757-7
62. Samer M, Hijazi O, Abdelsalam EM, El-Hussein A, Attia YA, Yacoub IH, Bernhardt H (2021) Life cycle assessment of using laser treatment and nanomaterials to produce biogas through anaerobic digestion of slurry. Environ Dev Sustain 23(10):14683–14696. https://doi.org/10.1007/s10668-021-01264-9
63. Sharifzadeh M, Sadeqzadeh M, Guo M, Borhani TN, Murthy Konda N, Garcia MC, Shah N (2019) The multi-scale challenges of biomass fast pyrolysis and bio-oil upgrading: review of the state of art and future research directions. Prog Energy Combust Sci 71:1–80. https://doi.org/10.1016/j.pecs.2018.10.006
64. Sharma A, Thakur M, Bhattacharya M, Mandal T, Goswami S (2019) Commercial application of cellulose nano-composites—A review. Biotechnol Rep 21:e00316. https://doi.org/10.1016/j.btre.2019.e00316
65. Singh DP, Dwevedi A (2019) Production of clean energy by green ways. *Solutions to Environmental Problems Involving Nanotechnology and Enzyme Technology*, 49–90. https://doi.org/10.1016/b978-0-12-813123-7.00002-5
66. Smith, & John, (2021) Laser technology in waste-biomass utilization. J Sustain Eng 25(2):45–62
67. Spreafico C, Russo D, Degl'Innocenti, R. (2021) Laser pyrolysis in papers and patents. J Intell Manuf 33(2):353–385. https://doi.org/10.1007/s10845-021-01809-9
68. Talib Hamzah H, Sridevi V, Seereddi M, Suriapparao DV, Ramesh P, Sankar Rao C, Pritam K (2022) The role of solvent soaking and pretreatment temperature in microwave-assisted pyrolysis of waste tea powder: analysis of products, synergy, pyrolysis index, and reaction mechanism. Biores Technol 363:127913. https://doi.org/10.1016/j.biortech.2022.127913
69. Tur E (2023) A comprehensive review on sustainability and environmental impact of laser powder bed fusion additively manufactured as-built Ti-6Al-4V parts. El-Cezeri Fen Ve Mühendislik Dergisi. https://doi.org/10.31202/ecjse.1325609
70. Tursunov O (2014) A comparison of catalysts zeolite and calcined dolomite for gas production from pyrolysis of municipal solid waste (MSW). Ecol Eng 69:237–243. https://doi.org/10.1016/j.ecoleng.2014.04.004
71. Tymchuk I, Nykyforov V, Maznytska O, Novokhatko O, Pasenko A, Malovanyy M (2021) Laser pretreating Of cyanobacteria biomass to produce lipids as a renewable energy source. Environ Eng Manage J 20(8):1255–1262. https://doi.org/10.30638/eemj.2021.116
72. Wang S, Gao L (2019) Laser-driven nanomaterials and laser-enabled nanofabrication for industrial applications. Ind Appl Nanomater 181–203. https://doi.org/10.1016/b978-0-12-815749-7.00007-4
73. Wang Z, Fu P, Chao X (2019) Laser absorption sensing systems: challenges, modeling, and design optimization. Appl Sci 9(13):2723. https://doi.org/10.3390/app9132723
74. Ye R, Chyan Y, Zhang J, Li Y, Han X, Kittrell C, Tour JM (2017) Laser-induced graphene formation on wood. Adv Mater 29(37). https://doi.org/10.1002/adma.201702211
75. Zhang S, Jiang SF, Huang BC, Shen XC, Chen WJ, Zhou TP, Yu HQ (2020) Sustainable production of value-added carbon nanomaterials from biomass pyrolysis. Nat Sustain 3(9):753–760. https://doi.org/10.1038/s41893-020-0538-1

Chapter 13
Bottlenecks and Opportunities in Applications of Irradiation-Based Technologies for Industrial Scale Waste Biomass Valorization

Paramjeet Baghel, Dibakar Rakshit, and Priyanka Kaushal

Abstract In light of the depletion of fossil fuel reserves and growing concerns regarding harm to the environment, it is imperative to shift towards a bioenergy-centric economy, which is more sustainable. As a feasible alternative to fuels and chemicals, biomass has been identified as a viable solution for it. Conversion of biomass to biofuels is done through various conventional heating techniques. Among these, irradiation-based technologies (gamma-ray radiation, electric field, microwave, and ultrasonication) lately attracted more attention on a global scale. In the context of mass production of the desired product (such as biofuels and value-added chemicals) is challenging because technology transfer from lab scale to industrial level is in the initial phases. Adopting these technologies requires consideration of many other aspects, such as economic and environmental sustainability. However, there are still several challenges for irradiation-based biorefineries. Silent characteristics such as rapid reaction, increased productivity, faster heating rates, better efficiency, and environmental friendliness make irradiation a more reliable technology. Acceptance of irradiation-based industry in the fossil-based industry is also challenging because of the feedstock supply, product composition, efficiency of resource recovery, quantity and cost of the product needed in the market demand, and techno-economic viability. This study thoroughly explains the techniques and concepts of these technologies, their current scenarios, bottlenecks, and future opportunities.

Keywords Irradiation-based technology · Value-added chemicals · Biomass · Biofuel · Pretreatment

---

P. Baghel · P. Kaushal (✉)
Clean Energy Laboratory, Centre for Rural Development and Technology, Indian Institute of Technology Delhi, Delhi 110016, India
e-mail: priyankak@iitd.ac.in

D. Rakshit
Department of Energy Science and Engineering, Indian Institute of Technology Delhi, Delhi 110016, India

© The Author(s), under exclusive license to Springer Nature Switzerland AG 2024
P. Verma and B. Kumar (eds.), *Sustainable Radiation Technologies in Waste-biomass Valorization*, Green Chemistry and Sustainable Technology,
https://doi.org/10.1007/978-3-031-63941-8_13

13.1 Introduction

The accelerated exhaustion of fossil resources is causing the worldwide energy crisis, thus resulting in a gradual rise in the price of crude oil. Moreover, the negative effects that are a direct result of the combustion of fossil fuels, such as the formation of acid rain and the phenomenon of global warming, are becoming an increasingly important concern in the records of human history [25]. In the last two decades, to replace fossil fuels, extensive research has been performed to find alternative energy sources [10, 28]. Alternative renewable sources offer clean energy like solar, hydropower, wind, geothermal, ocean energy, and bioenergy. The bioenergy boom in recent years has delivered a promising clean energy future [8, 29, 55]. Thus, the conversion of lignocellulosic feedstock into biofuels (solid, liquid, and gas) and different chemicals gained more attention for second-generation biofuels and lignocellulosic biomass-based biorefineries [48].

Biomass is defined as an organic material that is made up of a long chain of hydrocarbons (composition containing C, H, O, and N) [33]. It comes from different sources like plants, crops, agricultural residue, and animal waste. Furthermore, cellulose, hemicellulose, and lignin make up the majority of lignocellulosic feedstocks [33, 42]. Biomass composition and characteristics change rapidly according to the location of origin. To convert these feedstocks into the desired product, a different process is required [12, 32, 33]. Chemical, thermal, and biological treatment and pretreatment are used for specific kinds of products [57]. Because of variation in the biomass composition, it behaves differently in the treatment process [18]. In the pretreatment process of biomass to produce a particular biofuel and chemical, a reduction in cellulose crystallinity and delignification is required [28]. This can be done by conventional and irradiation-based methods [2, 36, 78]. Pretreatment leads to high surface area and active sites for the desired reaction for the process [2, 66, 75]. Irradiation-based methods are more selective than conventional methods and environmentally friendly [37, 51]. These methods also have short residence time, which makes them a cost-effective process.

This chapter gives full insight into irradiation-based methods used in pretreatment and direct conversion of biomass to final products. The basic principles of methods and difficulties in large-scale operation has been discussed. The recent research advancement and future scope of irradiation-based methods are also discussed.

13.2 Biomass Feedstock

Biofuels and value-added chemicals may be produced from a variety of biomass sources. In the quest for bioenergy resources, biomass plays a significant role. Biomass is mainly referred to as organic material and is composed of carbon, hydrogen, nitrogen, and oxygen at an elemental level [33]. As a result, biomass is a promising source of energy. The potential of biomass resource availability is

Table 13.1 Biomass feedstock estimation for the production of biofuel and value-added chemicals

Biomass feedstock	Major feedstock producers	Estimated feedstock (billion ton/year)	References
Wheat straw	India, China, France Russia, USA	1.10 by 2025	[3, 5, 61]
Rice straw	India, Vietnam, Indonesia, Bangladesh	1.27 by 2025	[3, 22, 61]
Commercial wood residue	Canada, Sweden, France	7.58 by 2050	[64]
Municipal solid waste	Germany, China, USA, Japan, Brazil	0.67 by 2025	[28]
Industrial Corn	Canada, Brazil, USA, Argentina, Mexico, China	4.82 by 2050	[64]
Industrial wood residue	Sweden, France, Canada	2.28 by 2050	[3, 64]
Forest residue	Canada, Sweden, China, Brazil, USA	6.00 by 2050	[3]

mentioned in Table 13.1. In the era of climate change and diminishing fossil fuel resources, biomass and its derived biofuels can be alternative solutions for fossil fuels [47]. Different types of potential biomass for biofuel and value-added chemical conversion are following.

13.2.1 Agricultural Residue and Wood-Based

Since ancient centuries, wood-based biomass has been used for cooking fuel purposes and heating the ambient for requirement. It is a traditional source of energy for people. Agricultural residue is left over after harvesting the crop. For example, rice straw, white straw, corn stover, etc. These materials are also called lignocellulosic biomass because of the high content of hemicellulose and cellulose [33]. As a result, these kinds of agricultural residues are easily processed for bioethanol and value-added chemicals.

13.2.2 Aquatic Plant-Based

It mainly includes algae, kelp, cattails, Salicornia, hyacinth, etc. These biomasses have been used for biofuel production in the last few decades [13]. Researchers and scientists found enormous potential for biofuel in aquatic plant-based biomass.

13.2.3 Animal Fiber

It includes wool, poultry waste, fish oil, silkworm Pupae, animal fats and oils, etc. [6]. The presence of triglycerides, lanolin, fatty acids, and lipids leads to biodiesel and other chemical conversion from animal-based biomass.

13.2.4 Industrial, Municipal, and Other Waste-Based

Different types of industries produce various kinds of waste for example. Sugarcane bagasse from sugar mills, sawdust, wood chips, paper, and logging residue from the timber industry. Municipal solid waste (MSW) and sewage sludge have been used to produce biofuel in recent years [32].

13.3 Different Techniques to Convert Biomass to Biofuel and Chemicals

The conversion of biomass to biofuel and chemicals is done via thermochemical conversion methods, biochemical conversion processes, and chemical conversion processes. The traditional way of harnessing energy from biomass is done by thermochemical (Pyrolysis, gasification, hydrothermal liquefaction) and biochemical processes (Fermentation and Anaerobic digestion) [33, 37, 58, 66]. Chemical conversion includes catalytic conversion and the use of some chemicals in traditional processes. New technology coupled with conventional technologies has been producing biofuel and chemicals. Such as irradiation techniques like Gamma-ray irradiation, electron beam irradiation, ultrasonication, and microwave irradiation are used to pretreat the biomass sample before biofuel conversion [37]. Pretreatment is the most essential step in liquid biofuel production. Because highly organized biomass makes it resistant to microbial, physical, and chemical attacks. Thus, these techniques help in the improvisation of the digestibility of carbohydrates, or polysaccharides present in the biomass for different processes. It also increases the enzyme accessibility of biomass for biofuel production [2, 43, 44]. According to biofuel and chemical production requirements, different biomass needs typical irradiation techniques. So, the appropriate selection of techniques for individual biomass plays a major role. The primary pretreatment procedures that make the residual solid biomass more accessible for later operation are size reduction, depolymerization, fractionation, and solubilization of the various components in the biomass [24, 37].

The pretreatment process is mostly determined by the chemical content of the biomass. Cellulose, hemicellulose, lignin, and extractives constitute biomass. The components of biomass are illustrated in Fig. 13.1.

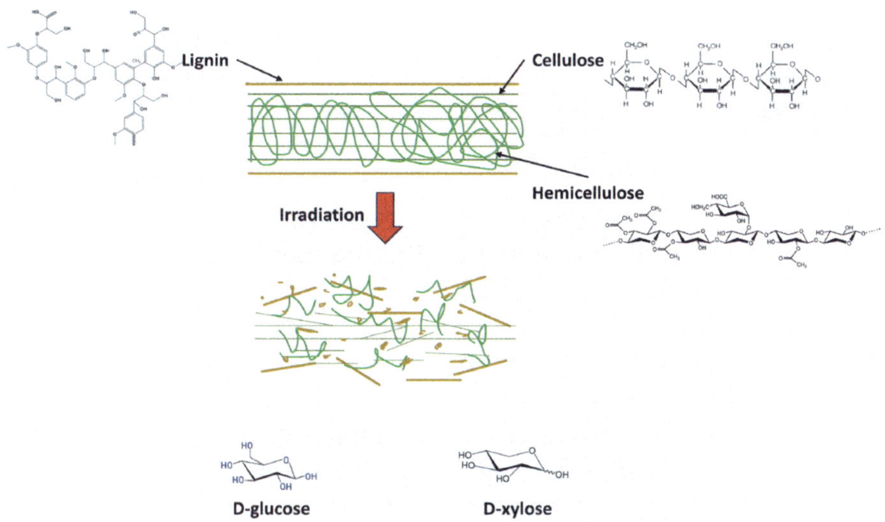

Fig. 13.1 Biomass feedstock decomposition after irradiation-based pretreatment producing sugar

A D-anhydroglucopyranose unit is interconnected via a β-(1–4) glycosidic bond to form the linear polymer cellulose. Hemicelluloses and amorphous polymers hold the microfibrils made up of these tightly packed cellulose chains together. Lignin covers these structures and connects them. An amorphous polymer called lignin gives plant cell walls stiffness and defends against microbial invasion [2].

A pretreatment procedure is required to remove the lignin and hemicellulose from the cellulose to improve access for enzymatic saccharification. An illustration of irradiation pretreatment is shown in Fig. 13.1. For the pretreatment of biomass to be successful, it must be highly digestible, there must be less sugar degradation, and there must be fewer inhibitors produced, which might potentially limit the efficiency of the fermentation process [68]. To rupture the cell structure of biomass that has a highly resistant component, intensive pretreatment conditions are required.

Industry-level chemicals (acid or alkali reagents) and biological pretreatment with heat are generally preferred because of proven technology. A different example of these pretreatments is illustrated in Fig. 13.2. In this method, microbes and enzymes easily decompose the chemical compounds present in the biomass for biofuel production [2]. Nowadays, physical pretreatment is also used to break the cell structure of the biomass according to the suitable product. Its prominent role is to reduce the size and increase the surface area so that it can be processed further. Microwave irradiation is directly applied to conventional technology and used for biofuel and value-chemical production. This method heats the biomass and produces the desired product in a single step. The conversion of several biomasses, including macro- and microalgae, agricultural waste, sewage sludge, and forestry waste, was investigated using microwave heating pyrolysis [63]. Compared to the other methods listed, it has been found that the irradiation pretreatment of biomass uses less energy.

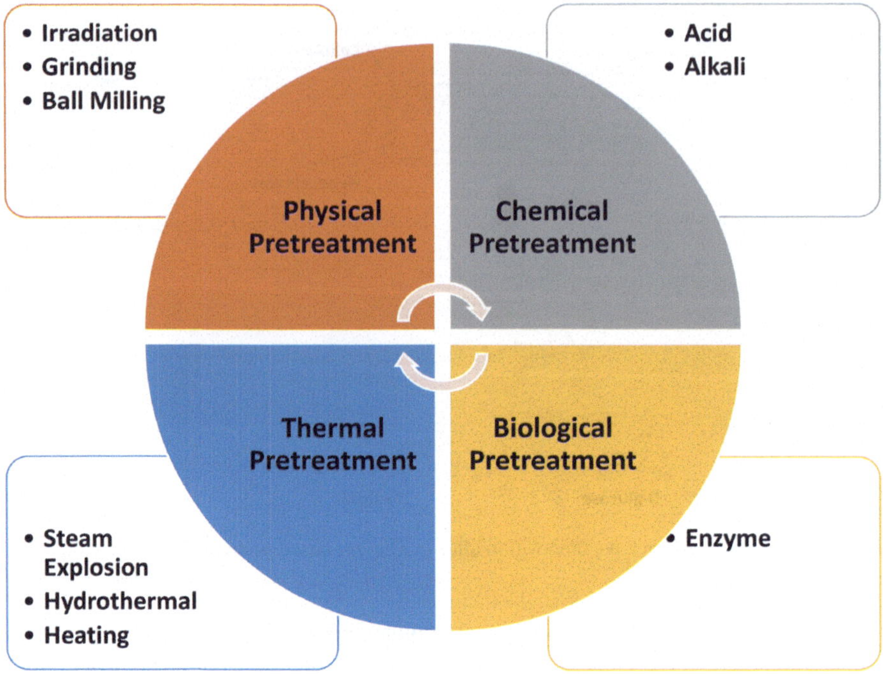

Fig. 13.2 Different pretreatment methods for conversion of biomass to biofuel and value-added chemicals

Additionally, this method is effective in producing desired output because it easier to manage the selective approach [28]. The next part goes into depth about the radiation pretreatment.

13.4 Working Principle and Mechanism of Irradiation-Based Techniques

Irradiation is regarded as an enticing physical technique for pretreating biomass, among other techniques. In the biomass irradiation process, high-energy radiations like electron beams, microwaves, γ-rays, and ultrasonic waves are used to treat biomass. Radiation has different effects on the biomass and process mechanisms depending on the method that is used.

13.4.1 Microwave Irradiation

Microwaves are a type of electromagnetic wave. Their wavelengths varies 1 mm to 1 m and their frequencies are between 300 MHz and 300 GHz [63]. The Federal Communications Commission (FCC) set aside two frequencies for use in the medical, scientific, and industrial fields. These frequencies are often used for heating with microwaves. The two frequencies that are most often used are 0.915 and 2.45 GHz. For the processing of materials, microwave furnaces with adjustable frequencies ranging from 0.9 to 18 GHz have recently been developed [49]. Induced rapid and selective heating is involved in the mechanism of the microwave irradiation technique. The process is generally integrated with the pyrolysis process for biofuel production. In the mechanism, irradiation works by polarity of molecules, which changes the orientation by induced dielectric heating. Microwave irradiation is absorbed by water molecules, cellulose, and hemicellulose effectively, leading to selective heating of the material. Due to the rotation of molecules in the direction of the electromagnetic field, friction occurs between the adjacent molecules. Thus, the rubbing motion generates heat from inside to outside. This also allows the biomass's rapid and volumetric heating, which is better than conventional heating [59, 72]. The traditional heating starts from inside to outside, whereas the microwave penetrates from the core of the material and heats it throughout as shown in Fig. 13.3. This also leads to heat loss to surroundings and makes this technique energy efficient.

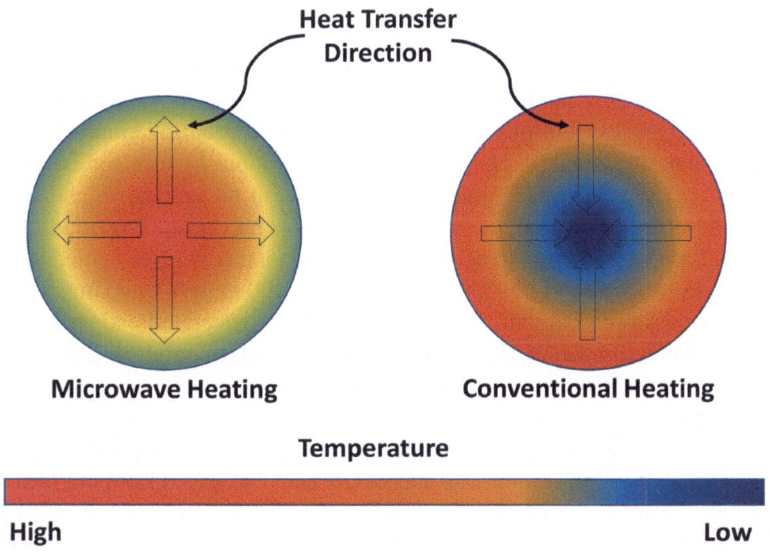

Fig. 13.3 Comparison of heat flow direction of microwave and conventional heating

13.4.2 Gamma-Ray Irradiation

This method is not directly used for biofuel conversion from biomass heating. This method is mostly involved with pretreatment of the biomass conversion. Gamma rays are a type of high-energy ionizing radiation that can easily pass through most materials [41, 45]. This radiation is made up of very large, high-frequency waves that depend a lot on the generation source. Recently, there has been a lot of interest in using this technology, especially to prepare biomass and fibers for the production of liquid biofuels [69]. Nuclides that are radioactive, like cobalt-60 and cesium-137, are often used in this step [51]. By breaking down the hydrogen bonds between molecules, the main goal of this irradiation is to make cellulose less ordered both inside and between molecules. Radiation from the sealed source will flow through the biomass structure during this phase, penetrating it. The biomass component receives the energy from gamma radiation through radiation collisions, which ionizes the atom by causing it to lose an electron [41]. The main component of biomass, cellulose polymers, split during radiation exposure, releasing a range of short- and long-lived radicals [41, 51]. During this process, the structure of the biomass changed significantly. As a result, this helps in further processes like the enzymatic saccharification process.

13.4.3 Electron Beam Irradiation

When it comes to the pretreatment of biomass to produce biofuels, this method is helpful. Regularly, it is utilized for biomass preparation for the enzymatic saccharification process. This method is famous for delignification, rupture of cellulose structure (decrease crystallinity), and increased sugar solubilizations [35, 52, 53, 74]. Significant alterations in cellulose are brought about by electron beam irradiation, such as a rise in surface area and a decrease in molecular weight and crystallinity. These structural changes make it easier for the various enzymatic fermentation [53]. Based on energy intensity and application, it may be classified into three distinct energy ranges: low (80–300 keV), medium (300–5 meV), and high (>5 meV) [37]. For biomass pretreatment, an electron cannon is generated and an accelerator accelerates an electron beam. As illustrated in Fig. 13.4, the high-intensity electron beam penetrates the biomass and distributes energy within the biomass. As a result, depolymerization of cellulose and heating within the material started. This causes the oxidation and production of different carbonyl groups in the biomass structure. The mechanical strength of biomass decreased as it was exposed to the beam. The crosslinking phenomenon of biomass components is also observed during irradiation [54]. In this way, the crystallinity of cellulose decreases, and a more amorphous structure is formed [26]. In addition, this process has several other benefits, such as low temperature, short treatment duration, and reduced production of unwanted inhibitory by-products compared to chemical pretreatment procedures.

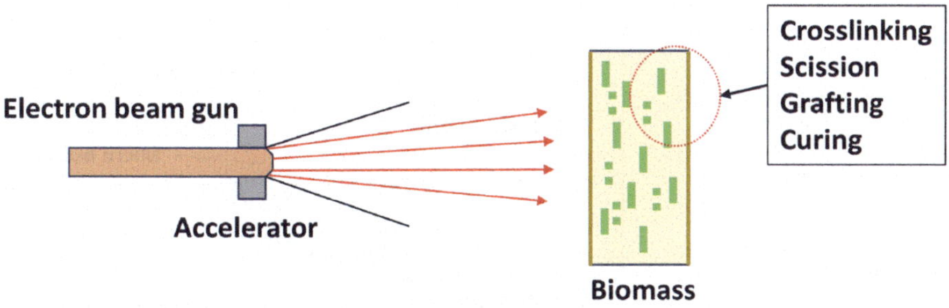

Fig. 13.4 Schematic representation of electron beam irradiation on biomass structure

13.4.4 Ultrasonication

High-frequency sound waves disrupt biomass structures, leading to further biofuel and chemical production processes. The disruption of biomass structure involves cavitation and localized heating through acoustic—waves through a medium [60]. Cavitation is a key phenomenon related to ultrasonication. The alternating low and high-pressure cycles are generated in the liquid medium during sonication. In the case of low-pressure cycles, tiny bubbles developed in the liquid. As these bubbles expand over the course of several cycles, the bubbles finally burst violently during high-pressure periods, causing shock waves and localized hot patches as shown in Fig. 13.5. These intense localized hot patches and pressure lead to heating inside the biomass material and treatment [60, 62]. Certain processes that convert biomass into fuel, such as pyrolysis and hydrolysis, benefit from the rapid and localized aspects of the heating process. It also improves the mass transfer within the biomass structure. The turbulence and microstreaming created by the sound waves in the liquid around the biomass particles improve the interaction between the biomass and reaction medium. This enhanced mass transfer has the potential to boost the conversion efficiency of biofuels through chemical processes. Thus, the reaction time is also reduced [50]. Sound waves can stimulate the dispersion of catalyst particles and their contact with biomass, resulting in more effective catalytic processes [1]. In this way, this process makes the industrial process more efficient (biodiesel production) and increases the yield of the product.

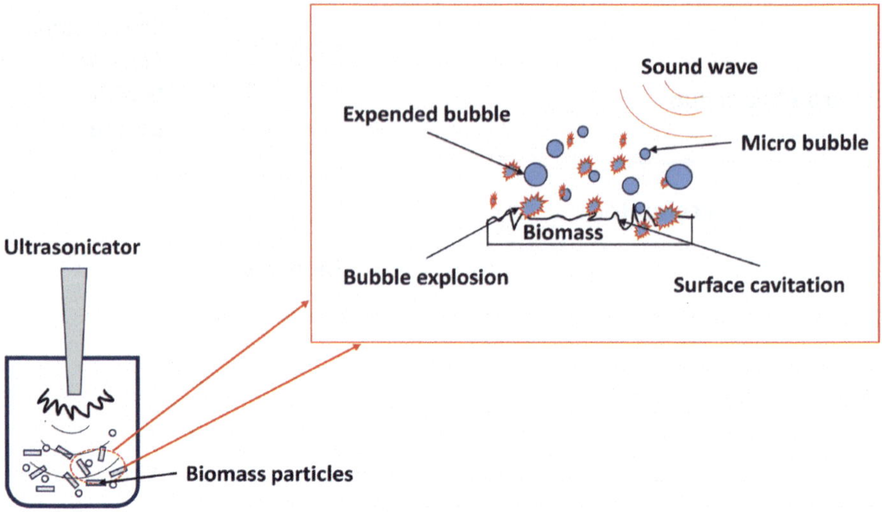

Fig. 13.5 Schematic illustration of biomass structure disruption through ultrasonication

13.5 Bottlenecks in Industrial Biomass Conversion Using Irradiation-Based Technologies

13.5.1 Biomass Heterogeneity

In the context of biomass, the term "heterogeneity" refers to the fact that different feedstocks might have varying structures and compositions. This aspect is a key contributor to the industrial application of irradiation-based biomass conversion. The diverse nature of the biomass also influences the process efficiency. Primary characteristics like biomass moisture and composition affect downstream processes and process optimization of irradiation-based biomass conversion [37, 51, 73]. According to geographical location and type, the composition of biomass changes rapidly. The composition of different feedstock is given in Table 13.2.

Softwood and hardwood, for instance, are treated differently due to their distinct compositions, which cause them to respond differently when exposed to certain irradiations. The sensitivity of biomass to irradiation depends upon the crystallinity of cellulose [51]. As a result, to achieve successful depolymerization and fragmentation, feedstock with a lower cellulose crystallinity requires less intense irradiation and vice versa. When it comes to feedstock, high lignin-based feedstock necessitates extensive irradiation and results in valuable chemical potential due to the complicated nature of lignin. Moisture content in biomass also affects the desired product yields [9, 34]. The water molecules present inside the biomass undergo decomposition into hydrogen and oxygen when exposed to radiation, which then triggers further reactions. Thus,

Table 13.2 Different compositions of various biomass

Biomass name	Components of biomass (%)			References
	Hemicellulose	Cellulose	Lignin	
Rice husks	28.70–40.00	15.50–26.00	12.00–29.30	[17]
Barley straw	24.00–33.00	36.00–43.00	6.30–13.10	[51]
Hybrid poplar	16.60–22.60	15.50–16.30	40.30–47.30	[39]
Sugarcane bagasse	25.00	42.00	20.00	[78]
Cornstalk	29.87	34.85	8.16	[71]
Rubber wood	28.43	39.56	27.16	[38]
Corn Stover's	20.90	32.70	25.40	[23]
Japanese ceder	13.80	33.50	52.70	[56]
Olive pomace	22.00	19.00	40.00	[4]
Coconut coir	44.20	32.80	22.10	[68]
Bamboo leaves	25.60	35.00	34.14	[31]
Banana waste	14.80	13.20	14.00	[40]
Oak	21.90	43.20	35.40	[77]
Sago palm bark	24.30	42.60	19.20	[21]
Pinus armandii Franch	17.80	24.10	48.40	[70]
Walnut shell	20.40	23.30	53.50	[18]
Miscanthus	21.30	52.10	18.60	[36]
Cocoa pods	27.70	32.30	21.44	[76]

the heterogeneity in feedstock affects the downstream processes following irradiation, such as microbial fermentation or enzymatic hydrolysis. Process optimization becomes a difficult task due to the heterogeneous nature of biomass. The parameters change very frequently for different biomass which can cause an overall cost hike in the process. Thus, the economic viability of irradiation-based biomass conversion at an industrial scale is often compromised [15, 51].

13.5.2 Reaction Selectivity

In the context of irradiation-based biofuel and chemical production from biomass, selectivity plays a crucial factor. Selectivity is defined as the desired reaction and product of the process. For optimizing the process for the desired product of biofuel and chemicals, reaction selectivity is essential. While irradiating the biomass feedstock, several reactions occur and produce an extensive range of products. So, to maximize the particular desired product of the reaction, selectivity of the reaction is necessary [24, 26, 51]. Otherwise, many unnecessary products are formed which are unsuitable for cost-effective processes. Furthermore, the product quality is also impacted by reaction selectivity.

13.5.3 Energy Intensity

The term "energy intensity" refers to the amount of energy that is applied to each unit of material throughout the irradiation process. This particular component is directly related to the cost-effectiveness, process efficiency, and environmental sustainability of the process of converting biomass. Energy distribution and efficiency problems typically arise throughout the process of moving from the laboratory to the industrial scale. The optimum level of energy intensity must be ensured to achieve consistency and efficiency in the process of converting biomass [15, 35, 51]. According to different feedstock, different intensity levels are required. The process becomes more complex since different irradiation techniques operate at different energy levels.

13.6 Recent Advances

Pretreatment is an essential step when converting biomass into biofuels such as ethanol and butanol. This step is responsible for the separation and structural changes in the three primary components (lignin, cellulose, and hemicellulose) of biomass. Production of by-products and low conversion of biomass components to desired products is a limitation of the typical chemical and enzymatic pretreatment procedures [67]. Numerous research have been done by researchers to extract energy via irradiation pretreatment over biomass for bioenergy. Most studies found that irradiation pretreatment might disrupt structure and reduce particle size, improving access to following procedures [9, 26, 37, 60]. It was evident that this irradiation-based pretreatment was able to improve the performance of different processes like enzymatic saccharification, extraction, esterification, thermochemical conversion, and fermentation [19, 65].

Current research studies show multiple advances in the technology for desired products. Xiang et al. performed a study on *Artemisia ordosica* to produce biofuels through enzymatic saccharification [74]. The authors used the 7.5 kGy dose of electron beam irradiation method to make material for the enzymatic process. When compared to the untreated biomass sample, the sugar production rose by 1136.75%. Ma et al. [54] demonstrated the degradation of cellulose in bamboo chips via electron beam irradiation. According to [24], the FAME composition of biodiesel changes after 0.7 kGy dose of electron beam irradiation. The reduction of $C = C$ double bond concentration in methyl was confirmed by GC–MS results. In this way, biofuel is sterilized and the growth of microbial life in storage tanks is prevented. Bak [9] employed electron beam irradiation to research rice straw (EBI). The glucose yield of 70.4% of the theoretical maximum was achieved after a water-soaking electron beam (WEBI). The findings demonstrate that WEBI surpassed untreated and EBI by 29.5 and 52.1%, respectively.

Nikolic et al. [60] reported the study of corn meal pretreatment via ultrasonication (Frequency 40 kHz). After pretreatment, bioethanol was prepared by saccharification and fermentation. The concentrations of glucose and ethanol rose by 6.82 and 11.15%, respectively, in comparison to the untreated sample. Rambahu et al. [62] conducted a study on Date seed oil to produce biodiesel through ultrasonication optimization. Maximum biodiesel generation of 97.3% was achieved under optimal conditions. Dutta et al. [20] produced 50.43 k/l Hydroxymethyl furfural (HMF) from banana peels using ultrasonication and subsequent hydrolysis. By using a normal approach, the delignification was only 45.78%; however, using ultrasound to demonstrate intensification, the percentage climbed to 92%. Lee et al. [50] checked the effect of ultrasonication irradiation on vegetable oil for biodiesel synthesis The authors found the synthesis process reaction time reduced to 30 min compared to the exiting process. Choedkiatsakul et al. [16] used an ultrasonication reactor to produce biodiesel from palm oil. The work showed that the reactor produced a high methyl ester yield within 5 min of reaction time compared to the mechanical stirrer reactor, which required 60 min. Zou et al. [80] used an ultrasound-assisted extraction method to extract the lipids from microalgae.

Liu et al. [53] compared gamma-ray irradiation to the other pretreatment methods for bioethanol production. The authors reported that the bioconversion of microcrystalline cellulose with a yield of 67% was achieved with an 891 kGy dose of irradiation. Following pretreatment, the procedures of saccharification and fermentation were employed to produce bioethanol. The structure of kraft lignin was modified by gamma and alpha ray with doses of 200 and 1200 kGy, respectively [41]. According to the authors, gamma radiation has a more significant effect on lignin crystalline structure, whereas alpha radiation primarily affects amorphous lignin. Free radicals and different chemical reactions were triggered in the material when irradiated. Wu et al. [73] elucidated the process of lignin breakdown using gamma irradiation at various dose ranges (0–1200 kGy). The authors reported more cracks and the creation of porous surfaces of irradiated samples compared to untreated. The molecular weight and thermal stability also decreased in irradiated samples. Chen et al. [15] produced biohydrogen using Laminaria japonica (microalgae). The samples were pretreated with gamma irradiation of doses (10–30 kGy) to disintegrate the structure and increase the soluble organic compound concentrations. The hydrogen production increased by approximately 71.4% as compared to untreated. Therefore, radiolysis of lignocellulosic biomass is an efficient method of pretreatment that may be used to extract cellulose and lignin [15].

Microwave technology is a promising technology in the pretreatment and conversion of lignocellulosic materials into renewable biofuel and value-added chemicals [11, 46]. Chen et al. [14] treated sorghum for ethanol production with diluted ammonia under microwave assistance. The study claimed that 48% of the original lignin was removed from sorghum with the help of microwave assistance. Kam et al. [34] demonstrated the one-step in situ transesterification process under pulsed microwave irradiation. The study reported that under optimal conditions, the FAME yield reached 96.18%. According to the authors finding, the biomass received high-energy infusions at brief intervals made it easier for the algal cell wall to be significantly disrupted, which in turn led to the release of a greater quantity of lipids. Zhang et al. [79] used microwave irradiation in the torrefaction process for solid biofuel production from sludge. The work showed that the calorific value of torrefied material is similar to coal. The by-product bio-oil contains compounds like phenols, cholesterol, and siloxanes. The biochar, bio-oil, and syngas production microwave-assisted heating is used in the reactors. Huang et al. [30, 31] conducted a microwave pyrolysis study on seven different biomasses (rice straw, sugarcane peel and bagasse, rice husk, coffee grounds, corn stover, and bamboo) at MW power of 250 W. The study provided comparative information to traditional pyrolysis which showed microwave pyrolysis was quicker and required less energy. Researchers also made different value-added chemicals from biomass using microwave irradiation like carbon nanoparticles, carbon nanotubes, nano biochar, phenols, lipids, catalysts, and activated biochars [7, 13, 27, 28].

13.7 Future Direction and Opportunities

The most important advantage of these methods is that they are environmentally friendly and do not use toxic chemicals. The irradiation processes have high selectivity in degrading biomass components, unlike chemical pretreatment, which may degrade cellulose and hemicellulose during the process [37]. In addition, this technique has a shorter duration (<10 min) in comparison to other methods, particularly in biological pretreatment, where lignin particles are removed from the biomass in seven days [66].

Despite the significant advantages of irradiation technologies, the utilization of irradiation in biomass pretreatment and its integration into physicochemical processes for industrial heating applications are slightly limited. In addition, the scientific community has rarely discussed the examination of how the electromagnetic interaction affects lignocellulosic biomass, which hinders a comprehensive understanding of this topic. Furthermore, the optimization of parameters (energy consumption and cost) is very important for the maximum yield of cellulose components and efficient process. Even though the advantages of using radiation heating in biomass pretreatment have been demonstrated, numerous issues remain unanswered. Hence, obstacles to the pretreatment of lignocellulosic biomass for large-scale biofuel generation persist, primarily revolving around the elements of capital spending, energy efficiency, and sustainability. The difficulties might include issues with energy usage and clarifying basic concepts dielectric properties of biomass and heating.

Additional research needs to be conducted to determine which irradiation-based technology is the most effective for use in commercial biorefineries. This is essential for the sustainable utilization of renewable resources that are presently available. Moreover it has potential to produce clean energy efficiently. It would be helpful to study energy balance and come up with products that are quite close to being ready for sale or commercialization. More research needs to be done on the procedure used to pretreat biomass that can be taken from the lab to commercial and industrial scenarios. Moreover, the incorporation of energy should be given careful consideration during the process of designing and developing increasingly sophisticated methods and procedures. Integration of irradiation with conventional technology is also important for a cost-effective process. There is a lot of proof that strong and thorough techno-economic assessments are very important for biorefineries to be successful. As a result, the economic equilibrium of biorefineries has the potential to be improved and enhanced.

13.8 Conclusion

A full discussion was conducted on the fundamental and characteristic aspects of several irradiation technologies that are used to treat biomass to produce a variety of value-added chemicals and biofuels. When compared to traditional pretreatment methods, the utilization of irradiation techniques offers a greater number of advantages. The absence of the use of chemicals is another reason why irradiation methods are considered to be environmentally beneficial. Numerous components that are present in various types of biomass play a key part in the process of converting biomass into the products that are needed. The literature also includes the introduction of an individual irradiation approach that targets specific components of biomass. It is possible to reduce the amount of time required for the reaction and enhance the amount of products that are needed by combining irradiation techniques with conventional techniques. The crystallinity of cellulose is the specific target of gamma rays, whereas the lignin component of biomass is the target of electron beams. The structure of biomass is disrupted by ultrasonication and microwaves, which also create a greater surface area for reactions to take place around. The microwave irradiation method is the one that is utilized in the pretreatment and direct conversion of biomass to desired products. This method is among the several irradiation methods. When all of the characteristics of a particular product are optimized, it will be conceivable in the future to implement these technologies on an industrial scale. The integration of many technologies results in the process being more cost-effective when applied on an industrial scale.

13.9 Credit Authorship Contribution Statement

Paramjeet Baghel: Methodology, Conceptualization, Investigation, Writing—original draft; **Dibakar Rakshit:** Resources, Supervision; **Priyanka Kaushal**: Resources, Writing—review & editing, Supervision.

13.10 Declaration

The authors have not used any AI tool during the writing of the chapter.

Acknowledgements The authors appreciatively acknowledge the financial and resources from the Centre for Rural Development and Technology, Indian Institute of Technology Delhi.

References

1. Adeoye AO, Quadri RO, Lawal OS, Emojevu EO (2024) Physicochemical characterization valorization of lignocellulosic waste (kola nut seed shell) via pyrolysis, and ultrasonication of its crude bio-oil for biofuel production. Clean Waste Syst 7:100138. https://doi.org/10.1016/J.CLWAS.2024.100138
2. Agbor VB, Cicek N, Sparling R, Berlin A, Levin DB (2011) Biomass pretreatment: fundamentals toward application. Biotechnol Adv 29(6):675–685. https://doi.org/10.1016/J.BIOTECHADV.2011.05.005
3. Ahmed N, Thompson S, Glaser M (2019) Global aquaculture productivity, environmental sustainability, and climate change adaptability. Environ Mana 63(2):159–172. https://doi.org/10.1007/S00267-018-1117-3/TABLES/3
4. Álvarez A, Cachero S, González-Sánchez C, Montejo-Bernardo J, Pizarro C, Bueno JL (2018) Novel method for holocellulose analysis of non-woody biomass wastes. Carbohyd Polym 189:250–256. https://doi.org/10.1016/J.CARBPOL.2018.02.043
5. Anand A, Pathak S, Kumar V, Kaushal P (2022) Biochar production from crop residues, its characterization and utilization for electricity generation in India. J Clean Prod 368:133074. https://doi.org/10.1016/J.JCLEPRO.2022.133074
6. Ao S, Changmai B, Vanlalveni C, Chhandama MVL, Wheatley AEH, Rokhum SL (2024) Biomass waste-derived catalysts for biodiesel production: recent advances and key challenges. Renew Energy 223:120031. https://doi.org/10.1016/J.RENENE.2024.120031
7. Baghel P, Kaushal P (2022) Rapid synthesis of carbon nanotubes from Prosopis Juliflora biochar using microwave irradiation. Mater Sci Eng, B 286:115987. https://doi.org/10.1016/J.MSEB.2022.115987
8. Baghel P, Sakhiya AK, Kaushal P (2022) Influence of temperature on slow pyrolysis of Prosopis Juliflora: an experimental and thermodynamic approach. Renew Energy 185:538–551. https://doi.org/10.1016/J.RENENE.2021.12.053
9. Bak JS (2014) Electron beam irradiation enhances the digestibility and fermentation yield of water-soaked lignocellulosic biomass. Biotechnol Reports 4(1):30–33. https://doi.org/10.1016/J.BTRE.2014.07.006
10. Balakrishnan P, Shabbir MS, Siddiqi AF, Wang X (2020) Current status and future prospects of renewable energy: a case study. Energy Sources Part A: Recov Utiliz Environ Effects 42(21):2698–2703. https://doi.org/10.1080/15567036.2019.1618983
11. Bhardwaj N, Kumar B, Verma P (2022) (2022) Microwave-assisted pretreatment using alkali metal salt in combination with orthophosphoric acid for generation of enhanced sugar and bioethanol. Biomass Convers Biorefinery 12:1069–1076. https://doi.org/10.1007/s13399-020-00640-1
12. Bridgwater AV (2012) Review of fast pyrolysis of biomass and product upgrading. Biomass Bienergy 38:68–94. https://doi.org/10.1016/j.biombioe.2011.01.048
13. Chaos-Hernández D, Reynel-Ávila HE, Bonilla-Petriciolet A, Villalobos-Delgado FJ (2023) Extraction methods of algae oils for the production of third generation biofuels – a review. Chemosphere 341:139856. https://doi.org/10.1016/J.CHEMOSPHERE.2023.139856

14. Chen C, Boldor D, Aita G, Walker M (2012) Ethanol production from sorghum by a microwave-assisted dilute ammonia pretreatment. Biores Technol 110:190–197. https://doi.org/10.1016/J.BIORTECH.2012.01.021
15. Chen Y, Yin Y, Wang J (2022) Biohydrogen production using macroalgal biomass of Laminaria japonica pretreated by gamma irradiation as substrate. Fuel 309:122179. https://doi.org/10.1016/J.FUEL.2021.122179
16. Choedkiatsakul I, Ngaosuwan K, Cravotto G, Assabumrungrat S (2014) Biodiesel production from palm oil using combined mechanical stirred and ultrasonic reactor. Ultrason Sonochem 21(4):1585–1591. https://doi.org/10.1016/J.ULTSONCH.2013.12.025
17. Daza Serna LV, Orrego Alzate CE, Cardona Alzate CA (2016) Supercritical fluids as a green technology for the pretreatment of lignocellulosic biomass. Biores Technol 199:113–120. https://doi.org/10.1016/J.BIORTECH.2015.09.078
18. de Caprariis B, De Filippis P, Petrullo A, Scarsella M (2017) Hydrothermal liquefaction of biomass: influence of temperature and biomass composition on the bio-oil production. Fuel 208:618–625. https://doi.org/10.1016/J.FUEL.2017.07.054
19. Di Y, Na R, Xia H, Wang Y, Li F (2023) Irradiation effects on characteristics and ethanol fermentation of maize starch. Int J Biol Macromol 246:125602. https://doi.org/10.1016/J.IJBIOMAC.2023.125602
20. Dutta A, Kininge MM, Priya, Gogate PR (2023) Intensification of delignification and subsequent hydrolysis of sustainable waste as banana peels for the HMF production using ultrasonic irradiation. Chem Eng Process - Process Intensif 183:109247. https://doi.org/10.1016/J.CEP.2022.109247
21. Ethaib S, Omar R, Mustapa Kamal SM, Awang Biak DR (2020) Comparison of sodium hydroxide and sodium bicarbonate pretreatment methods for characteristic and enzymatic hydrolysis of sago palm bark. Energy Sources Part A: Recov Utiliz Environ Effects. https://doi.org/10.1080/15567036.2020.1753857
22. Gadde B, Menke C, Wassmann R (2009) Rice straw as a renewable energy source in India, Thailand, and the Philippines: overall potential and limitations for energy contribution and greenhouse gas mitigation. Biomass Bioenergy 33(11):1532–1546. https://doi.org/10.1016/J.BIOMBIOE.2009.07.018
23. Gong L, Xu ZY, Dong JJ, Li H, Han RZ, Xu GC, Ni Y (2019) Composite coal fly ash solid acid catalyst in synergy with chloride for biphasic preparation of furfural from corn stover hydrolysate. Biores Technol 293:122065. https://doi.org/10.1016/J.BIORTECH.2019.122065
24. Grabowski P, Tomkielski D, Szajerski P, Gwardiak H (2019) Changes of biodiesel composition after electron beam irradiation. J Radioanal Nucl Chem 319(3):727–736. https://doi.org/10.1007/S10967-018-6364-5/FIGURES/9
25. Haites E (2018) Carbon taxes and greenhouse gas emissions trading systems: what have we learned? Climate Policy 18(8):955–966. https://doi.org/10.1080/14693062.2018.1492897
26. Henniges U, Hasani M, Potthast A, Westman G, Rosenau T (2013) Electron beam irradiation of cellulosic materials-opportunities and limitations. Materials 6:1584–1598. https://doi.org/10.3390/ma6051584
27. Hidalgo P, Navia R, Hunter R, Camus C, Buschmann A, Echeverria A (2023) Carbon nanotube production from algal biochar using microwave irradiation technology. J Anal Appl Pyrol 172:106017. https://doi.org/10.1016/J.JAAP.2023.106017
28. Hoang AT, Nižetić S, Ong HC, Mofijur M, Ahmed SF, Ashok B, Bui VTV, Chau MQ (2021) Insight into the recent advances of microwave pretreatment technologies for the conversion of lignocellulosic biomass into sustainable biofuel. Chemosphere 281:130878. https://doi.org/10.1016/J.CHEMOSPHERE.2021.130878
29. Hovelius K, Hansson PA (1999) Energy- and exergy analysis of rape seed oil methyl ester (RME) production under Swedish conditions. Biomass Bienergy 17(4):279–290. https://doi.org/10.1016/S0961-9534(99)00047-1
30. Huang YF, Chiueh PT, Kuan WH, Lo SL (2016) Microwave pyrolysis of lignocellulosic biomass: heating performance and reaction kinetics. Energy 100:137–144. https://doi.org/10.1016/J.ENERGY.2016.01.088

31. Huang YF, Chiueh PT, Lo SL (2016) A review on microwave pyrolysis of lignocellulosic biomass. Sustain Environ Res 26(3):103–109. https://doi.org/10.1016/J.SERJ.2016.04.012
32. Iroba KL, Baik OD, Tabil LG (2017) Torrefaction of biomass from municipal solid waste fractions I: temperature profiles, moisture content, energy consumption, mass yield, and thermochemical properties. Biomass Bioenergy 105:320–330. https://doi.org/10.1016/j.biombioe.2017.07.009
33. Yogalakshmi KN, Sivashanmugam P, Kavitha S, Kannah Y, Varjani S, Adish Kumar S, Kumar G (2022) Lignocellulosic biomass-based pyrolysis: a comprehensive review. Chemosphere 286:131824. https://doi.org/10.1016/J.CHEMOSPHERE.2021.131824
34. Kam YL, Agutaya JKCN, Quitain AT, Ogasawara Y, Sasaki M, Lam MK, Yusup S, Assabumrungrat S, Kida T (2023) In-situ transesterification of microalgae using carbon-based catalyst under pulsed microwave irradiation. Biomass Bienergy 168:106662. https://doi.org/10.1016/J.BIOMBIOE.2022.106662
35. Karthika K, Arun AB, Melo JS, Mittal KC, Kumar M, Rekha PD (2013) Hydrolysis of acid and alkali presoaked lignocellulosic biomass exposed to electron beam irradiation. Biores Technol 129:646–649. https://doi.org/10.1016/J.BIORTECH.2012.12.048
36. Kashcheyeva EI, Gismatulina YA, Budaeva VV (2019) Pretreatments of non-woody cellulosic feedstocks for bacterial cellulose synthesis. Polymers 11(10):1645. https://doi.org/10.3390/POLYM11101645
37. Kassim MA, Khalil HPSA, Serri NA, Kassim MHM, Syakir MI, Aprila NAS, Dungani R, Kassim MA, Khalil HPSA, Serri NA, Kassim MHM, Syakir MI, Aprila NAS, Dungani R (2016) Irradiation pretreatment of tropical biomass and biofiber for biofuel production. Radiat Effects Mater. https://doi.org/10.5772/62728
38. Khan AS, Man Z, Bustam MA, Nasrullah A, Ullah Z, Sarwono A, Shah FU, Muhammad N (2018) Efficient conversion of lignocellulosic biomass to levulinic acid using acidic ionic liquids. Carbohyd Polym 181:208–214. https://doi.org/10.1016/J.CARBPOL.2017.10.064
39. Kim H, Ahn Y, Kwak SY (2016) Comparing the influence of acetate and chloride anions on the structure of ionic liquid pretreated lignocellulosic biomass. Biomass Bienergy 93:243–253. https://doi.org/10.1016/J.BIOMBIOE.2016.07.022
40. Kumar A, Jha A (2017) Host–pathogen interaction. In: Anticandidal Agents, pp 7–10. https://doi.org/10.1016/B978-0-12-811311-0.00002-8
41. Kumar A, Kumar GA (2023) Modification of lignin properties using alpha particles and gamma-rays for diverse applications. Radiat Phys Chem 202:110562. https://doi.org/10.1016/J.RADPHYSCHEM.2022.110562
42. Kumar Sakhiya A, Anand A, Aier I, Baghel P, Vijay VK, Kaushal P (2020) Sustainable utilization of rice straw to mitigate climate change: a bioenergy approach. Mater Today: Proc 46:5366–5371. https://doi.org/10.1016/j.matpr.2020.08.795
43. Kumar B, Bhardwaj N, Verma P (2019) Pretreatment of rice straw using microwave-assisted $FeCl_3$-H_3PO_4 system for ethanol and oligosaccharides generation. Bioresour Technol Reports 7:100295. https://doi.org/10.1016/j.biteb.2019.100295
44. Kumar B, Verma P (2019) Optimization of microwave-assisted pretreatment of rice straw with FeCl3 in combination with H3PO4 for improving enzymatic hydrolysis. Adv Plant & Microb Biotechnol 41–48. https://doi.org/10.1007/978-981-13-6321-4_7
45. Kumar B, Bhardwaj N, Agrawal K, Chaturvedi V, Verma P (2020) Current perspective on pretreatment technologies using lignocellulosic biomass: an emerging biorefinery concept. Fuel Process Technol 199:106244. https://doi.org/10.1016/j.fuproc.2019.106244
46. Kumar B, Bhardwaj N, Verma P (2020) Microwave assisted transition metal salt and orthophosphoric acid pretreatment systems: generation of bioethanol and xylo-oligosaccharides. Renew Energy 158:574–584. https://doi.org/10.1016/j.renene.2020.05.006
47. Kumar B, Verma P (2020) Biomass-based biorefineries: an important architype towards a circular economy. Fuel 288:119622 Elsevier
48. Kumar B, Verma P (2021) Life cycle assessment: blazing a trail for bioresources management. Energy Convers Manag X 10:100063 Elsevier

49. Lauf RJ, Bible DW, Johnson AC, Everleigh CA (1993) Two to 18 GHz broadband microwave heating systems. Microw J 36(11):24–30. https://go.gale.com/ps/i.do?p=AONE&sw=w&issn=01926225&v=2.1&it=r&id=GALE%7CA14649530&sid=googleScholar&linkaccess=fulltext
50. Lee SB, Lee JD, Hong IK (2011) Ultrasonic energy effect on vegetable oil based biodiesel synthetic process. J Ind Eng Chem 17(1):138–143. https://doi.org/10.1016/J.JIEC.2010.12.012
51. Liu Y, Guo L, Wang L, Zhan W, Zhou H (2017) Irradiation pretreatment facilitates the achievement of high total sugars concentration from lignocellulose biomass. Bioresour Technol 232:270–277. https://doi.org/10.1016/J.BIORTECH.2017.01.061
52. Liu Y, Zhou H, Wang L, Wang S, Fan L (2016) Improving Saccharomyces cerevisiae growth against lignocellulose-derived inhibitors as well as maximizing ethanol production by a combination proposal of γ-irradiation pretreatment with in situ detoxification. Chem Eng J 287:302–312. https://doi.org/10.1016/J.CEJ.2015.10.086
53. Liu Y, Zhou H, Wang S, Wang K, Su X (2015) Comparison of γ-irradiation with other pretreatments followed with simultaneous saccharification and fermentation on bioconversion of microcrystalline cellulose for bioethanol production. Bioresour Technol 182:289–295. https://doi.org/10.1016/J.BIORTECH.2015.02.009
54. Ma X, Zheng X, Zhang M, Yang X, Chen L, Huang L, Cao S (2014) Electron beam irradiation of bamboo chips: degradation of cellulose and hemicelluloses. Cellulose 21(6):3865–3870. https://doi.org/10.1007/S10570-014-0402-4/FIGURES/2
55. Mullen CA, Boateng AA, Goldberg NM, Lima IM, Laird DA, Hicks KB (2010) Bio-oil and biochar production from corn cobs and stover by fast pyrolysis. Biomass Bioenergy 34(1):67–74. https://doi.org/10.1016/J.BIOMBIOE.2009.09.012
56. Muranaka Y, Nakagawa H, Hasegawa I, Maki T, Hosokawa J, Ikuta J, Mae K (2017) Lignin-based resin production from lignocellulosic biomass combining acidic saccharification and acetone-water treatment. Chem Eng J 308:754–759. https://doi.org/10.1016/J.CEJ.2016.09.117
57. Nair LG, Agrawal K, Verma P (2022) An insight into the principles of lignocellulosic biomass-based zero-waste biorefineries: a green leap towards imperishable energy-based future. Biotechnol Genet Eng Rev 1–51
58. Nair LG, Agrawal K, Verma P (2022) An overview of sustainable approaches for bioenergy production from agro-industrial wastes. Energy Nexus 100086
59. Nair LG, Agrawal K, Verma P (2023) Organosolv pretreatment: an in-depth purview of mechanics of the system. Bioresour Bioprocess 10(1). Springer
60. Nikolić S, Mojović L, Rakin M, Pejin D, Pejin J (2010) Ultrasound-assisted production of bioethanol by simultaneous saccharification and fermentation of corn meal. Food Chem 122(1):216–222. https://doi.org/10.1016/J.FOODCHEM.2010.02.063
61. Olsen RL, Hasan MR (2012) A limited supply of fishmeal: Impact on future increases in global aquaculture production. Trends Food Sci Technol 27(2):120–128. https://doi.org/10.1016/J.TIFS.2012.06.003
62. Rambabu K, Bharath G, Hai A, Show PL, Banat F, Manickam S (2023) Enhancing biodiesel production from waste date seed oil through ultrasonic-assisted optimization: a sustainable approach using non-edible feedstocks. Chem Eng Process - Process Intensif 194:109601. https://doi.org/10.1016/J.CEP.2023.109601
63. Ren X, Shanb Ghazani M, Zhu H, Ao W, Zhang H, Moreside E, Zhu J, Yang P, Zhong N, Bi X (2022) Challenges and opportunities in microwave-assisted catalytic pyrolysis of biomass: a review. Appl Energy 315:118970. https://doi.org/10.1016/J.APENERGY.2022.118970
64. Ruth M, Mai T, Newes E, Aden A, Warner E, Uriarte C, Inman D, Simpkins T, Argo A (2013) Projected biomass utilization for fuels and power in a mature market. Transp Energy Futur Ser
65. Shenbagamuthuraman V, Kasianantham N (2023) Microwave irradiation pretreated fermentation of bioethanol production from Chlorella vulgaris Biomasses: Comparative analysis of response surface methodology and artificial neural network techniques. Bioresour Technol 390:129867. https://doi.org/10.1016/J.BIORTECH.2023.129867

66. Sindhu R, Binod P, Pandey A (2016) Biological pretreatment of lignocellulosic biomass – an overview. Bioresour Technol 199:76–82. https://doi.org/10.1016/J.BIORTECH.2015.08.030
67. Singh R, Krishna BB, Kumar J, Bhaskar T (2016) Opportunities for utilization of non-conventional energy sources for biomass pretreatment. Bioresour Technol 199:398–407. https://doi.org/10.1016/J.BIORTECH.2015.08.117
68. Subhedar PB, Babu NR, Gogate PR (2015) Intensification of enzymatic hydrolysis of waste newspaper using ultrasound for fermentable sugar production. Ultrason Sonochem 22:326–332. https://doi.org/10.1016/J.ULTSONCH.2014.07.005
69. Varella LYZ, Mesquita AZ, de Rezende DB, Mourão IG, Cren ÉC, Lopes NP, Silva WG, Alonso TC (2024) Experimental evaluation of the use of gamma radiation in sugarcane bagasse (Saccharum spp.) as a pretreatment for bioethanol production. Radiat Phys Chem 111611. https://doi.org/10.1016/J.RADPHYSCHEM.2024.111611
70. Wang S, Lin H, Zhang L, Dai G, Zhao Y, Wang X, Ru B (2016) Structural characterization and pyrolysis behavior of cellulose and hemicellulose isolated from Softwood Pinus armandii Franch. Energy Fuels 30(7):5721–5728. https://doi.org/10.1021/ACS.ENERGYFUELS.6B00650/ASSET/IMAGES/LARGE/EF-2016-006509_0006.JPEG
71. Wang Z, Cheng Q, Liu Z, Qu J, Chu X, Li N, Noor RS, Liu C, Qu B, Sun Y (2019) Evaluation of methane production and energy conversion from corn stalk using furfural wastewater pretreatment for whole slurry anaerobic co-digestion. Bioresour Technol 293:121962. https://doi.org/10.1016/J.BIORTECH.2019.121962
72. Wu C, Budarin VL, Gronnow MJ, De Bruyn M, Onwudili JA, Clark JH, Williams PT (2014) Conventional and microwave-assisted pyrolysis of biomass under different heating rates. J Anal Appl Pyrol 107:276–283. https://doi.org/10.1016/j.jaap.2014.03.012
73. Wu X, Chen L, Chen J, Su X, Liu Y, Wang K, Qin W, Qi H, Deng M (2019) The effect of 60Co γ-irradiation on the structure and thermostability of alkaline lignin and its irradiation derived degradation products. Waste Biomass Valoriz 10(10):3025–3035. https://doi.org/10.1007/S12649-018-0300-3/TABLES/5
74. Xiang Y, Xiang Y, Wang L (2017) Electron beam irradiation to enhance enzymatic saccharification of alkali soaked Artemisia ordosica used for production of biofuels. J Environ Chem Eng 5(4):4093–4100. https://doi.org/10.1016/J.JECE.2017.07.058
75. Xiao R, Chen X, Wang F, Yu G (2010) Pyrolysis pretreatment of biomass for entrained-flow gasification. Appl Energy 87(1):149–155. https://doi.org/10.1016/j.apenergy.2009.06.025
76. Yetri Y, Hoang AT, Mursida, Dahlan D, Muldarisnur, Taer E, Chau MQ (2020). Synthesis of activated carbon monolith derived from cocoa pods for supercapacitor electrodes application. Energy Sources Part A: Recov Utiliz Environ Effects. https://doi.org/10.1080/15567036.2020.1811433
77. Yu J, Paterson N, Blamey J, Millan M (2017) Cellulose, xylan and lignin interactions during pyrolysis of lignocellulosic biomass. Fuel 191:140–149. https://doi.org/10.1016/J.FUEL.2016.11.057
78. Zadeh ZE, Abdulkhani A, Aboelazayem O, Saha B (2020) Recent insights into lignocellulosic biomass pyrolysis: a critical review on pretreatment, characterization, and products upgrading. Processes 8(7):799. https://doi.org/10.3390/PR8070799
79. Zhang C, Ho SH, Chen WH, Eng CF, Wang CT (2021) Simultaneous implementation of sludge dewatering and solid biofuel production by microwave torrefaction. Environ Res 195:110775. https://doi.org/10.1016/J.ENVRES.2021.110775
80. Zou X, Xu K, Chang W, Qu Y, Li Y (2021) Rapid extraction of lipid from wet microalgae biomass by a novel buoyant beads and ultrasound assisted solvent extraction method. Algal Res 58:102431. https://doi.org/10.1016/J.ALGAL.2021.102431

Chapter 14
Radiation-Based Technologies for Application in Industry, Human Health, and Environment Protection: A Techno-Economic Perspective

Sridevi Veluru, Husam Talib Hamzah, S. Nikhil, Jampana Vishnu Varma, and Geddada S. G. D. Saketh

Abstract Radiation technology provides adaptable resources that can significantly contribute to sustainable development. Because it uses less energy and produces less trash, it is usually more economical and better for the environment than the alternatives. This is crucial for all nations, especially those struggling financially. To prevent the contamination of the environment, we can use radiation to eliminate emerging organic pollutants (EOPs) in soil and water. When administered properly, radiation can significantly improve human and environmental health and well-being while bolstering global security. At high enough doses, ionizing radiation can kill us or harm our cells and organs, making it an unsuitable therapeutic option for many disorders, including cancer. Energy generation, industry, research, and medical diagnoses and treatment of illnesses are just a few of the numerous positive uses for this form of radiation when used correctly, at appropriate dosages, and with appropriate protective gear. To combat the potential dangers of ionizing radiation to humans, communities, and ecosystems, the International Atomic Energy Agency (IAEA) provides regulatory bodies and lawmakers with a comprehensive set of worldwide safety standards. Nevertheless, the regulation of radiation sources and their protection falls under the purview of individual nations. Creating novel biodegradable "active" packaging materials to address plastic waste is an excellent illustration of this trend. This chapter helps fill gaps in our knowledge of radiation-based technologies' technical, economic, and ecological uses in many areas of human endeavour, such as basic research, manufacturing, farming, and healthcare.

Keywords Radiation technology · Food safety · Environmental remediation · Techno-economic analysis · IAEA

S. Veluru (✉) · H. T. Hamzah · S. Nikhil · J. V. Varma · G. S. G. D. Saketh
Department of Chemical Engineering, College of Engineering, Andhra University, Visakhapatnam, India
e-mail: vellurusridevi@yahoo.co.in

© The Author(s), under exclusive license to Springer Nature Switzerland AG 2024
P. Verma and B. Kumar (eds.), *Sustainable Radiation Technologies in Waste-biomass Valorization*, Green Chemistry and Sustainable Technology,
https://doi.org/10.1007/978-3-031-63941-8_14

14.1 Introduction

Industrial and economic growth are both influenced by the state of nuclear science and technology. Their numerous benefits include better materials, more efficient manufacturing processes, and a cleaner environment. They help accomplish the UN Sustainable Development Goals, such as Goal 9, which aims to enhance infrastructure, innovation, and industry [6]. Regarding improving people's quality of life through radiation research and technology, the International Atomic Energy Agency backs its 168 member states 100% [25].

The use of chemicals has been on the rise in recent decades due to rising industrial, agricultural, and human activities [10, 50]. There has been a rise in the chemical load in the soil and aquatic environment due to human society's development and the wrong understanding of personal safety. Agricultural uses of untreated sewage are common in many developing nations, leading to inadequate or untreated wastewater that introduces numerous contaminants into the food chain [15, 24, 26]. Soil and water pollution threatens human and environmental health; hence, effective technology to remove these contaminants is essential. There are no regulations to monitor or limit the emissions of emerging organic pollutants (EOPs) despite these contaminants being harmful to ecosystems and people [75]. Chemicals that can affect the endocrine system, pharmaceuticals and personal care items, and long-term organic pollutants are the three main categories of EOPs [2, 49]. Potentially regulated EOPs include those that pose risks to human and environmental health, are persistently released into the environment, and cannot be fully converted or eliminated by even the most sophisticated wastewater treatment plants (WWTPs) [4].

Irradiated polymers are increasingly finding creative new homes in medical applications, like hydrogel sheets for burn and wound management and radiotherapy for cancer treatment, thanks to advancements in the radiation processing sector [25]. According to Lacroix, Governments are enacting new legislation that holds industries accountable for their plastic consumption and trash production. As a result, researchers are increasingly concentrating on environmentally friendly packaging materials. One potential way to develop new and safe product materials and help achieve the environmental goal of minimizing food packaging waste is by irradiating natural polymers. Processing innovations for specific polymers are lowering environmental impact while increasing production.

Many researchers have tried to assess the possibility of industrial energy needs concerning electrical and thermal energy. Ravi Kumar et al. investigated a variety of industrial applications of solar thermal collectors, including food processing, building, drying, dehydration, refrigeration, HVAC, and photovoltaic (PV) desalination systems [64].

Scientific groups actively seek and develop novel ingredients to meet the increasing demand for healthier food options. Despite their potential environmental harm, agro-industrial leftovers are a rich source of bioactive chemicals, and recovering them can help ensure economic and environmental sustainability [46]. One green method that can enhance bioactive component extraction is ionizing radiation [52]. Updates in the field of radiation processing, environmentally friendly packaging, the possibility of increased heat and electrical energy demands in the industry, and the impact of radiation technology on bioactive compound extraction are all topics covered in this chapter, which helps to clarify AOPs role in water remediation.

14.2 Advanced Oxidation Process (AOP) Using Ultra Violet Radiation

Water purification is a common use of the advanced oxidation process, which relies on the breakdown of hydrogen peroxide when exposed to ultraviolet light. Because of their chemical stability, stubborn organic pollutants are difficult to cure using current conventional processes. One of the most environmentally beneficial ways to remove these contaminants is using modern oxidation technology. Reactive species, such as ozone, hydroxyl radicals, and hydrogen peroxide, decompose pollutants that are biologically persistent when exposed to ambient temperature and pressure. As illustrated in Fig. 14.1, two distinct kinds of AOTs exist. Photochemical and non-photochemical processes are subcategories of the two main groups [71]. A highly reactive form of the hydroxyl radical is at the heart of AOT's fundamental mechanism. Ozone and hydroxyl radicals are examples of reactive substances. When compared to well-known oxidizing agents like potassium permanganate, these hydroxyl radicals are more reactive, unstable, and non-selective. Hydroxyl radicals are highly effective in breaking organic contaminants into inert carbon dioxide and water. Even though it only lasted briefly, the hydroxyl radical could attack organic contaminants by forming a double bond, transferring electrons, and abstracting hydrogen.

14.2.1 Classification of AOPs

This general term "advanced oxidation processes" encompasses several different approaches. Most of these techniques involve using a powerful oxidizing chemical, catalyst, and irradiation, typically in ultraviolet light. Titanium dioxide and ultraviolet light, Fenton's reactions, and hydrogen peroxide with ultraviolet light are the three most used methods in AOP for producing hydroxyl radicals, according to the literature [4]. There are two main categories into which AOPs fall: heterogeneous and homogenous. The energy consumption is a standard metric for further subdividing the latter [66].

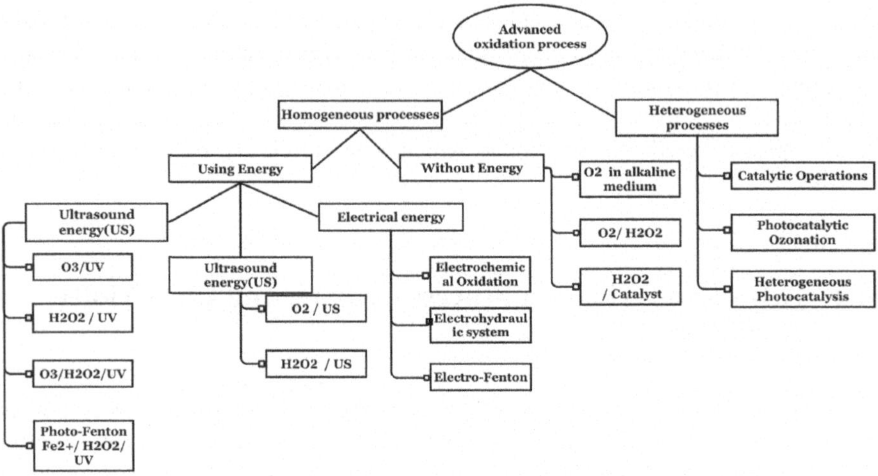

Fig. 14.1 Various AOTs based on wastewater treatment technologies [71]

14.2.1.1 Homogenous Advanced Oxidation Processes

UV-based homogeneous AOPs are often used to break down compounds that absorb UV light in the right spectrum. These chemicals are ideal for photodegradation because they absorb ultraviolet light at lower wavelengths. Toxic organic chemicals in wastewater can be successfully oxidized and destroyed using advanced water treatment methods such as photolytic ozonation (O_3/UV) [71]. Illuminating water-based systems already high in ozone is the usual practice when using 254 nm UV light. Researchers discovered that O_3/UV is far more effective than UV photolysis and ozonation at destroying chlorophenols (CPs). Leitus and his colleagues [54] discovered that ozonation alone isn't always more efficient than photolytic ozonation. The photolysis of hydrogen peroxide and other related propagation processes produce hydroxyl radicals in the H_2O_2/UV process, which is another example of an advanced oxidation mechanism. As demonstrated in the following equation, the photolysis of hydrogen peroxide takes place after the application of UV radiation:

$$H_2O_2 + h\upsilon \rightarrow 2HO \tag{14.1}$$

More extended periods of UV exposure and/or a reasonably high concentration of H_2O_2 are necessary for this procedure.

14.2.1.2 Heterogeneous Advanced Oxidation Processes

Heterogeneous AOPs degrade substances with the help of catalysts. In this context, "heterogeneous" means impurities in the liquid phase, and the catalyst is typically solid. The latter causes chemical reactions to proceed more quickly because electron–hole pairs are present. Photogenerated electrons undergo oxidation, while photogenerated holes undergo reduction. OH radicals are produced in aqueous solutions due to the oxidation of water molecules absorbed by the catalyst. As a rule, the reduction reaction occurs in aerobic conditions and uses oxygen as the target species; this produces a superoxide radical [71].

14.2.1.3 Applications of AOPs

Research has demonstrated that AOPs outperform individual agents (such as UV, ozone, or hydrogen peroxide) in terms of effectiveness. The inexpensive cost of H_2O_2 and/or ozone, necessary for producing hydroxyl radicals, makes AOPs an attractive option for wastewaters with low COD [78]. It is now possible to convert previously resistant materials to degradation into compounds that require additional biological treatment. Two of AOPs' most essential uses are treating refractory organic compounds and disinfecting treated wastewater [43]. Disinfecting wastewater with AOPs is possible because the free radicals produced by ozone are more potent oxidants than ozone alone, rendering bacteria and refractory organic compounds in the wastewater susceptible to oxidation [56]. Unfortunately, getting high concentrations is not feasible due to the short half-life of hydroxyl free radicals. When dealing with deficient concentrations, it is necessary to calculate, using the CT concept, significantly longer detention times for microbes disinfection [66]. For this reason, hydroxyl radicals are not typically employed for disinfection but rather to oxidize trace elements of refractory organic compounds present in highly treated effluents. Hydroxyl radicals, once formed, can damage organic compounds by transferring electrons, adding radicals, or abstracting hydrogen [14].

14.3 Food Industry

Food is a fundamental necessity for human survival [44]. When food is left out for too long, it starts to spoil. This means preservation is necessary to increase the time these foods can be stored without compromising their nutritional value or quality[32]. Salting, burning, blanching, and other preservation techniques were standard in ancient times [9]. Many new methods of food preservation have emerged with the current era's technological advancements [31].

Modern concerns in the food sector extend to assurance of product quality, protection from contamination, availability of food, and overall food safety [67]. These matters are paramount, as are concerns about environmental sustainability, food waste valorization, and waste minimization. Most projections put the world's population at 11 billion by 2050. Worldwide, there has been a corresponding rise in the size of the economic "middle class" in recent decades. Opportunities for food firms to expand internationally and serve this demographic are emerging. Modern shoppers prefer year-round variety in portion-controlled packaging. In the future, many of the food processing methods used now may be considered outdated. What's more, we haven't even begun to scratch the surface of the technological possibilities in this field. Still, innovations in food processing bode well for the future of "fresh" and "chemical-free" food options for shoppers all year round. This chapter section aims to present a synopsis of the food industry's use of food irradiation.

14.3.1 Food Irradiation Technology

Radiowaves, incandescent lights, television transmissions, microwaves, ultraviolet radiation, and cosmic rays are all electromagnetic spectrum components, including food irradiation technologies [34]. The three main methods for irradiating food are gamma rays, X-rays (emitted by sources like linear accelerators or X-ray tubes), and electron beams. The first two methods use radioactive isotopes like Cs-137 (cesium-137) or Co-60 (cobalt-60). All three of these irradiation methods apply to food irradiation on a global scale [61].

Electromagnetic waves are known as irradiation to process food to decrease damage and deterioration [35]. Because it attacks DNA directly, irradiation kills harmful microbes and prevents them from reproducing and living in food. A broad definition of irradiation is a beam of light whose intensity varies with wavelength and is inversely related to frequency as it passes through various food materials. The penetrating power increases as the wavelength decreases. The dosage is an important consideration when irradiating food. Because various doses of radiation serve distinct functions [76], consuming an irradiated food product will result in material degradation since the dosage is too high. Once ionizing radiation has penetrating properties, a substance can absorb some or all of the energy from the radiation. The relative absorption rate is expressed as a unit of gray or kilo gray (Gy or kGy) [48].

14.3.1.1 Irradiation Principles

Preserving food using ionizing radiation, a form of high energy, is possible. Because anything it comes into contact with has the potential to be ionized. When these meals are exposed to irradiation sources like X-rays, gamma rays, and electron beams, their components undergo ionization, excitation, and transformation [5]. The process by which live cells lose their resistance to environmental stimuli is called excitation. When macromolecules are ionized, they are reduced to free radicals. Inhibition of DNA synthesis, disruption of microbial cell division, and biological impacts result from changes in the components of living cells [62]. This impact prevents the growth of microorganisms in food items.

14.3.1.2 Sources of Irradiation

Particles of gamma radiation, Nuclear fission produces caesium-137 as a byproduct and neutron bombardment of cobalt-59 produces the radioisotope cobalt-60, which in turn produces gamma irradiation [22]. Irradiation of food is the usual application for cobalt-60. Stainless steel encases a metal source in two layers. It is common practice to place the source in a water-absorbing pool or a protected container while it is not in use. Since the supplier is replenished or retained with the decomposing source, it generates no waste. Because it is insoluble in water, cobalt-60 poses a low threat of environmental contamination through leaking into water systems; commercial-scale facilities use it more than any other source to gamma irradiate foods [22]. In compliance with the Nonviolent Transport of Radioactive Materials Regulations established by the International Atomic Energy Act, specialized vehicles are needed to convey cobalt-60, which stipulate stringent safety requirements and rigorous testing for vehicles that carry radioactive materials. Environmental pollution is a concern because cesium-137 is water soluble, unlike cobalt-60 [6]. Table 14.1 shows the widespread use of its applications in food.

Table 14.1 Results of gamma-ray testing on different foods and their properties

Food products	Dose	Findings	References
Brown rice	0, 2, 4, 6, 8 and 10 kGy	Antioxidant activity, total phenolic content, and the rehydration ratio/ browning index all went up, β-carotene content and enhanced irradiation dose resulted in microbiological stability	[38]
Kimchi seasoning mixture	0–10 kGy	Microbe populations declined, acidity and pH levels stabilized, sugar reduction and headspace gas changes slowed, and colour values did not alter as irradiation dose increased	[41]
Pomegranate	1, 3 and 5 kGy	Make pomegranates last longer: reduce the amount of microbes on pomegranate aryl and lower polyphenol oxidase activity	[8]
Wheat	0–5 kGy	Decrease in rheological characteristics and dynamic module: increase in antioxidant activity (by a large margin)	[12]

Electron beam: A device that can be turned on and off to produce electrons accelerated to 99% of the speed of light is used to create it. Although electron beams have a lower price tag and more throughput, they aren't very consistent in their doses and can only penetrate a few centimetres deep. Because of this, it is employed to handle foods that are not very thick [22]. Foods can be subjected to the electron beam, as illustrated in Table 14.2.

X-rays: The bremsstrahlung-conversion process involves bombarding dense target materials with high-energy accelerated electrons, which produces the Continuous Energy Spectra (CES). "Fellows" [22]. Because of its low atomic number and high melting point, tantalum is more suitable for industrial, large-area, high-power targets than tungsten, which is more workable and has a lower threshold energy for induced reactions [7]. It is possible to switch off X-rays like electron beams when they are not in use and do not require a radioactive source. Despite their excellent dose uniformity and deep penetration, X-rays are an extremely costly irradiation source since they only convert 8% of the incident energy into X-rays [22]. So far, X-rays can be seen in Table 14.3. In the future, X-rays will be utilized extensively due to the rapid advancement of technology.

14.3.1.3 Effect of Irradiation on Food Quality: Benefits and Drawbacks

Irradiation has several uses in food, one of which is preserving the material's original properties without subjecting it to excessive heat. Irradiation can also kill some bacteria and other microbes that inhabit food. It only takes one operation to irradiate packed, frozen, and fresh foods without chemical additives [36]. Low operating costs, controllability of the automatic process, limited energy use, and comparison of nutrition changes to alternative preservation methods are all benefits of irradiation [33]. One of the significant problems with irradiation is that it kills so many germs that the food is no longer fit for human consumption. Another issue is that people can't tell whether any remaining harmful bacteria are in the food. Destroying harmful germs after they have contaminated food makes the food unsafe to eat [37]. Concern about radioactive fallout, nutritional value loss, analytical methods for determining if food has been irradiated insufficiently, and the evolution of microbes resistant to radiation are all outcomes of the irradiation process. There are several benefits and drawbacks to irradiating food, as demonstrated in Table 14.4.

14.3.1.4 Irradiation Legality

The IAEA aims to apply irradiation for food safety and quality. Verifying that food irradiation technology complies with all applicable laws and regulations is a crucial and basic step in the process. Nearly forty nations have established rules on food irradiation that all must follow. All member nations' stated national standards as well as the international CODEX Alimentarius standards formed the basis for these

Table 14.2 Electron beam analysis of food items and their properties

Food products	Dose	Findings	References
Fish gelatin film with bamboo leaves antioxidants	5 and 7 kGy	Decreasing the pace of antioxidant release can potentially enhance the film's physical properties	[30]
Rice starch	0, 1, 2, 4, 8 and 10 kGy	At low levels, lipase activity dropped in rice and had no discernible effect on cooked rice quality; however, lipase activity degraded starch molecules at larger doses	[58]
Mango	0, 5 kGy	Preventing illnesses that manifest after harvest; stimulation of respiration; increased production of hydrogen peroxide and carbon dioxide gas; preservation of firmness; reduction of the rate of rise in total soluble solids; and no change in ethylene production or vitamin C content during storage	[72]
The waxy maize starch and its films	2–30 kGy	There was a decrease in the starch's molecular weight, branching, and thermal characteristics. There were more α-1,6-glucosidic linkages at low doses than α-1,4-glucosidic bonds, and the solubility and mechanical characteristics were good	[79]
Egg white protein	1.0, 1.5, 2.0 and 2.5 kGy	Due to hydrophobic amino acid residues and S–S bond breaking, the antioxidant activity of egg white protein rises, leading to oligomerization, and the microstructure of egg white protein changes to a "honeycomb" microstructure	[51]

Table 14.3 Characteristics and applications of X-rays to different food products

Food products	Dose	Findings	References
Strawberry	0–1 kGy	At onekGy, it can slow down rotting and weight loss in fruit, postpone discolouration, and improve sensory qualities while storing	[77]
Spinach leaf	Combination of 0.3 kGy, 1% citric acid	Including 1% citric acid in X-ray irradiation resulted in a more effective antibacterial impact without compromising product quality	[40]
Sliced cheese	0, 0.2, 0.4, 0.6, 0.8 kGy	The cheese quality is unaffected by increasing the dose of irradiation, which can inactivate dangerous bacteria such as *Escherichia coli, Salmonella typhimurium,* and *Listeria monocytogenes*	[59]
Lettuce	Combination of 0.05–0.3 kGy, 0.5% gallic acid	There is no adverse effect on lettuce quality from the combination of gallic acid at 0.5% and X-rays at 0.3 kGy, which efficiently inactivate bacteria such as *Escherichia coli, Salmonella typhimurium,* and *Listeria monocytogenes*	[39]
Rice	0–1.5 kGy	Essential oregano oil enhances the effectiveness of microbicides when exposed to radiation	[11]

Table 14.4 Pros and cons of irradiation on food

Benefits	Drawbacks
Foods would have a longer shelf life due to the variance in irradiation and other treatments	When irradiation enters a cell, it disrupts the nucleus, which has a biological impact
Removing viral activity requires a combination of heating and irradiation	The production of oxygen-centred reactive radicals during water radiolysis can directly affect irradiation. The meal has potentially dangerous microbes because water is a significant component
Irradiation is a direct method for controlling pests in spices	Lipid oxidation occurs as a result of free radicals produced by irradiation
Radiation may decrease the infectious nature of an organism	Thiamine levels were 16% lower in irradiated poultry thaninnon-irradiated chicken after onekGy of radiation exposure
Irradiated food usually retains most of its nutritional value	There was no evidence that food irradiation posed a higher risk of mycotoxin development

regulations. The worldwide regulatory standard for irradiated goods is food label no. 69/1999 [1].

14.4 Applications of Ionizing Radiation in Materials Processing

14.4.1 Electron Accelerators

Radiation processing technology uses a range of linear accelerators powered by microwave sources, such as direct current models and single resonant cavity models. The several kinds of accelerators are defined by the different methods of producing accelerated electrons. While resonators directly accelerate particles by harnessing the electrical component of electromagnetic waves, accelerators directly accelerate particles by harnessing high-voltage power sources. These two methods are the main ones for producing electrical fields. MW accelerators (GHz), low radiofrequency (100–200 MHz), and continuous beam (DC) have traditionally formed the basis of radiation processing [65]. In industrial applications, transformer accelerators can attain high beam power while being very efficient. These kinds of accelerators may push electron velocities up to 5 meV [63]. Typically, the energy of transformer accelerators ranges from 0.15 to 1 meV. Resonance techniques are required to accelerate electrons to energies between 5 and 10 meV. It is crucial for processing dense materials or objects with a more significant volume using radiation [63]. Electron acceleration methods utilizing electrically-coupled electromagnetic waves in the microwave frequency range of 1,300–3,000 MHz are created in a number of small, concentrically-arranged resonators [17].

Radiation modification of polymers is a potential industrial application, but these devices are prohibitively expensive and have a low energy efficiency of only 10% [70]. System resonances powered by frequencies about 100 MHz provide better options. Rather than using pricey microwave power sources like klystrons or magnetrons, these systems employ inexpensive generators that use robust vacuum tubes. There is a considerable improvement (up to 50%) in the accelerator's efficiency. Modern electron accelerators are multi-function devices capable of producing beams of varying intensities and energy (Table 14.5). In a particular application, the type of electron accelerator that is most suitable to employ is determined by the amount of electron energy that is required, which is influenced by the density and structure of the materials that are being targeted. The beam power, which is an indication of the overall processing capability, is yet another approach to classify it [70].

Table 14.5 Maximum allowable performance of electron accelerators for radiation processing

Parameter	Direct DC	UHF 100–200 MHz	Linac 1.3–9.3 GHz
Beam average current	<2 A	<100 mA	< 30 mA
Electrons energy	0.05–5 meV	0.3–10 meV	2–10 meV
Beam power (max)	~500 kW	700 kW	150 kW
Electrical efficiency	60–80%	20–50%	10–20%

14.5 Applications of Ionizing Radiation in Sterilization

The primary objective of a standard sterilization process is the removal of any harmful microorganisms that may contaminate the product or, perhaps, the user's health. These days, the radiation-based sterilization treatments that are utilized the most frequently are X-rays, gamma rays, ultraviolet light, and electron beams [17]. Factors related to the product or package's attributes determine the optimal radiation-based sterilization method. It contains many one-time-use medical devices and several other items that play an essential role in the medical field, such as cosmetics, hydrogels, packaging for medical purposes, food items, pharmaceuticals, and implants and transplants [16]. Ionizing radiation sterilization of surgical transplants is a socially significant issue. As a result of advancements in restorative surgery, there is a growing demand for tissue transplants, including sclera, cornea, bone, tendon, skin, amniotic grafts, and heart valves. Many patients have their disability eliminated or significantly reduced, and their treatment period is drastically cut in half thanks to biostatic transplants, particularly bone grafts [16]. Ionizing radiation breaks down nucleic acid structures by interacting with materials to produce free electrons and unstable ions. Ionizing radiation also generates reactive OH radicals in biological materials, further damaging RNA, DNA, and other macromolecules in cells, ultimately leading to cell death. When it comes to ionizing radiation, more-5 giant genomes of microbes are more susceptible than micromolecules. Because their genomes are smaller than those of other harmful species, such as bacteria, viruses require higher radiation dosages. Viruses cannot fix damage to their nucleic acids due to their lack of enzymes. Initial investigations of the re-sterilization of N95 masks were conducted by researchers during the epidemic using both gamma radiation and electron beam [16]. According to the findings of the study, the effectiveness of respirators' filtration is diminished when they are subjected to radiation because the electric charge that normally exists within the PP fibres is lost. Therefore, decontaminating Filtering Facepiece Respirators (FFRs) with Electron beam (EB) radiation is a significant nuisance, regardless of the levels that are utilized. By recovering the electric charge of the treated or processed masks, it is necessary to maintain the filtration function of the moulds [29]. Only then can their use in the mode above be contemplated. Because the structural and functional properties of the filtering composites are unaffected by radiation dosages. [28, 29] also noted a decline in the N95 respirator's filtration efficacy, at around 63.6%. But charge regeneration might make the recycled respirator work better. Radiation sterilization of packing materials is an area that has seen a lot of recent advancement in study. Many medical device manufacturing processes include EB irradiating empty packaging, such as cartons, cans, boxes, and bottles [3]. The use of medicines in radiation therapy is growing. Sterilization of cosmetic preparations and their components also involves heating, albeit on a much smaller scale compared to pharmaceuticals. Medications in an aqueous solution, where their radiation resistance is often low ($\leq$0.01–0.1 mol dm^{-3}), are the primary cause of radiation sterilization's limited efficiency [17].

14.6 Environmental Applications

The quantity and variety of environmental pollutants have skyrocketed due to the exponential growth of human settlements, industrial output, and population. Natural contamination occurs due to random events; manufactured contamination, on the other hand, is a contemporary worldwide problem. Toxic substances significantly threaten human health, ecological damage, environmental degradation, and global warming. Power plants, industries, and transportation are the primary sources of air pollution, primarily as particulates and acidic gases or gases that emit greenhouse gases [77] As cities and industries have grown, so has the volume of sewage from homes, farms, and businesses. The excessive concentration of sewage in the water has depleted their capacity for self-purification. As a result of fast industrialization, solid waste is increasing, leading to landfill overflow and water and soil pollution with harmful substances. Governments worldwide have passed new laws to control and reduce emissions of various pollutants. In the fight against environmental degradation, recycling and recovery of waste are essential steps. Reversible damage to Earth's life support systems has significantly higher environmental costs than effective environmental regulation, which comes at a far lower economic cost. However, many novel strategies and techniques are necessary to meet these demands [40]. One innovative aspect of the technology's application was the creation of new, very powerful electron accelerators for processing large streams of gaseous or liquid contaminants online. The accelerators were installed incredibly efficiently to treat wastewater and off-gas. Nevertheless, additional developments in accelerator technology are necessary to execute these procedures to make this production technique competitive because machines operating in demanding industrial sectors necessitate robust power, efficient energy usage, and a high level of reliability [40].

14.6.1 EB Technology for Wastewater Treatment

A city in Guangdong Province, China, known as Jiangmen, treated its actual dyeing wastewater utilizing EB procedures. An estimated 30,000 m^3/d of effluent was produced. Continuous monitoring of effluent water quality has shown that EB technology has the potential to be a state-of-the-art treatment system that greatly improves the effectiveness of COD and colour removal. Even after raising the influent COD from 180 to 280 mg/L, the EB unit's effluent COD remained within the 20–50 mg/L range. The colour would have faded approximately ten times faster if the dyeing wastewater had been treated. The EB service unit was included in the overall running cost of the wastewater treatment system, which varied between 2.0 and 2.5 ¥/m^3 [55]. Although recalcitrant wastewater treatment in industry is an extremely effective and economically complex procedure, evidence gathered from monitoring for almost a year of operation suggests that EB technology could be useful in this situation [74]. Irradiation effectively breaks down many underground pollutants, including pesticides, aromatic hydrocarbons, chlorinated organic compounds, polyfluorylalkyl substances (PFASs), and gasoline additives. Bao et al. (2022) [73] reviewed recent findings and current knowledge regarding the use of ionizing radiation to clean up contaminated soils and groundwater. The writers provided a comprehensive overview of the fields studying these technologies and their practical uses and experimental methods.

14.6.2 Sludge Hygienization

The pace of urbanization on a global scale is accelerating. Today, almost half of the world's population calls a significant city or metropolitan area home [19]. Municipalities face a substantial disposal challenge when it comes to city sewage. The nutrients, fuel, and water included in human stream waste are enormous. Within the modern era of circular economies [47], which prioritizes these materials, the idea of WWTPs as "resource recovery facilities" has essentially replaced it. After decades of research, the radiation sludge hygienization [41] technology was successfully deployed in Vadodara, India [18]. Municipal sewage sludge must first kill harmful bacteria before being released into the environment. Using high energy (10 meV) electron beam irradiation, investigators examined the efficiency of inactivating various microbiological pathogens and indicator species in sewage sludge that had undergone aerobic and anaerobic digestion [17]. EB irradiation has the potential to damage indicator organisms as well as diseases caused by bacteria and viruses [68].

The first comparative cost–benefit analysis of utilizing EB technology for sludge hygienization against EB technique for cogeneration of biogas and sludge hygienization was carried out [18], and the results are shown in Table 14.6. In contrast to the Vadodara plant, which uses water suspension irradiation, several plants in India employ gamma irradiation to irradiate solar-powered dried sludge. Exposure to gamma rays eliminates germs, reduces odour, kills weeds, and breaks down compounds in the sludge. A gamma radiation treatment of 10 kGy on average may be enough to sanitize the sludge since the study found that it reduced the bacteria burden by four log cycles. The use of an inoculation medium with sterile sludge resulted in increasing the quantity of plant growth-promoting bacteria (PGPB) by 100 times (Bio-Gold). Applying 1 tonne of Bio-Gold per hectare to each crop reduces soil contamination, speeds up growth for potatoes, tomatoes, and wheat, and uses 20% less chemical fertilizer. Bio-Gold standards have been proposed as a result of the research. The facilities listed above were built and are operational right now. They can process 100 tonnes per day in Ahmedabad (2019) and Indore (2022), and each one has a 1.5 MCi capacity. Greater output with reduced waste has resulted from the first-ever broad use of inoculation and irradiation (Varshney 2022). An initial economic assessment, with prices denoted in $ as of 2019, compares the utilization of Electron Beam (EB) methods solely for sludge hygienization to the integration of EB techniques for both biogas cogeneration and sludge hygienization. This analysis focuses on operational costs, with no consideration given to investment costs [18].

Table 14.6 Cost analysis of effluent processing unit for E beam technology

Throughput-250,000 m^3/annually, sludge outcome: 1500 tons dry mass annually		
Parameter	Method-1 (sludge hygenization)	Method-2 (sludge hygenization and biogas generation)
Accelerator information	100 kW, 2 meV	100 kW, 2 meV, 2 meV + biogas production
Income		
From avoiding sludge disposal cost	1500 tons@109.86$/ton = 164,790$	1500 tons@109.86$/ton = 164,790$
From bio solid based fertilizer	1500 tons@103.92 $/ton = 155,880$	1277.5 tons@103.92 $/ton = 132,737$
From electricity production		350 kW × 8000 h × 0.14$/kWh = 392,000$
(note: biogas produced = 1,022,000 m^3/annually is converted to electricity and heat by co-generator, generated power is 350 kW)		
Expenses		
Power consumption		
(1) E beam accelerator	130 kW × 8000h × 0.14$ = 145,600 $	130 kW × 8000h × 0.14$ = 145,600 $
(2) By processing unit equipment	70 kW × 8000h × 0.14$ = 78,400 $	70 kW × 8000h × 0.14$ = 78,400 $
(3) Heat generation	10kW × 8000h × 0.14$ = 11,200 $	
(4) Grass silage		1055 tons × 11.01 $/ton = 11,615$
Total cost for power consumption	235,200$/annually	235,615$/annually
Net income and savings	85,470$	289,122$

14.7 New Applications

In a variety of economic sectors all around the world, radiation processing is gaining widespread recognition and continues to experience steady developments [27]. Examples of current radiation techniques include photo-, electro-, and thermal-induced reductions of CO_2 and the modification of catalysts for these processes by radiation. Developing innovative materials that may be used to manufacture mixed matrix membranes for CO_2 capture is an essential part of radiation technology [69]. Chemical reactions involving carbonates, washing of non-ferrous metal ores, rock breakdown, gas phase synthesis, concrete transformation, cracking of petroleum products, lignin, and paraffin, turning natural gas into motor fuel, cleaning up waste gas from metalworking, and the thermal process [80]. The study of radiochemical and radiothermal technologies talks about coal-fired power plants that are already in use and those that will be built in the future [60].

14.8 The Effects of Ionizing Radiation on the Extraction of Bioactive Compounds from Agro-Industrial Residues

Scientific groups actively seek and develop novel ingredients to meet the increasing demand for healthier food options. Recovering bioactive molecules from agro-industrial leftovers, which would otherwise harm the environment, can help ensure long-term economic and ecological viability [52, 57]. Using ionizing radiation to enhance the extraction of bioactive substances is a safe and environmentally friendly option. It is possible to extract bioactive molecules from such waste materials and the bioactive qualities that are associated with them by the application of ionizing radiation [66]. Extracts made from irradiated waste have better bioactive characteristics, which makes them advantageous to the food industry. This reduces the negative impact on the environment and produces high-value products [52].

Both traditional and non-conventional technologies are capable of extracting these chemicals from their respective components. The limits of traditional methodologies are inspiring the development of new and interesting ways. Some examples of these methods are molecular distillation, pressurized low-polarity water extraction (PLOUGH)(, pressurized liquid extraction (PLE) enzyme-assisted extraction (EAE), microwave-assisted extraction (MAE) and supercritical fluid extraction (SFE) [52]. These approaches demand the use of expensive, exceptionally pure, and safe substances, and the extraction procedure is time-intensive. Their assessment of antioxidant activity, total flavonoids, total phenol concentration, and selectivity [13]. In comparison to ME, UAE, HPE, and SC-CO2, it was shown that MAE was the most efficient method for extracting antioxidants from *Citrus sinensis* (L.) Osbeck peel. MAE had higher levels of phenols, flavonoids, and individual flavonoids than the other methods. Purifying, isolating, and further characterizing the extracted components is necessary because the resultant extracts contain complex combinations of natural substances. A number of different chromatographic techniques, such as high-performance liquid chromatography (HPLC), thin-layer chromatography (TLC), flash chromatography, column chromatography, and others, are utilized in order to separate and purify the molecules [20].

14.8.1 Applications of Bioactive Compounds Extracted from Agro-Industrial Residues

Functional foods can be created by incorporating extracts from irradiation residues, which include enhanced characteristics, into food products. This meets the growing need for dietary bioactives that enhance human health and mitigate the risk of diseases [21, 45, 53]. Research into fortifying meals with one or more bioactive substances to enhance or rectify a food product's potential biological activity has shifted focus to creating functional foods. In Fig. 14.2, we can see the process of bioactive component

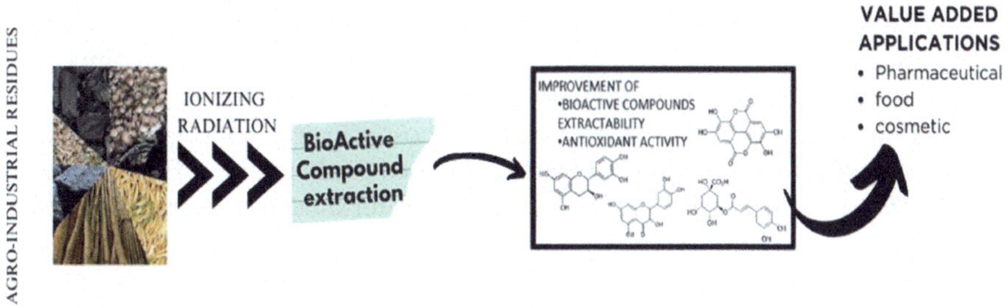

Fig. 14.2 Extraction, improvement, and added value applications of bioactive compounds from agricultural wastes

extraction from green waste. Bread, pastries, biscuits, and yoghurt are some of the most critical nutritional matrices, and there have been multiple research on how to incorporate bioactive chemicals from agro-industrial residues into them [23, 42]. An increase in the bioactive components of the new product was achieved through the utilization of *Myrciaria dubia* (Kunth) McVaugh peel as a yoghurt fortifier. This was accomplished without a significant alteration in the product's nutritional or fatty acid profile [42]. Incorporating epicarp extracts from *Punica granatum* L. into a traditional Brazilian pastry was a smart move [42]. The antioxidant activity of the novel formulations was higher and maintained throughout 14 days of storage. Simultaneously, the cake made using the new recipes had a moist texture and an attractive rose shade, which may influence customers to buy it.

14.9 Conclusions

One promising method for cleaning wastewater is using advanced oxidation technologies. Physical or chemical processes, or a combination of the two, comprise the bulk of wastewater treatment procedures. Among the most cutting-edge techniques, AOPs rely on a simple idea: to break down substances resistant to oxidation by producing and employing a hydroxyl free radical (HO*), a potent oxidant. Due to their superior effectiveness in pollutant elimination, AOPs have shown promise in several water treatment industries. The primary forces propelling technological advancement are ionizing radiation sources. When it comes to conventional application sectors, such as materials processing, food irradiation, sterilization, particularly those involving polymers, the most easily observable recent regulatory or technological improvements are those that include polymers. Perhaps e−/X systems will be an essential tool in these growing fields of application, especially in sterilization, and play a significant role. More intricate scenarios are associated with ecological applications, necessitating large amounts of electricity with excellent electrical efficiency and cheap accelerators. To get more significant concentrations of phenols, flavonoids, and individual flavonoids, the extraction method used was MAE. It is necessary to isolate, purify, and further describe the resulting extracts because they contain complex mixes of natural components. These natural bioactive compounds can play a vital role in future biorefinery applications.

Radiation technology is a sustainable and cost-effective solution, consuming less energy and generating minimal waste compared to alternatives. Heterogeneous Advanced Oxidation Processes (AOPs) use catalysts for substance degradation. Electromagnetic waves, or irradiation, preserve food by reducing damage and deterioration. This method boasts advantages like low operating costs, process controllability, and minimal energy usage, outperforming alternative preservation techniques. Utilizing linear accelerators powered by microwave sources, such as direct current and single resonant cavity models, radiation processing technology stands out. Electron Beam (EB) technology, a state-of-the-art system, shows promise in significantly improving Chemical Oxygen Demand (COD) and colour removal effectiveness. Ionizing radiation enhances the extraction of bioactive substances, offering a safe and environmentally friendly approach.

14.10 Conflicts of Interest

The authors declare no conflict of interest.

14.11 Declaration

No AI Tool has been used during the compilation of MS.

References

1. Abdel-Wahab M, Zubizarreta E, Polo A, Meghzifene A (2017) Improving quality and access to radiation therapy—an IAEA perspective. Semin Radiat Oncol 27(2):109–117. https://doi.org/10.1016/j.semradonc.2016.11.001
2. Ahmad HA, Ahmad S, Cui Q, Wang Z, Wei H, Chen X, Ni S-Q, Ismail S, Awad HM, Tawfik A (2022) The environmental distribution and removal of emerging pollutants, highlighting the importance of using microbes as a potential degrader: a review. Sci Total Environ 809:151926. https://doi.org/10.1016/j.scitotenv.2021.151926
3. Al-Hadyan K, Alsbeih G, Alharbi N, Judia S, Alghamdi M, Almousa A, Alsharif I, Bakheet R, Al-Romaih K, Al-Mozaini M, Al-Ghamdi S, Moftah B, Alhmaid R (2021) Effect of gamma irradiation on filtering facepiece respirators and SARS-CoV-2 detection. Sci Rep 11:19888. https://doi.org/10.1038/s41598-021-99414-6
4. Al-Maqdi KA, Elmerhi N, Athamneh K, Bilal M, Alzamly A, Ashraf SS, Shah I (2021) Challenges and recent advances in enzyme-mediated wastewater remediation-a review. Nanomaterials (Basel, Switzerland) 11(11). https://doi.org/10.3390/nano11113124

5. Alp M, Parihar V, Limoli C, Cucinotta F (2015) Irradiation of neurons with high-energy charged particles: an in silico modeling approach. PLoS Comput Biol 11:e1004428. https://doi.org/10.1371/journal.pcbi.1004428
6. Amano Y (2017) Radiation technology for a more prosperous and sustainable future
7. Ashraf S, Sood M, Professor A, Bandral J, Professor A, Jammu, Chatha J, Kashmir I, Trilokia, Kashmir J, Manzoor I, Trilokia M, Manzoor M (2019) Food irradiation: A review 7:131–136
8. Ashtari M, Khademi O, Soufbaf M, Afsharmanesh H, Askari MA (2019) Effect of gamma irradiation on antioxidants, microbiological properties and shelf life of pomegranate arils cv. 'Malas Saveh'. Sci Hortic 244:365–371. https://doi.org/10.1016/j.scienta.2018.09.067
9. Aurina K, Sari A (2022) Ohmic heating: a review and application in food industry. In: Proceedings of the 2nd international conference on smart and innovative agriculture (ICoSIA 2021), vol 19, pp 107–113. https://doi.org/10.2991/absr.k.220305.016
10. Barbosa Ferreira M, Sales Solano AM, Vieira dos Santos E, Martínez-Huitle CA, Ganiyu SO (2020) Coupling of anodic oxidation and soil remediation processes: a review. Materials 13(19). https://doi.org/10.3390/ma13194309
11. Begum T, Follett P, Hossain F, Christopher L, Salmieri S, Lacroix M (2020) Microbicidal effectiveness of irradiation from gamma and X-ray sources at different dose rates against the foodborne illness pathogens *Escherichia coli*, *Salmonella Typhimurium* and *Listeria monocytogenes* in rice. LWT 132:109841. https://doi.org/10.1016/j.lwt.2020.109841
12. Bhat N, Wani I, Hamdani A (2020) Effect of gamma-irradiation on the thermal, rheological and antioxidant properties of three wheat cultivars grown in temperate Indian climate. Radiat Phys Chem 176:108953. https://doi.org/10.1016/j.radphyschem.2020.108953
13. Boudhrioua N (2016) Comparison of the efficiency of different extraction methods on antioxidants of maltease orange peel. Int J Food Nutr Sci 3:1–13. https://doi.org/10.15436/2377-0619.16.789
14. Candal RJ (2014) Advanced oxidation technologies. In: Advanced oxidation technologies. https://doi.org/10.1201/b16657
15. Cheng W, Li J, Wu Y, Xu L, Su C, Qian Y, Zhu Y-G, Chen H (2016) Behavior of antibiotics and antibiotic resistance genes in eco-agricultural system: a case study. J Hazard Mater 304:18–25. https://doi.org/10.1016/j.jhazmat.2015.10.037
16. Chmielewska D, Werner Ł, Gryczka U, Migdal W (2022) Effect of electron beam irradiation on filtering facepiece respirators integrity and filtering efficiency. Nukleonika-Orig Edn 67:23–33. https://doi.org/10.2478/nuka-2022-0002
17. Chmielewski AG (2023) Radiation technologies: the future is today. Radiat Phys Chem 213:111233. https://doi.org/10.1016/j.radphyschem.2023.111233
18. Chmielewski AG (2023) Radiation technologies: the future is today. Radiat Phys Chem 213. https://doi.org/10.1016/j.radphyschem.2023.111233
19. Chmielewski A, Sudlitz M, Han B, Pillai S (2021) Electron beam technology for biogas and biofertilizer generation at municipal resource recovery facilities. Nukleonika-Orig Edn 66:213–219
20. Duistermaat JJ, Kolk JAC (2000) Proper Actions 8:93–130. https://doi.org/10.1007/978-3-642-56936-4_2
21. Faustino M, Veiga M, Sousa P, Costa E, Silva S, Pintado M (2019) Agro-food byproducts as a new source of natural food additives. Molecules 24:1056. https://doi.org/10.3390/molecules24061056
22. Fellows P (nd) Food processing
23. Fierascu R, Fierascu I, Avramescu S, Sieniawska E (2019) Recovery of natural antioxidants from agro-industrial side streams through advanced extraction techniques. Molecules 24:4212. https://doi.org/10.3390/molecules24234212
24. Fu Q, Malchi T, Carter LJ, Li H, Gan J, Chefetz B (2019) Pharmaceutical and personal care products: from wastewater treatment into agro-food systems. Environ Sci Technol 53(24):14083–14090. https://doi.org/10.1021/acs.est.9b06206
25. Gil L (2017) Culture meets nuclear in brazil. IAEA Bull 58(1):16–17

26. Goswami RK, Mehariya S, Verma P, Lavecchia R, Zuorro A (2021) Microalgae-based biorefineries for sustainable resource recovery from wastewater. J Water Process Eng 40:101747. https://doi.org/10.1016/j.jwpe.2020.101747
27. Güven O, Han B (2022) The prospects for radiation technology in mitigating carbon footprint. Radiat Phys Chem 198:110282. https://doi.org/10.1016/j.radphyschem.2022.110282
28. Huang M, Hasan K, Pillai SD, Pharr M, Staack D (2020) Since January 2020 Elsevier has created a COVID-19 resource centre with free information in English and Mandarin on the novel coronavirus COVID- 19. The COVID-19 resource centre is hosted on Elsevier Connect, the company's public news and information, January 2020
29. Huang M, Hasan MK, Pillai SD, Pharr M, Staack D (2023) Electron beam technology for re-processing of personal protective equipment. Radiat Phys Chem 202:110557. https://doi.org/10.1016/j.radphyschem.2022.110557
30. Huang T, Fang Z, Zhao H, Xu D, Yang W, Yu W, Zhang J (2020) Physical properties and release kinetics of electron beam irradiated fish gelatin films with antioxidants of bamboo leaves. Food Biosci 36:100597. https://doi.org/10.1016/j.fbio.2020.100597
31. Indiarto R (2020) Post-harvest handling technologies of tropical fruits: a review. Int J Emerg Trends Eng Res 8(7):3951–3957. https://doi.org/10.30534/ijeter/2020/165872020
32. Indiarto R, Nurannisa RL (2020) Quality characteristics of dairy products using ohmic heating technology: a review. Int J Emerg Trends Eng Res 8(7):3461–3467
33. Indiarto R, Nurhadi B, Tensiska T, Subroto E, Istiqamah YK (2019) Effect of liquid smoke on microbiological and physico-chemical properties of beef meatballs during storage. Food Res 4:522–531. https://doi.org/10.26656/fr.2017.4(2).341
34. Indiarto R, Pratama AW, Sari TI, Theodora HC (2020) Food irradiation technology: a review of the uses and their capabilities. International Journal of Engineering Trends and Technology 68(12):91–98. https://doi.org/10.14445/22315381/IJETT-V68I12P216
35. Indiarto R, Qonit M (2020) A review of irradiation technologies on food and agricultural products. Int J Sci Technol Res 9:4411–4414
36. Indiarto R, Rezaharsamto B (2020) A review on ohmic heating and its use in food. Int J Sci Technol Res 9:485–490
37. Indiarto R, Rezaharsamto B (2020) The physical, chemical, and microbiological properties of peanuts during storage: a review. Int J Sci Technol Res 9:1909–1913
38. Jan A, Sood M, Younis K, Islam R (2020) Brown rice based weaning food treated with gamma irradiation evaluated during storage. Radiat Phys Chem 177:109158. https://doi.org/10.1016/j.radphyschem.2020.109158
39. Jeon M-J, Ha J-W (2020) Bactericidal and synergistic effects of X-ray irradiation and gallic acid against foodborne pathogens on lettuce. Food Microbiol 92:103584. https://doi.org/10.1016/j.fm.2020.103584
40. Jeon M-J, Ha J-W (2020) Synergistic bactericidal effect and mechanism of X-ray irradiation and citric acid combination against food-borne pathogens on spinach leaves. Food Microbiol 91:103543. https://doi.org/10.1016/j.fm.2020.103543
41. Jeong S-G, Yang JE, Park J, Ko S, Choi I, Kim H, Chun H, Kwon M-J, Park HW (2020) Gamma irradiation improves the microbiological safety and shelf-life of kimchi seasoning mixture. LWT 134:110144. https://doi.org/10.1016/j.lwt.2020.110144
42. Jos M, Barros L, Genena AK, Barreira CM (2020) Characterization and application of pomegranate epicarp extracts as functional ingredients in a typical brazilian pastry product
43. Khalid M, Park Y, Karri R, Walvekar R (2023) Advanced oxidation processes for micropollutant remediation. https://doi.org/10.1201/9781003247913
44. Khan A, Khan SU, Khan S, zia-ul-islam S, Naimatullah KB, Khan M (2018) Nutritional complications and its effects on human health. J Food Sci Nutr 01(01):17–20. https://doi.org/10.35841/food-science.1.1.17-20
45. Kumar P, Duhan S, Duhan J (2018) Agro-industrial wastes and their utilization using solid state fermentation: a review. Bioresour Bioprocess 5. https://doi.org/10.1186/s40643-017-0187-z
46. Kumar B, Verma P (2021) Life cycle assessment: blazing a trail for bioresources management. Energy Convers Manag X 100063. Elsevier

47. Li X, Jin W, Yan D, Tan Z, Ruiyuan L, Xiao R, Wang J, Wang J, Zhou L (2020) Electron beam irradiation effect on the physicochemical characteristics of municipal excess sludge. Radiat Phys Chem 180:109246. https://doi.org/10.1016/j.radphyschem.2020.109246
48. Lima F, Vieira K, De Santos MA, De Souza P (2018) Effects of radiation technologies on food nutritional quality. https://doi.org/10.5772/intechopen.80437
49. Liu B, Zhang S-G, Chang C-C (2020) Emerging pollutants-part II: treatment. Water Environ Res: Res Publ Water Environ Fed 92(10):1603–1617. https://doi.org/10.1002/wer.1407
50. Liu J, Ren S, Cao J, Tsang DCW, Beiyuan J, Peng Y, Fang F, She J, Yin M, Shen N, Wang J (2021) Highly efficient removal of thallium in wastewater by $MnFe_2O_4$-biochar composite. J Hazard Mater 401:123311. https://doi.org/10.1016/j.jhazmat.2020.123311
51. Liu X, Liu J, Zhang W, Han S, Zhang T, Liu B (2019) Electron beam irradiation-induced structural changes increase the antioxidant activities of egg white protein. LWT 111:846–852. https://doi.org/10.1016/j.lwt.2019.05.066
52. Madureira J, Barros L, Cabo Verde S, Margaça FMA, Santos-Buelga C, Ferreira ICFR (2020) Ionizing radiation technologies to increase the extraction of bioactive compounds from agro-industrial residues: a review. J Agric Food Chem 68(40):11054–11067. https://doi.org/10.1021/acs.jafc.0c04984
53. Maria S, Shirahigue LD, Ceccato-antonini SR (2020) Agro-industrial wastes as sources of bioactive compounds for food and fermentation industries
54. Al Mayyahi A, Al-asadi HAA (2018) Advanced oxidation processes (AOPs) for wastewater treatment and reuse: a brief review2(3):18–30
55. Meeroff DE, Bloetscher F, Shaha B (2020) Economics of wastewater/biosolids treatment by electron beam technology. Radiat Phys Chem 168:108541. https://doi.org/10.1016/j.radphyschem.2019.108541
56. Miklos D, Remy C, Jekel M, Linden K, Drewes J, Hübner U (2018) Evaluation of advanced oxidation processes for water and wastewater treatment – a critical review. Water Res 139:118–131. https://doi.org/10.1016/j.watres.2018.03.042
57. Nair LG, Agrawal K, Verma P (2022) An overview of sustainable approaches for bioenergy production from agro-industrial wastes. Energy Nexus 100086
58. Pan L, Xing J, Zhang H, Luo X, Chen Z (2020) Electron beam irradiation as a tool for rice grain storage and its effects on the physicochemical properties of rice starch. Int J Biol Macromol 164:2915–2921. https://doi.org/10.1016/j.ijbiomac.2020.07.211
59. Park J-S, Ha J-W (2018) X-ray irradiation inactivation of *Escherichia coli* O157:H7, *Salmonella enterica* Serovar Typhimurium, and *Listeria monocytogenes* on sliced cheese and its bactericidal mechanisms. Int J Food Microbiol 289:127–133. https://doi.org/10.1016/j.ijfoodmicro.2018.09.011
60. Pavlov YS, Petrenko VV, Alekseev PA, Bystrov PA, Souvorova OV (2022) Trends and opportunities for the development of electron-beam energy-intensive technologies. Radiat Phys Chem 198:110199. https://doi.org/10.1016/j.radphyschem.2022.110199
61. Pillai S (2016) Introduction to electron-beam food irradiation. Chem Eng Prog 112:36–44
62. Pillai S, Shayanfar S (2016) Electron beam technology and other irradiation technology applications in the food industry. Top Curr Chem 375:6. https://doi.org/10.1007/s41061-016-0093-4
63. Postek MT, Poster DL, Vládar AE, Driscoll MS, LaVerne JA, Tsinas Z, Al-Sheikhly MI (2018) Ionizing radiation processing and its potential in advancing biorefining and nanocellulose composite materials manufacturing. Radiat Phys Chem 143:47–52. https://doi.org/10.1016/j.radphyschem.2017.09.015
64. Ravi Kumar K, Krishna Chaitanya NVV, Sendhil Kumar N (2021) Solar thermal energy technologies and its applications for process heating and power generation – a review. J Clean Prod 282:125296. https://doi.org/10.1016/j.jclepro.2020.125296
65. Rosa L (2019) Solar heat for materials processing: a review on recent achievements and a prospect on future trends. ChemEngineering 3:83. https://doi.org/10.3390/chemengineering3040083

66. Saravanan A, Deivayanai VC, Kumar PS, Rangasamy G, Hemavathy RV, Harshana T, Gayathri N, Alagumalai K (2022) A detailed review on advanced oxidation process in treatment of wastewater: mechanism, challenges and future outlook. Chemosphere 308:136524. https://doi.org/10.1016/j.chemosphere.2022.136524
67. Shayanfar S, Pillai S (2015) Future trends in electron beam technology for food processing. In: Electron beam pasteurization and complementary food processing technologies, pp 295–308. https://doi.org/10.1533/9781782421085.3.295
68. Siwek M, Edgecock T (2020) Application of electron beam water radiolysis for sewage sludge treatment—a review. Environ Sci Pollut Res 27:1–25. https://doi.org/10.1007/s11356-020-10643-0
69. Sun Y, Chmielewski AG (2017) Applications of ionizing radiation, vol 1. http://tl-irmp.eu/wp-content/uploads/2018/01/APPLICATIONS-OF-IONIZING-RADIATION-IN-MATERIALS-PROCESSING-Volume-2.pdf, http://tl-irmp.eu/download/book-application-of-ionizing-radiation-in-materials-processing/
70. Sun Y, Chmielewski AG, Jądrowej ICiT (2017) Applications of ionizing radiation in materials processing, vol 1. TA – TT, Institute of Nuclear Chemistry and Technology Warszawa. https://doi.org/LK, https://worldcat.org/title/989768781
71. Tijani JO, Fatoba OO, Madzivire G, Petrik LF (2014) A review of combined advanced oxidation technologies for the removal of organic pollutants from water. Water Air Soil Pollut 225(9):2102. https://doi.org/10.1007/s11270-014-2102-y
72. Truc N, Uthairatanakij A, Srilaong V, Laohakunjit N, Jitareerat P (2020) Effect of electron beam radiation on disease resistance and quality of harvested mangoes. Radiat Phys Chem 180:109289. https://doi.org/10.1016/j.radphyschem.2020.109289
73. Wan X-M, Lei M, Chen T (2020) Review on remediation technologies for arsenic-contaminated soil. Front Environ Sci Eng 14:24. https://doi.org/10.1007/s11783-019-1203-7
74. Wang S, Wang J, Chen C, He S, Hu J, Zhang Y (2022) First full-scale application of electron beam technology for treating dyeing wastewater (30,000 m3/d) in China. Radiat Phys Chem 196:110136. https://doi.org/10.1016/j.radphyschem.2022.110136
75. What is "emerging"? (2006) Environ Sci & Technol 40(23):7105. https://doi.org/10.1021/es062982z
76. Wojcik A, Harms-Ringdahl M (2019) Radiation protection biology then and now. Int J Radiat Biol 95:1–23. https://doi.org/10.1080/09553002.2019.1589027
77. Yoon Y-S, Ameer K, Song B-S, Kim J-K, Park H-Y, Lee K-C, Eun J-B, Park J-H (2020) Effects of X-ray irradiation on the postharvest quality characteristics of 'maehyang' strawberry (Fragaria × ananassa). Food Chem 325:126817. https://doi.org/10.1016/j.foodchem.2020.126817
78. Zhi D, Lin Y, Jiang L, Zhou Y, Huang A, Yang J, Luo L (2020) Remediation of persistent organic pollutants in aqueous systems by electrochemical activation of persulfates: a review. J Environ Manag 260:110125. https://doi.org/10.1016/j.jenvman.2020.110125
79. Zhou X, Ye X, He J, Wang R, Jin Z (2020) Effects of electron beam irradiation on the properties of waxy maize starch and its films. Int J Biol Macromol 151:239–246. https://doi.org/10.1016/j.ijbiomac.2020.01.287
80. Zwolińska E, Sun Y, Chmielewski AG, Pawelec A, Bułka S (2020) Removal of high concentrations of NOx and SO_2 from diesel off-gases using a hybrid electron beam technology. Energy Reports 6:952–964. https://doi.org/10.1016/j.egyr.2020.04.008

SPRINGER NATURE

GPSR Compliance

The European Union's (EU) General Product Safety Regulation (GPSR) is a set of rules that requires consumer products to be safe and our obligations to ensure this.

If you have any concerns about our products, you can contact us on ProductSafety@springernature.com

In case Publisher is established outside the EU, the EU authorized representative is:

Springer Nature Customer Service Center GmbH
Europaplatz 3
69115 Heidelberg, Germany

The manufacturer's authorised representative in the EU is Springer Nature Customer Service Centre GmbH, Europaplatz 3, 69115 Heidelberg, Germany. If you have any concerns regarding our products, please contact ProductSafety@springernature.com

Printed and bound by CPI Group (UK) Ltd, Croydon, CR0 4YY
01/10/2026
02225904-0013